Praise for *God's Politics*

"We need a religious revival. And we may get one. . . . The revival Jim Wallis predicts."

—Nick Coleman, *Minneapolis Star-Tribune*

"Wallis provides a refreshing alternative voice to the polarizing rhetoric currently popular. Those who find themselves uncomfortable in either camp will welcome his approach, even if they disagree with some aspects of his solution."

—*Library Journal*

"More than any other name, people mentioned the Rev. Jim Wallis as the next Niebuhr. The liberal evangelical's book, *God's Politics,* shows how the Religious Right doesn't own evangelical Protestantism, much less American Christianity. The *Sojourners* magazine editor chides liberals for failing to understand how religion and politics mix. And he rebukes conservatives for equating Christianity with red-state views on abortion and other social questions."

—William McKenzie, *Dallas Morning News*

"*God's Politics* is part argument for why the wind needs to be changed and part manual for how to change it. The book reads like a sermon by a minister who has learned that the best way to keep his congregation from falling asleep is to break up the theology with anecdotes and provide plenty of lists for parishioners to scribble down in the margins of their bulletins."

—Amy Sullivan, *Washington Monthly*

"One suggestion in the post-election reality for igniting the fires of discussion would be to encourage every church, synagogue, and mosque study group, every Sunday School class, every youth group, and every secular book club to read and discuss *God's Politics.* It is a primer for life at the intersection of religion and politics in 2005 and will most assuredly instigate lively and productive discussion that may even lead to action."

—*National Catholic Reporter*

"Wallis's theology will be welcomed by many, and ought to be read by all who have an interest in politics and religion in this country."

—Dan Wakefield, *Boston Globe*

"I found this book intriguing, compelling and often convincing, to the point that it caused me to re-examine longstanding positions."

—Alex Coffin, *Charlotte Observer*

"As a leading evangelical Christian on a national and global scale, Wallis is attempting to reshape the dialogue over 'moral values' that has polarized the United States in the wake of the November 2004 elections."

—Ed Conroy, *San Antonio Express-News*

"Wallis's thoughtful theological critique of the Bush Administration is making an impact."

—Ron Ferguson, *Glasgow Herald* (Scotland)

"Wallis is a progressive evangelical Christian, and no, that's not an oxymoron. He believes the country is ready for a new and better conversation about faith and values, and he wants to break the cultural zeitgeist that equates religion with the right wing."

—Eleanor Clift, *Newsweek* online

"Canadians in general and Christians in particular will find this book a welcome and forceful challenge to what's most objectionable about George W. Bush and Republican politics in the United States. And, if we ponder as we read, we discover many things that are objectionable about current Canadian politicians and politics, too."

—Jim Romahn, *Toronto Star*

"As for blue and red religion, the most noted attempt to counter the common wisdom comes from Jim Wallis. *God's Politics* has the subject in the subtitle: *Why the Right Gets It Wrong and the Left Doesn't Get It*. Wallis has written a sermon on the moral imperative to build a progressive movement."

—Ellen Goodman, *Boston Globe*

GOD'S POLITICS

ALSO BY JIM WALLIS:

Faith Works: How Faith Based Organizations are Changing Lives, Neighborhoods, and America

Living God's Politics: A Guide to Putting Your Faith into Action

Peacemakers: Christian Voices from the New Abolitionist Movement

Revive Us Again: A Sojourner's Story

The Rise of Christian Conscience: The Emergence of a Dramatic Renewal Movement in the Church Today

The Call to Conversion: Recovering the Gospel for These Times

The Soul of Politics: A Practical and Prophetic Vision for Change

Who Speaks for God?

GOD'S POLITICS

Why the Right Gets It Wrong and the Left Doesn't Get It

JIM WALLIS

HarperOne
An Imprint of HarperCollins*Publishers*

To Luke Wallis

For how the probing questions, startling insights, and emerging social conscience of a six-year-old son helped to shape this book

HarperCollins books may be purchased for educational, business, or sales promotional use. For information please write: Special Markets Department, HarperCollins Publishers, 10 East 53rd Street, New York, NY 10022.

HarperCollins Web site: http://www.harpercollins.com

FIRST HARPERCOLLINS PAPERBACK EDITION PUBLISHED IN 2006

Library of Congress Cataloging-in-Publication Data is available.
ISBN: 978-0-06-083447-0
ISBN-10: 0-06-083447-1

12 13 14 15 16 RRD(H) 11

CONTENTS

Without a vision, the people perish.
—Proverbs 29:18

Write the vision; make it plain upon tablets, so he may run who reads it. For still the vision awaits its time; it hastens to the end—it will not lie. If it seems slow, wait for it; it will surely come, it will not delay.
—Habakkuk 2:2–3

ACKNOWLEDGMENTS

I have one of the best jobs in the world because I get to know some of the most faith-inspired people everywhere who are doing some of the most important things anywhere. They care for the poor and vulnerable and challenge those who neglect and oppress the marginalized; they often risk their own security by seeking to reduce our conflicts and end our wars; they reconcile divided peoples, gangs, and nations—they are healers, activists, service providers, peacemakers, and justice seekers who actually do what most other people just believe in. I hear their stories, know their struggles, understand their pains, and feel their hopes. Their visions and dreams always help shape mine, and I give them all great credit for helping to inspire this book.

More particularly, I want to thank the remarkable staffs of Sojourners and Call to Renewal (and both organization's boards of directors), the men and women with whom I work to make a difference in this world. Every day, they influence the ways I think, speak, and act; this book was influenced by them too.

From my staff, most especially, I want to thank my collaborator and friend, Duane Shank, without whose feedback, editing, research, fact-checking, constant dialogue, and consistently wise suggestions this book would not have been possible. He gets the biggest thanks for this book. Since the launch of *God's Politics,* Deanna Murshed has done a magnificent job in coordinating the book tour from Washington, D.C., which quickly became a "movement tour" with a deluge of requests for my new special assistant Rich Bland to schedule. Their calm competence has been remarkable. Helena Brantley has been the main media coordinator of the book tour from our offices, working closely with HarperSanFrancisco, and has handled the barrage with grace and efficiency.

And I can't say enough about my whole team at HarperSanFrancisco and their colleagues at HarperCollins in New York City. It is a gift to have editors and publishers who are not only so good at what they do but who also really believe in your work. Mark Tauber, HarperSanFrancisco's vice president and deputy publisher, was my agent when I started writing this book. He successfully presented it to Harper Collins, and was later hired by them. It's not often that an author's agent becomes his publisher! Mark's belief in this book and in me has helped to carry us together though the whole process. Steve Hanselman, the Harper publisher who initially accepted the book, was also the one who told me why he believed this book could "break through," and how much he felt personally committed to its message. As has often been the case, Steve's publishing judgment proved prophetic. Also supportive from the start was my first editor at Harper, my longtime friend and publishing advisor Roy Carlisle. He has helped me editorially with all my books, and the breakthough of *God's Politics* is an accomplishment that belongs also to him. My new editor for *God's Politics* is Eric Brandt, one of the most skillful people I have ever worked with. Eric has a grace and calmness about him that we all have greatly benefited from, especially when Harper moved the book's launch date up by almost four months! After the election and the discussion it spurred, Harper wanted the book out as soon as possible, and Elaine Merrill, senior production editor, shepherded the book through a very rushed production. I am very grateful for Claudia Riemer Boutote, senior publicity director, and publicist Laina Adler, whose careful strategy and tireless efforts contributed invaluably to getting the word out and helped to make this a bestseller across the country. My warm thanks also go to Jeff Hobbs, director of sales; Nina Olmsted, vice president of national accounts; Margery Buchanan, director of marketing operations; and to my old friend, Mickey Maudlin, Harper's new editorial director. Finally, thanks to Louise Miller, vice president and director of publishing operations, who makes sure that supply keeps up with the demand for the book. I mention all these people because of the enormous demands put on a publishing staff when a book helps to spark a national conversation like God's Politics has already done. Visiting with them during the San Francisco stop of my book tour made me again appreciate how we really are all a team on this project.

On a very personal level, without the patient, steady, and loving support of my wife, Joy Carroll, and the creative partnership we have found together, the book could not have been written. And my two sons Luke (six years old) and Jack (two) have this way of keeping their dad both human and sane. Time with them is the best part of my busy life. My family anchors my life and provides both the spiritual discipline and nourishment that helps sustain all the rest of work and activity. Coaching T-ball with 15 first graders on Friday nights provides just the balance I need!

Only in the re-reading of the manuscript did I realize how many stories there were from Luke and our ongoing conversations. Luke's probing questions, often startling insights, and emerging social conscience provide me a regular source of eye-opening experiences. His response to my work and activity is already serving to deepen my own understanding of it—so much so that I have decided to dedicate this book to Luke.

Thank you and bless you all.

—*Jim Wallis*

INTRODUCTION

Why Can't We Talk about Religion and Politics?

WHY CAN'T WE TALK about religion and politics? These are the two topics you are not supposed to discuss in polite company. Don't break up the dinner party by bringing up either of these subjects! That's the conventional wisdom. Why? Perhaps these topics are too important, too potentially divisive, or raise the issues of core values and ultimate concerns that make us uncomfortable.

Sojourners magazine, where I serve as editor, commits the offense, in every single issue, of talking about faith, politics, and culture. Yet our subscriber and on-line lists are growing, especially among a younger generation. I am also on the road a lot, speaking almost every week to very diverse audiences of people. I hear and feel the hunger for a fuller, deeper, and richer conversation about religion in public life, about faith and politics. It's a discussion that we don't always hear in America today. Sometimes the most strident and narrow voices are the loudest, while more progressive, prophetic, and healing faith often gets missed. But the good news is about how all that is changing—really changing.

Abraham Lincoln had it right. Our task should not be to invoke religion and the name of God by claiming God's blessing and endorsement for all our national policies and practices—saying, in effect, that God is on our side. Rather, Lincoln said, we should pray and worry earnestly whether we are on God's side.

Those are the two ways that religion has been brought into public life in American history. The first way—God on our side—leads inevitably to triumphalism, self-righteousness, bad theology, and, often, dangerous foreign policy. The second way—asking if we are on God's side—leads to much healthier things, namely, penitence and even repentance, humility, reflection, and even accountability. We need much more of all these, because these are often the missing values of politics.

Of course, Martin Luther King Jr. did it best. With his Bible in one hand and the Constitution in the other, King persuaded, not just pronounced. He reminded us all of God's purposes for justice, for peace, and for the "beloved community" where those always left out and behind get a front-row seat. And he did it—bringing religion into public life—in a way that was always welcoming, inclusive, and inviting to all who cared about moral, spiritual, or religious values. Nobody felt left out of the conversation. I try to do that too in this book.

The night before I wrote this introduction, I was speaking in Denver, Colorado. A young man was waiting patiently in a greeting line after my talk. Finally we shook hands and he told me that he was an agnostic (not religiously affiliated). But he said that he cared deeply about the moral issues at stake in his country. Then he told me he had been "spiritually inspired" by the evening. And he thanked me for making him feel included. "I just wanted to give you some feedback from outside the community," the sincere agnostic said. It was an encouraging word to me.

The values of politics are my primary concern in this book. Of course, God is not partisan; God is not a Republican or a Democrat. When either party tries to politicize God, or co-opt religious communities for their political agendas, they make a terrible mistake. The best contribution of religion is precisely not to be ideologically predictable nor loyally partisan. Both parties, and the nation, must let the prophetic voice of religion be heard. Faith must be free to challenge both right and left from a consistent moral ground.

God's politics is therefore never partisan or ideological. But it challenges everything about our politics. God's politics reminds us of the people our politics always neglects—the poor, the vulnerable, the left behind. God's politics challenges narrow national, ethnic, economic, or cultural self-interest, reminding us of a much wider world and the creative human diversity of all those made in the image of the creator. God's politics reminds us of the creation itself, a rich environment in which we are to be good stewards, not mere users, consumers, and exploiters. And God's politics pleads with us to resolve the inevitable conflicts among us, as much as is possible, without the terrible cost and consequences of war. God's politics always reminds us of the ancient prophetic prescription to "choose life, so that you and your children may live," and challenges all the selective moralities that would choose one set of lives and issues over another.

This book challenges both the Right and the Left—offering a new vision for faith and politics in America. To do that, we will enter into a new conversation of personal faith and political hope. I hope you will join the discussion.

The 2004 election was one of those where, no matter who won, almost half the population was going to feel absolutely crushed (the half that lost in an almost fifty-fifty electorate). We are now deeply divided, the media say, between "red states" (Republican) and "blue states" (Democrat). So how do the people in the red and blue states talk to each other, especially people of faith? I believe there are more common concerns and more potential common ground in both the red and blue states than many think. The new senator from Illinois, Barack Obama, proclaimed in his widely praised Democratic Convention keynote speech, "We have an awesome God in the blue states!"

In the days following November 2, the "moral values voter" became the defining story line of the election. A flawed national exit poll question has created an enormous and important political debate in America, and one that will be with us far beyond the 2004 election. When asked to choose the most important issue that influenced their vote, 22 percent (a slight plurality) chose "moral values" from a list that also included the war in Iraq, terrorism, health care, education, taxes, and the economy/jobs. And 80 percent of those who chose moral values voted for George W. Bush. That poll result has sparked a firestorm in the media and in Washington's political circles about who gets or doesn't get the moral values issue. The conventional wisdom claims that the Republicans do and the Democrats don't get it, that the moral values responders simply meant voters who are against abortion and gay marriage, and that religious conservatives won the election for George Bush, which was Karl Rove's strategy all along. That perception has both thrilled and emboldened the religious Right, whose leaders and organizations are now flexing their political muscles by making demands on an administration they claim to have helped re-elect.

I have no doubt that the issues of abortion and gay marriage played an important role in the political outcome. But it was not necessarily the religious Right that won this election. Demographic information in the same exit polls showed that nationally, 52 percent of Catholics voted for Bush, 5 percent more than in 2000; 44 percent of Latinos, 9 percent higher; and 11 percent of African Americans, 2 percent higher. In what proved to be the key state, Ohio, 55 percent of Catholics went for Bush, 5 percent higher than in 2000; and 16 percent of African Americans, 7 percent more than four years ago. And conservative moral values likely played a part in those increases.[1]

But, of course, if I were a Christian who cared deeply about peace I would have likely checked the war in Iraq (one of the choices) instead of moral values, and if I were the Catholic coordinator of a food pantry I would have checked the closest thing I could find to poverty, which would have been the economy or health care.

The single moral values question was a whole *different kind* of choice than the rest of the "issues," ignoring the moral values inherent in those other concerns. Putting an ambiguous moral values choice in a list of specific issues skewed the results. Another way of looking at the poll is to group the issues together. In that case, war/peace values received 34 percent and economic values 33 percent.

A week after the election another poll, conducted by Zogby International, confirmed that when a broader and more specific list of moral issues was presented, the results were quite different. When asked "Which moral issue most influenced your vote?" 42 percent chose the war in Iraq while 13 percent said abortion, and 9 percent said same-sex marriage. When asked to choose the most urgent moral problem in American culture, 33 percent selected greed and materialism; 31 percent, poverty and economic justice; 16 percent, abortion; and 12 percent, same-sex marriage. The "greatest threat to marriage" was identified as infidelity by 31 percent, rising financial burdens by 25 percent, and same-sex marriage by 22 percent. But regardless of the validity of various polls and poll questions, values are now the hot topic in American politics.

I welcome the discussion of "moral values." And I believe the values debate should be the future of American politics. Of course, the questions are, which values and whose values? How narrowly or widely will values be defined and how partisan will the discussion be? Will the moral values debate cut both ways in politics, challenging both the political Left and the political Right, both Republicans and Democrats? Will values be used as wedges to further divide us or bridges to bring us together—to find common ground by moving to higher ground? We must now ask two questions. Where is the real debate in the moral values conversation (because there are real differences in America on the values issues)? And where can we find common ground (because there is also much that we share in common, which could be built upon)? This book is about both of those questions.

Religion was *indeed* a big factor in this election with moral values talk in the air the entire campaign. On the Republican side, George W. Bush talked comfortably and frequently about his personal faith and ran on what his conservative religious base called the "moral issues." On the Democratic side, Senator John Kerry invoked the New Testament story of the Good Samaritan, talked about the importance of loving our neighbors, and said that faith without works is dead—but only began talking that way at the very end of his campaign. Critics of the Kerry campaign say he got religion too little and too late, while critics of the Bush campaign say he used religion in one of the most partisan ways ever seen. But the results of the election point to how incredibly important the cultural and political definitions of religion and moral values really are.

Almost a year before the election, I wrote in *Sojourners* and in an op-ed piece for the *New York Times* that too many Democrats still wanted to restrict religion to the private sphere and were very uncomfortable with the language of faith and values even when applied to their own agenda. And that

Republicans wanted to narrowly restrict religion to a short list of hot-button social issues and obstruct its application to other matters that would threaten their agenda.

Well, after a year of political campaigning we ended up at about the same place. While some Democrats are now realizing the importance of faith, values, and cultural issues, a strong group of secular fundamentalists still fights to keep moral and spiritual language out of the liberal discussion. And while some Republicans would like to see an expanded application of faith, the religious fundamentalists still want to restrict religious values to gay marriage and abortion. And a very smart group of Republican strategists effectively appealed to both the faith and the fears of an important conservative religious constituency.

The religious Right fought to keep the primary focus on gay marriage and abortion (two important issues) and some of their leaders even said that good Christians could only vote for President Bush. But many other Christians disagreed. Sojourners led a petition and ad campaign titled "God Is Not a Republican. Or a Democrat." We insisted that poverty is a religious issue, pointing to thousands of verses in the Bible on the poor. We said that the environment—protection of God's creation—is a religious concern. And because millions of Christians in America believe the war in Iraq was not a just war we insisted that war is also a serious religious and theological matter. We spoke of a "consistent ethic of life," which addressed all the threats to human life and dignity, challenged the agendas of both parties, and revealed that no consistent pro-life candidate was running for president this year. More than 100,000 people signed the petition, and ads were placed in more than fifty newspapers around the country, including the *New York Times* and *USA Today.* (The complete text of the ad is at the end of this introduction.)

In this election, one side talked about the number of unborn lives lost to abortion each year, while the other pointed to the one hundred thousand civilian casualties in Iraq. But both are life issues—according to the pope, for example, who opposes both John Kerry's views on abortion and George Bush's war policy. Some church leaders challenged both candidates on whether just killing terrorists would really end terrorism and called for a deeper approach. And two hundred theologians, many from leading evangelical institutions, warned that a "theology of war emanating from the highest circles of government is also seeping into our churches." Clearly, God is not a Republican or a Democrat, and the best contribution of religion is precisely not to be ideologically predictable or loyally partisan but to maintain the moral independence to critique both the Left and the Right.

Syndicated *Washington Post* columnist E. J. Dionne said days after the election, "What's required is a sustained and intellectually serious effort by religious moderates and progressives to insist that social justice and inclusion are 'moral values' and that war and peace are 'life issues.' As my wife and I prepared our three kids for school the day after the day after, we shared our outrage that we in Blue America are cast as opponents of 'family values' simply because we don't buy the Right Wing's agenda. No political faction can be allowed to assert a monopoly on the family."[2]

My vision—a progressive and prophetic vision of faith and politics—was not running in this election. George Bush and John Kerry were, and John Kerry lost—not progressive religion. Neither candidate championed the poor as a "moral value," or made the war in Iraq a clearly religious matter. And neither advocated a "consistent ethic of human life" beyond single-issue voting, or a serious "pro-family" agenda without being anti-gay. The ways in which both parties' visions are morally and politically incomplete must now be taken up by people of faith. That can best be done by reaching into both the conservative Christian communities who voted for George Bush and more liberal Christian communities who voted for John Kerry.

We've now begun a real debate in this country over what the most important religious issues are in politics, and that discussion will continue well past the 2004 elections. It's time to spark a public conversation in this country over what the "moral values" in politics should be—and how broadly and deeply they should be defined. Religion doesn't fall neatly into Right and Left categories. If there were ever candidates running with a strong set of personal moral values *and* a commitment to social justice and peace, they could build many bridges to the other side. Personal and social responsibility are *both* at the heart of religion, and the two together could make a very powerful and compelling political vision for the future of our bitterly divided nation.

In a deeply polarized country, when half the people do feel crushed, the need for some kind of political healing and reconciliation becomes clear and that could be one of the most important roles for the religious community. But that shouldn't mean compromise of deeply held convictions on any side. Instead, building on our most deeply held values might be the best way to move forward and create some new political opportunities. In the spirit of America's greatest religious leader, the Reverend Martin Luther King Jr., the religious community could help a divided nation find common ground by moving to higher ground.

In George Bush's second term, he could return to the promise of his early and forgotten "compassionate conservatism," offering more than just "faith-based initiatives" but a serious plan for dramatic poverty reduction (both at home and internationally) with the resources to back it up. Now, having won the election, Bush could safely take some of John Kerry's advice for significantly involving the international community in helping to achieve both security and elections in Iraq and begin to withdraw the American occupation, whose continued presence will only bring disaster. And with the death of Yassar Arafat, the American president could join his English cousin, Tony Blair, in pushing both Israeli and Palestinian leaders toward a fair two-state peace solution, which more than any other accomplishment would lead to greatly reduced tensions in the world, and a serious reduction of terrorism. But if he doesn't move in those directions, George Bush may find that a significant part of his opposition will come from other Christians and people of faith who are acting out of their deepest "moral values."

In their preelection statement, "Faithful Citizenship," the Catholic bishops wrote: "Politics in this election year and beyond should be about an old idea with a new power—the common good. The central question . . . should be 'How can we—all of us, especially the weak and vulnerable—be better off in the years ahead? How can we protect and promote human life and dignity? How can we pursue greater justice and peace?"[3]

To start, the way forward is a dialogue on those questions among people of faith who will then address them to policy makers. We should hold ourselves and both major political parties accountable to the challenge of the biblical prophet Micah, to "do justice, love kindness, and walk humbly with your God." This book is about how we might begin that critical task.

Here is a quick tour of this book. The privatizing of faith has weakened its impact on critical public issues and opened the door for a right-wing "Christian politics," which both narrows and distorts a biblical agenda. We will show how God is personal, but never private, and how the witness of the biblical prophets and Jesus must be recovered for these times and courageously applied to a whole range of moral and political issues. If we make "prophetic faith" a public issue as it has been before at critical times in history, we might literally change the political wind on matters of great importance.

But protest is not enough anymore, and truly prophetic religion must always have better alternatives to offer. Those alternatives will not fit into the traditional categories of left and right, liberal and conservative, but could open up a whole new option for politics. And the real issue in "bringing God into politics" is not *whether* but *how;* and we will critique both the Republicans and the Democrats on how they deal with matters of faith.

Is fear the best foundation for foreign policy, or is there a better response, even a moral response, to terrorism? How do we judge matters of war and peace by theology and faith, and not just by politics? Is there a theology of empire emerging in America, and how dangerous is that? The prophet Micah's vision of how to find real security in this world is a clear alternative to Washington's proponents of Pax Americana. That wiser strategy understands how dramatic reductions in global poverty and effective responses to crises such as the HIV/AIDS pandemic are both morally required *and* crucial to our own security. And in those struggles, the world's religious communities could provide the "tipping point" for change. In an era aflame with war, the gospel vocation of peacemaking has never been more important. But peacemaking must now be turned into realistic strategies for practically resolving conflicts in a violent world. And perhaps the most important test case will be the Middle East.

Because the Scriptures spend so much time on the poor, we will too. This is the most important "political" issue in the Bible, and it must be ours as well. We will show how the poor are not only trapped in poverty, but in the ideological debate *about* poverty. How can they be set free? The prophet Isaiah's promise of just, secure, and healthy communities could be very appealing to people across the political spectrum. That prophetic vision reminds us that budgets are moral documents, revealing our true priorities, and must be judged morally, not just economically. What is the promise and danger of so-called faith-based initiatives, and how can religion function as the conscience of the state? We'll ask what the prophet Amos might think about our contemporary corporate scandals.

In societies becoming more and more multiracial, what can religious faith teach us about the vision and practice of reconciliation? Is there a consistent ethic upholding the sacredness of life that crosses political boundaries? And how do we really strengthen the family and community ties that bind us together and help us raise our children while rejecting the mean-spirited scapegoating that has proved so divisive?

Who can provide the critical leadership for all this? Well, as a dear departed colleague and incredible street organizer used to remind me, "We are the ones we have been waiting for."

This book will explore how people concerned about social change and hungry for spiritual values can actually combine those two quests. Too often politics and spirituality have been separated, polarized, and even put into competition with one another. We have been buffeted by private spiritualities that have no connection to public life and a secular politics showing disdain for religion or even spiritual concerns. That leaves spirituality without social consequences and a politics with no soul. And political discourse that is disconnected from moral values quickly degenerates. We will ask how we might change our public life with the values that many of us hold most dear. This book is about how to connect a genuinely "prophetic" spirituality to the urgent need for social justice. This is the connection the world is waiting for.

God Is Not a Republican. Or a Democrat.

> "It is the responsibility of every political conservative, every evangelical Christian, every pro-life Catholic, every traditional Jew, every Reagan Democrat, and everyone in between to get serious about re-electing President Bush."
>
> —Jerry Falwell, *The New York Times*, July 16, 2004

> "I think George Bush is going to win in a walk. I really believe I'm hearing from the Lord it's going to be like a blowout election in 2004. The Lord has just blessed him. . . . It doesn't make any difference what he does, good or bad."
>
> —Pat Robertson, *AP/Fox News*, January 2, 2004

These leaders of the Religious Right mistakenly claim that God has taken a side in this election, and that Christians should only vote for George W. Bush.

We believe that claims of divine appointment for the President, uncritical affirmation of his policies, and assertions that all Christians must vote for his re-election constitute bad theology and dangerous religion.

We believe that sincere Christians and other people of faith can choose to vote for President Bush or Senator Kerry—*for reasons deeply rooted in their faith.*

We believe all candidates should be examined by measuring their policies against the complete range of Christian ethics and values.

We will measure the candidates by whether they enhance human life, human dignity, and human rights; whether they strengthen family life and protect children; whether they promote racial reconciliation and support gender equality; whether they serve peace and social justice; and whether they advance the common good rather than only individual, national, and special interests.

We are not single-issue voters.

We believe that poverty—caring for the poor and vulnerable—is a religious issue. Do the candidates' budget and tax policies reward the rich or show compassion for poor families? Do their foreign policies include fair trade and debt cancellation for the poorest countries? (Matthew 25:35–40, Isaiah 10:1–2)

We believe that the environment—caring for God's earth—is a religious issue. Do the candidates' policies protect the creation or serve corporate interests that damage it? (Genesis 2:15, Psalm 24:1)

We believe that war—and our call to be peacemakers—is a religious issue. Do the candidates' policies pursue "wars of choice" or respect international law and cooperation in responding to real global threats? (Matthew 5:9)

We believe that truth-telling is a religious issue. Do the candidates tell the truth in justifying war and in other foreign and domestic policies? (John 8:32)

We believe that human rights—respecting the image of God in every person—is a religious issue. How do the candidates propose to change the attitudes and policies that led to the abuse and torture of Iraqi prisoners? (Genesis 1:27)

We believe that our response to terrorism is a religious issue. Do the candidates adopt the dangerous language of righteous empire in the war on terrorism and confuse the roles of God, church, and nation? Do the candidates see evil only in our enemies but never in our own policies? (Matthew 6:33, Proverbs 8:12–13)

We believe that a consistent ethic of human life is a religious issue. Do the candidates' positions on abortion, capital punishment, euthanasia, weapons of mass destruction, HIV/AIDS—and other pandemics—and genocide around the world obey the biblical injunction to choose life? (Deuteronomy 30:19)

We also admonish both parties and candidates to avoid the exploitation of religion or our congregations for partisan political purposes.

By signing this statement, we call Christians and other people of faith to a more thoughtful involvement in this election, rather than claiming God's endorsement of any candidate.

This is the meaning of responsible Christian citizenship.

PART I

Changing the Wind

CHAPTER 1

Take Back the Faith

Co-opted by the Right, Dismissed by the Left

MANY OF US FEEL that our faith has been stolen, and it's time to take it back. In particular, an enormous public misrepresentation of Christianity has taken place. And because of an almost uniform media misperception, many people around the world now think Christian faith stands for political commitments that are almost the opposite of its true meaning. How did the faith of Jesus come to be known as pro-rich, pro-war, and only pro-American? What has happened here? And how do we get back to a historic, biblical, and *genuinely* evangelical faith rescued from its contemporary distortions? That rescue operation is even more crucial today, in the face of a deepening social crisis that cries out for more prophetic religion.

Of course, nobody can steal your personal faith; that's between you and God. The problem is in the political arena, where strident voices claim to represent Christians when they clearly don't speak for *most* of us. It's time to take back our faith in the public square, especially in a time when a more authentic social witness is desperately needed.

The religious and political Right gets the public meaning of religion mostly wrong—preferring to focus only on sexual and cultural issues while ignoring the weightier matters of justice. And the secular Left doesn't seem to get the meaning and promise of faith for politics at all—mistakenly dismissing spirituality as irrelevant to social change. I actually happen to be conservative on issues of personal responsibility, the sacredness of human life, the reality of evil in our world, and the critical importance of individual character, parenting, and strong "family values." But the popular presentations of religion in our time (especially in the media) almost completely ignore the biblical vision of social justice and, even worse, dismiss such concerns as merely "left wing."

It is indeed time to take back our faith.

Take back our faith from whom? To be honest, the confusion comes from many sources. From religious right-wingers who claim to know God's political views on every issue, then ignore the subjects that God seems to care the most about. From pedophile priests and cover-up bishops who destroy lives and shame the church. From television preachers whose extravagant lifestyles and crass fund-raising tactics embarrass more Christians than they know. From liberal secularists who want to banish faith from public life and deny spiritual values to the soul of politics. And even from liberal theologians whose cultural conformity and creedal modernity serve to erode the foundations of historic biblical faith. From New Age philosophers who want to make Jesus into a nonthreatening spiritual guru. And from politicians who love to say how religious they are but utterly fail to apply the values of faith to their public leadership and political policies.

It's time to reassert and reclaim the gospel faith—especially in our public life. When we do, we discover that faith challenges the powers that be to do justice for the poor, instead of preaching a "prosperity gospel" and supporting politicians who further enrich the wealthy. We remember that faith hates violence and tries to reduce it and exerts a fundamental presumption against war, instead of justifying it in God's name. We see that faith creates community from racial, class, and gender divisions and prefers international community over nationalist religion, and we see that "God bless America" is found nowhere in the Bible. And we are reminded that faith regards matters such as the sacredness of life and family bonds as so important that they should never be used as ideological symbols or mere political pawns in partisan warfare.

The media like to say, "Oh, then you must be the religious Left." No, not at all, and the very question is the problem. Just because a religious Right has fashioned itself for political power in one utterly predictable ideological guise does not mean that those who question this political seduction must be their opposite political counterpart. The best public contribution of religion is precisely *not* to be ideologically predictable or a loyal partisan. To always raise the moral issues of human rights, for example, will challenge both left- and right-wing governments that put power above principles. Religious action is rooted in a much deeper place than "rights"—that place being the image of God in every human being.

Similarly, when the poor are defended on moral or religious grounds, it is certainly not "class warfare," as the rich often charge, but rather a direct response to the overwhelming focus on the poor in the Scriptures, which claim they are regularly neglected, exploited, and oppressed by wealthy elites, political rulers, and indifferent affluent populations. Those Scriptures don't simply endorse the social programs of the liberals or the conservatives, but they make it clear that poverty is indeed a religious issue, and the failure of political leaders to help uplift the poor will be judged a moral failing.

It is precisely because religion takes the problem of evil so seriously that it must always be suspicious of too much concentrated power—politically *and* economically—either in totalitarian regimes or in huge multinational corporations that now have more wealth and power than many governments. It is indeed our theology of evil that makes us strong proponents of both political and economic democracy—not because people are so good, but because they often are not and need clear safeguards and strong systems of checks and balances to avoid the dangerous accumulations of power and wealth.

It's why we doubt the goodness of *all* superpowers and the righteousness of empires in any era, *especially* when their claims of inspiration and success invoke theology and the name of God. Given the human tendencies of military and political power for self-delusion and deception, is it any wonder that hardly a religious body in the world regards the ethics of unilateral and preemptive war as "just"? Religious wisdom suggests that the more overwhelming the military might, the more dangerous its capacity for self- and public deception. If evil in this world is deeply human and very real, and religious people believe it is, it just doesn't make spiritual sense to suggest that the evil all lies "out there" with our adversaries and enemies, and none of it is "in here" with us—embedded in our own attitudes, behaviors, and policies. Powerful nations dangerously claim to "rid the world of evil" but often do enormous harm in their self-appointed vocation.

The loss of religion's prophetic vocation is terribly dangerous for any society. Who will uphold the dignity of economic and political outcasts? Who will question the self-righteousness of nations and their leaders? Who will question the recourse to violence and rush to wars, long before any last resort has been unequivocally proven? Who will not allow God's name to be used to simply justify ourselves, instead of calling us to accountability? And who will love the people enough to challenge their worst habits, coarser entertainments, and selfish neglects?

Prophetic religion always presses the question of the common good. Indeed, the question, "Whatever became of the common good?" must be a constant religious refrain directed to political partisans whose relentless quest for power and wealth makes them forget the "commonwealth" again and again. That common good should always be constructed from the deepest wells of our personal *and* social responsibility and the absolute insistence to never separate the two.

I am always amazed at the debate about poverty, with one side citing the need for changes in personal behaviors and the other for better social programs, as if the two were mutually exclusive. Obviously, both personal and social responsibility are necessary for overcoming poverty. When this absurd bifurcation is offered by ideological partisans on either side, I am quickly convinced that both sides must never have lived or worked anywhere near poverty or poor people. That there are behaviors that further entrench and even cause poverty is indisputable, as is the undeniable power of systems and structures to institutionalize injustice and oppression. Together, personal and social responsibility creates the common good. Because we know these realities as *religious* facts, taught to us by our sacred Scriptures, religious communities can teach them to those still searching more for blame than solutions to pressing social problems.

But recovering the faith of the biblical prophets and Jesus is not just about politics; it also shapes the way we live our personal and communal lives. How do we live a faith whose social manifestation is compassion and whose public expression is justice? And how do we raise our children by those values? That may be the most important battle of spiritual formation in our times, as I am personally discovering as a new father. Our religious congregations are not meant to be social organizations that merely reflect the wider culture's values, but dynamic countercultural communities whose purpose is to reshape both lives and societies. That realization perhaps has the most capacity to transform both religion and politics.

We contend today with both religious and secular fundamentalists, neither of whom must have their way. One group would impose the doctrines of a political theocracy on their fellow citizens, while the other would deprive the public square of needed moral and spiritual values often shaped by faith. In a political and media culture that squeezes everything into only two options of left and right, religious people must refuse the ideological categorization and actually build bridges between people of goodwill in both liberal and conservative camps. We must insist on the deep connections between spirituality and politics while defending the proper boundaries between church and state that protect religious and nonreligious minorities and keep us all safe from state-controlled religion. We can demonstrate our commitment to pluralistic democracy and support the rightful separation of church and state without segregating moral and spiritual values from our political life.

Neither religious nor secular fundamentalism can save us, but a new spiritual revival that ignites deep social conscience could transform our society. Movements do change history, and the strongest ones are those with a spiritual foundation. Most important of all is the spiritual power of hope, which may be the only thing that can finally overcome our too characteristic cynicism. Hope versus cynicism is the key moral and political choice of our time. This book is about the politics of hope.

The 2004 Election

Every week of the 2004 campaign, I did interviews with reporters who started the conversation by saying, "I am doing a story about religion and the election." And religion turned out to be one of the critical factors in the election.

That was demonstrated by a National Public Radio story in mid-September 2004 about a swing voter in West Virginia. Now in her seventies, this woman had voted in every election since she was twenty-one years old.

But this time, she felt more conflicted than ever. She told the reporter she thought the war in Iraq was a mistake and was turning into a real mess. "We shouldn't have gone to Iraq," she said. "I feel Bush took us into that." But, she said, she liked the way he talks about his Christianity and brings his faith into what he's doing. On the other hand, "We have lost so many jobs in West Virginia," she said, and that caused her to lean away from the president again. But then she said she was with him on gay marriage and abortion. Her conflicts exemplify both the policy and cultural issues that defined this campaign.

When reporters started talking about the religious issues of this election being abortion and gay marriage, I often corrected the narrow perception that reduces all Christian ethics and values to one or two hot-button social issues. I talked about how poverty, the environment, the war in Iraq, and our response to terrorism were also key religious and moral values questions that were important to people of faith. That wider perspective always made sense to the reporters, and their stories took on a broader view of the issues at stake.

The coverage of religion and politics has begun to change. Some journalists honestly admitted to me that they used to cover only what members of the religious Right had to say about politics, because they were the loudest voices; but this year the media saw more moderate and progressive religious voices having much more visibility and impact and, therefore, some members of the press wanted to present a more balanced coverage.

Sojourners' successful petition/ad campaign—"God is Not a Republican. Or a Democrat"—probably helped a lot. It suggested that endorsing political candidates is a fine thing, but ordaining them is not—the way that some leaders of the religious Right named George W. Bush as "God's candidate" in this election and proclaimed that real Christians could vote only for him. Just making it clear that good people of faith would be voting for both George Bush and John Kerry in this election for reasons deeply rooted in their faith was an important statement. It also directly challenged the single-issue voting that comes from shrinking all our religious and moral values down to only one or two issues, and said that all candidates should be examined by measuring their policies against the complete range of Christian ethics and values. On that wider and deeper list of religious and moral values were poverty, the environment, war, truth-telling, human rights, our response to terrorism, and a "consistent ethic of human life" that included abortion, but also capital punishment, euthanasia, weapons of mass destruction, HIV/AIDS and other pandemics, and genocide around the world. From the ad's August placement in the *New York Times* on the first day of the Republican National Convention to its appearance the weekend before the election in *USA Today,* with placement in more that fifty local city and college newspapers in between (often by local churches and student groups), the God is not a Republican ør a Democrat campaign created vital and rigorous dialogue across the country within the religious community and beyond and became a significant media discussion.

Differences in how the campaigns and candidates treated the issue of religion in the 2004 election year were very stark, with the Republican Convention appearing at times to be a "praise service," according to religion writer Amy Sullivan, especially before the prime-time television coverage. The Bush campaign's chief political strategist, Karl Rove, made no secret of his intent to reach out aggressively to conservative religious voters. But in doing so, the Bush re-election team seriously overstepped the proper boundaries of church and state by suggesting religious "duties" that included turning over congregational membership lists to local Republican parties. That offended even members of Bush's own religious base like Richard Land of the Southern Baptists, who said such partisan activities were "inappropriate" and that he was "appalled"—an honest and honorable criticism.[1] In mailings to churches in some states, the Republican National Committee suggested that liberals (read Democrats) would ban the Bible and accept gay marriage if they were to win.

I have never seen such outrageous behavior by a political party in trying to manipulate religion for its own agenda while so disrespecting the faith of millions of other believers who disagree with the Republican political agenda. What do such tactics say about the Republicans' respect for the black churches, when the African American vote was again almost 90 percent for the Democrats? Is something wrong with their faith? Do black churches ban the Bible? The Republicans virtually claim to own religion. And the Democrats still don't seem to know how to take back the faith.

If the Republicans overstepped in their religious outreach, the Democrats understepped in their effort to be more religion friendly than they have in the recent past. Listening more carefully to religious voices both inside and outside the Kerry campaign would have provided more strategic help and public capacity in speaking directly to the important issues of religion in politics and seeking to broaden their definition in this election campaign.

To be fair, the Democrats, both at their convention and in their campaign, did try to offer a new open door to the religious community in important ways, and Kerry began to talk about how his own faith influenced his values. But Kerry could have done much more to speak to religious audiences, talk to the religious press, and redefine the religious issues at stake in this campaign by moving away from just abortion and the Eucharist, and including poverty and war. The Democrats should be much more willing to use moral and religious language in defense of economic fairness and justice. But they shouldn't make the same mistake the Republicans have made in trying to co-opt religious leaders and communities for their political agenda. Nor should they suggest that religious people have an exclusive hold on the issue of morality, thus disrespecting millions of Americans who have deep moral concerns about the direction of their country but no religious affiliation. The issue here is not religiosity per se, but rather the moral compass a political leader or party brings to public life. Religion is often a critical factor creating that compass, and therefore is an appropriate campaign discussion, but faith is certainly not the only issue. But, as a frustrated Democrat lamented to me after the election, "My party is still afraid of the 'G' word."

The Democrats' fledgling attempts to reach out to the religious community and counter the now infamous "God Gap" between the two parties have been steps in the right direction. But with its loss in November, there is little question now that the Democratic Party should move much more deliberately to embrace religious communities and concerns, to use moral and religious language to argue for social reform, and to learn from the lessons of progressive religious movements in American history as they advance their agenda for the future. In large part, the desire to affirm progressive religion is coming from spiritually devout Democratic elected officials, who feel they have been religiously disrespected not only by Republicans but even by those from within their own political ranks.

What did progressive religion have to say about this election?

Religious and political conservatives often raise the issues of abortion and gay marriage. I have clearly disagreed with the Democrats on abortion, believing that Christians can be both progressive and pro-life. I've urged the Democrats to be much more respectful and welcoming of pro-life Democrats. Someday, a smart Democrat will figure out how both pro-choice and pro-life people could join together in concrete measures to dramatically reduce the abortion rate by focusing on teen pregnancy, adoption reform, and real support for low-income women. That would be so much better than both sides using the issue as a political football and political litmus test during elections, and then doing little about it afterward. I also have strongly affirmed the critical importance of strengthening marriage and family and of supporting parents in the most difficult and important task in our society, but have also supported gay civil rights and legal protection for same-sex couples.

If the Democrats could take the opportunity of a political defeat to really reassess their language and style, the way they morally frame public policy issues, and their cultural disconnect with too many Americans including many people of faith, they could transform the political discourse. But it will require a serious reassessment. And if they are further willing to re-examine their positions on some of the cultural/moral issues the Republicans beat them with in 2004, they could virtually change the political landscape. If the Democrats could be persuaded by both good political sense and sound moral values to moderate some of their positions by becoming anti-abortion without criminalizing an agonizing and desperate choice, and being pro-family without being anti-gay, they would change politics in America by giving permission to millions of voters who would naturally vote for them except for the cultural and moral divide they feel with Democratic language and policies.

But there were two issues in the 2004 election year that most tugged at my heart, worry my Christian conscience, and compel me to faithful citizenship and discipleship. The first is poverty, the second is war.

As the Bush administration began, I joined a small group of religious leaders to meet with the president-elect in Austin, Texas. To his credit, George W. Bush invited both those who had voted for him and against him. We encouraged him to commit himself to a concrete and measurable goal in the battle against poverty—such as cutting child poverty by half in ten years, as the

British Labour government under Tony Blair had pledged. I thought a Republican president, in the name of compassionate conservatism, could make new progress on the critical issue of poverty, much like Nixon's going to China. I told him he should surprise everybody with an aggressive antipoverty agenda. I supported the president's faith-based initiative, much to the chagrin of Democratic friends, but from the beginning of the Bush presidency many of us have had a very consistent message: significant resources must be committed to serious poverty reduction, not just in a faith-based initiative but especially in budget decisions, tax policies, and spending priorities.

Two years later, a statement organized by Call to Renewal and signed by thirty-four Christian leaders across the theological and political spectrum concluded, sadly, that the president had failed the test of resources and priorities, which made our continuing support for his faith-based initiative increasingly untenable. Without the resources and policies to seriously reduce poverty, the faith-based initiative became words without backing, faith without works. A faith-based initiative could have been done differently, with the resources and policies to back it up, but this one has turned out to be a big disappointment, with policy failures such as the denial of child tax credits to low-income families that would have brought the biblical prophets to the White House lawn.

Other priorities were just more important to the Bush administration than poverty reduction. Tax cuts that mostly benefited the wealthy were more important, the war in Iraq was more important, and homeland security was more important—all without the key recognition of how poverty, despair, family instability, and social disintegration undermine our national security. A budget based on a windfall of benefits for the wealthy and harsh cuts for poor families and children is an unbiblical budget. The good people who have run the White House faith-based office were clearly not the ones making policy and budget decisions for the Bush administration. One result of the lack of White House leadership has been the steady rise in the number of people, families, and children living in poverty in each of the last three years, according to the 2003 U.S. Census report. And that is a religious issue.

In his speech to the 2004 Republican Convention, the president spoke about many important issues—education reform and opportunity, health care security, job training, support for low-income families and neighborhoods. There were new and promising directions in his notion of "an ownership society," which focuses on things such as tax credits, educational equality, and home ownership for lower-income families as an alternative to relying on only entitlement programs.

In an August 2004 article in the *New York Times Magazine,* conservative writer David Brooks laid out a vision for "progressive Republicanism" that has a clear role for the positive action of government to make work actually work for low-income families, with a whole range of wage supplements and wealth creation for poor working families.[2] There were signs of such a vision in the Bush speech. But the president failed to deal with how his central domestic priority, making permanent his tax cuts that most benefit the wealthy, will simply not allow such positive government initiatives—because of a lack of resources. The Brooks vision will never be possible if Republicans stick to their characteristic anti-government ideology that is so reluctant to spend money to reduce poverty. George W. Bush has not changed that mentality, but rather submitted to it. Until it changes, the poor will continue to suffer.

From what I have seen and heard of George W. Bush (including in small meetings and personal conversations I've had with the president) I believe his faith to be both personal and real. And I also believe that he has a heart genuinely concerned for poor people. But I think the president is often guilty of bad theology. On the issue of poverty, George Bush believes in a God of charity, but not a God of justice. And after September 11, George Bush's theology became much worse and much more dangerous.

The heart and passion of President Bush's speech and of the whole Republican Convention was a ringing defense of the Bush administration's war on terrorism and the war in Iraq; and an attack on John Kerry as too weak, indecisive, and unfit to command, bolstered by the Swift Boat attack ads on John Kerry's Vietnam record.

In the furious August 2004 debate on that topic, the press eventually began to scrutinize the credibility and accuracy of those attacks on Kerry's military service (after the damage had already been done), but mostly stayed away from the most controversial question about Vietnam—whether the war was fundamentally and basically wrong, and characterized by the regular commission of war crimes. That's what the young and decorated naval officer John Kerry said when he testified to Congress after he came home from the war. I was a young antiwar organizer then. I still say today, thirty years later, that what Kerry said about Vietnam was true then and is still true now; and it was John Kerry's finest political hour.

But the election showed that our country is still polarized over Vietnam and is now polarized again over another war. There is no disagreement in America about the need to protect our families, our nation, and the world against terrorism and that this vicious and, yes, evil terrorist violence must be defeated. But whether that goal and our national security were advanced or whether they were seriously damaged by the war in Iraq is indeed the real and divisive question. Nobody was willing to take the word of a "madman," as the president has caricatured his war opponent, but many of us, including almost every major Christian body in the world, believed this "war of choice" to be unnecessary and unjust.

Even as an opponent of the war, I found the most moving part of Bush's convention speech to be the stories of his times with military families who had lost their precious loved ones. Those losses are heartbreaking for all of us (as the loss of Iraqi lives should be too). Spending time with those who have lost a brother, a son, a daughter, a mother or father, a husband or wife has been a heartbreaking experience for me too. But the most heartwrenching question is whether those deaths were, tragically, unnecessary.

Congress voted to give the president the authority to go to war, and Bush misused and abused that authority in why, when, and how he took America to war. Now we are in a real mess and George W. Bush should not be allowed to get away with the deceptions, incompetence, and consequences of this awful war. The war in Iraq was wrong from the start. A Christian president ignored the conviction of the vast majority of the world's Christians that the war in Iraq was wrong and that there is a better way—a more effective and morally consistent way to fight terrorism.

Bush's war in Iraq is the beginning of a long-term strategy of preemptive war and mostly unilateral American foreign policy that is both mistaken and terribly dangerous for the future. My older son is six years old, and my other son is almost two. If America's present course continues, they will be facing endless war in their lives, perhaps even still in Iraq and who knows where else. Many Americans believe there is a better way, including many people of faith.

The Political Problem of Jesus

I do lots of radio shows, especially talk radio—everything from conservative evangelical Christian and Catholic radio to black gospel radio, National Public Radio, drive-time talk radio, liberal Air America, and left-wing Pacifica radio. I was doing an interview with Air America one day when host Al Franken asked me about Jesus. He told me he was a Jew but not practicing or particularly devout, yet was sincerely puzzled over how some people could think Jesus could ever support tax cuts for the rich while the poor saw their meager resources slashed so dramatically. He just didn't get it. A lot of people don't.

I had lunch one day with the head of a major nonprofit housing provider for the poor, who put the question to me again, saying, "I am a reader of the New Testament, and I just don't understand how a right-wing economic agenda can be squared with the clear teachings of Jesus on wealth and poverty in Matthew, Mark, Luke, and John."

That's the problem with the economic and political agenda of the religious Right—most people know what Jesus said about these things, whether they are Christians or not. And the conformity of many conservative evangelical leaders to the political Right and its agenda that favors the wealthy over the poor and middle class just doesn't make any sense to them. They know that Jesus was not pro-rich, pro-war, and only pro-American, as we described at the beginning of this chapter. So why are so many conservative evangelicals oblivious to the teaching of Jesus, they wonder. Why do "family values" groups support the Republican right-wing economic agenda when it hurts so many low-income families? And how can some even claim that God is pro-war? Most people just don't get it, because they know that Jesus was on the side of the poor and the cause of peace. The politics of Jesus is a problem for the religious Right.

Jesus said, "Blessed are the poor," and opened up his own ministry by proclaiming, "The Spirit of the Lord is upon me, because he has anointed me to bring good news to the poor. He has sent me to proclaim release to the captives and recovery of sight to the blind, to let the oppressed go free, to proclaim the year of the Lord's favor" (which was a direct biblical reference to the Jubilee Year in the Hebrew Scriptures where, periodically, the debts of the poor were cancelled, slaves were set free, and land was redistributed for the sake of equity). People such as U2's lead singer, Bono, see the contemporary relevance of such Scripture for issues such as global poverty and the HIV/AIDS pandemic in Africa, and so do many of his young fans—so why don't others see it? In Matthew's twenty-fifth chapter, Jesus speaks of the hungry, the homeless, the stranger, prisoners, and the sick and promises he will challenge all his followers on the judgment day with these words, "As you have done to the least of these, you have done to me." James Forbes, the pastor of Riverside Church in New York City, concludes from that text that, "Nobody gets to heaven without a letter of reference from the poor!" How many of America's most famous television preachers could produce the letter?

In a world of violence and war, the words of Jesus, "Blessed are the peacemakers, for they shall be called the children of God," are not only challenging, they are daunting. The hardest saying of Jesus and perhaps the most controversial in our post–September 11 world must be: "Love your enemies, pray for those who persecute you." Let's be honest: How many churches in the United States have heard sermons preached from either of these Jesus texts in the years since America was viciously attacked on that world-changing September morning in 2001? Shouldn't we at least have a debate about what the words of Jesus mean in the new world of terrorist threats and pre-emptive wars?

Jesus knows no national boundaries or national preferences. The body of Christ is an international one, and the allegiance of Christians to the church must always supersede their national identities. The words of Jesus stand as a virtual roadblock to any nation's pretension to rationalize and sanctify the preference for war. Jesus's instruction to be peacemakers leads either to nonviolent alternatives to war or, at least, a rigorous application of the church principles of just war.

Christ commands us to not only see the splinter in our adversary's eye but also the beams in our own, which often obstruct our own vision. To name the face of evil in the brutality of terrorist attacks is good theology, but to say they are evil and we are good is bad theology that can lead to dangerous foreign policy. Self-reflection should provide no excuses for terrorist violence, but it is crucial to defeating the terrorists' agenda. Christ instructs us to love our enemies, which does not mean a submission to their hostile agendas or domination, but does mean treating them as human beings also created in the image of God and respecting their human rights as adversaries and even as prisoners. And Christ calls us to confession and humility, which does not allow us to say that if persons and nations are not in support of all of our policies, they must be with the evildoers.

The words of Jesus are either authoritative for Christians, or they are not. And they are not set aside by the very real threats of terrorism. They do not easily lend themselves to the missions of nation-states that would usurp the prerogatives of God. The threat of terrorism does not overturn Christian ethics.

Also at issue here is the politics of fear. Jesus says, "Be not afraid," an attitude that could undermine the entire basis of our current foreign policy. Effective campaigns of fear easily co-opt anxious people—believers and unbelievers alike—and could lead our nation and our world to decades of pre-emptive, unilateral, and virtually endless war, despite the clear warnings of Jesus's own words.

The issue here is not partisan politics, and there are no easy political solutions. The governing party has increasingly struck a religious tone in an aggressive foreign policy that seems much more nationalist than Christian, while the opposition party has offered more confusion than clarity. In any election we choose between very imperfect choices. Yet it is always important to examine what is at stake prayerfully and theologically.

This examination among evangelicals became clear in the 2004 Evangelical Call to Civic Responsibility, an unprecedented call to social action from the National Association of Evangelicals. In contrast to the Jerry Falwell and Pat Robertson era, evangelicals are now showing moral leadership in the fight against global poverty, HIV/AIDS, human trafficking, and sustainability of God's earth.

These changes represent both a reaction against overt partisanship and a desire to apply Christian ethics to a broader set of issues. Many people of faith have grown weary of the religious Right's attempts to narrow the moral litmus test to abortion and gay marriage. For example, when likely voters were asked in a 2004 poll whether they would rather hear a candidate's position on poverty or on gay marriage, 75 percent chose poverty.

Only 17 percent chose gay marriage.[3] Any serious reading of the Bible points toward poverty as a religious issue, and candidates should always be asked by Christian voters how they will treat "the least of these." Stewardship of God's earth is clearly a question of Christian ethics. Truth telling is also a religious issue that should be applied to a candidate's rationales for war, tax cuts, or any other policy, as is humility in avoiding the language of "righteous empire," which too easily confuses the roles of God, church, and nation.

War, of course, is also a deeply theological matter. The near unanimous opinion of religious leaders worldwide that the Iraq war failed to fit "just war" criteria is an issue for many Christians, especially as the warnings from religious leaders have proved prophetically and tragically accurate. The "plagues of war," as the pope has referred to the continuing problems in Iraq, are in part a consequence of a "Christian president" simply not listening to the counsel of religious leaders who tried to speak to the White House. What has happened to the "consistent ethic of life," suggested by Catholic social teaching, which speaks against abortion, capital punishment, poverty, war, and a range of human rights abuses too often selectively respected by pro-life advocates?

The religious Right's grip on public debates about values has been driven in part by a media that continues to give airtime to the loudest religious voices, rather than the most representative, leaving millions of Christians and other people of faith without a say in the values debate. But this is starting to change as progressive and prophetic faith voices are speaking out with a confidence and moral urgency not seen for twenty-five years. Mobilized by human suffering in many places, groups motivated by religious social conscience (including many evangelicals not defined by the religious Right) have hit a new stride in efforts to combat poverty, destructive wars, human rights violations, pandemics like HIV/AIDS, and genocide in places like Sudan.

In politics, the best interest of the country is served when the prophetic voice of religion is heard—challenging both Right and Left from consistent moral ground. The evangelical Christians of the nineteenth century combined revivalism with social reform and helped lead movements for abolition and women's suffrage—not to mention the faith-based movement that directly preceded the rise of the religious Right, namely the American civil rights movement led by the black churches.

The truth is that most of the important movements for social change in America have been fueled by religion—progressive religion. The stark moral challenges of our time have once again begun to awaken this prophetic tradition. As the religious Right loses influence, nothing could be better for the health of both church and society than a return of the moral center that anchors our nation in a common humanity. If you listen, these voices can be heard rising again.

CHAPTER 2

A Lack of Vision

Too Narrow or None at All

IT WAS ONE OF THOSE warm spring days in the nation's capital when the fresh promise of new possibilities seems, just for a moment, to defy the entrenched ways of Washington. Surrounded by the impressive vista of monuments and museums on the Mall, I stood behind a rough lectern on a makeshift stage, looking into the eyes of one thousand low-income people—mostly single mothers who had been on welfare. My job was to speak, and my topic was hope. In a city where the currency is power, these poor Americans seemed a bit out of place. Not used to having much clout in their political system, you could tell they were feeling the energy that comes from just being together. They had come on buses from urban and rural communities to lobby the Congress for a new welfare reform bill—one that would effectively help people like themselves escape poverty and move to self-sufficiency. I told them a story.

I remember another group of people who wanted to change things meeting in a high mountain town in Mexico, two thousand miles from Washington, D.C. Two hundred fifty Christian leaders from fifty countries (mostly from the Southern Hemisphere) were gathered for a whole week to ask how they could learn to do a new kind of "advocacy." Having spent years doing service to the poor in their own countries, and now engaged in effective community development projects, they still saw the poor losing ground. So they had come from Latin America, Africa, and Asia to ask how they together might help change the rules of global trade and transform international economic practices enough to give poor countries and their people a fair chance to break the bonds of misery and deprivation. I told them the same story.

I've also told the story at Harvard University, where I teach part-time. In my class at the Kennedy School of Government, the students wrestled with the question of what to do with their lives after graduation—trying to sort out the differences between career and vocation. Harvard graduate students are being groomed to run the systems of power that the poor mothers in Washington and the Christian leaders from the global south want to change. But they too are looking for reasons to hope that some transformation might be possible.

I've told the same story at countless public gatherings and town meetings in hundreds of communities across the length and breadth of the United States, where people of faith, conscience, goodwill, and fragile hopes want to make a difference but are searching for ways to do it. Maybe you are like some of the people I've talked to.

Here's the story.

I urged the moms on the Mall not to waste any valuable time while they were in Washington. I wanted them to be able to quickly recognize the members of Congress whom they had come to see. They're the ones, I told them, who walk around town with their fingers held high in the air, having just licked them and put them up to see which way the wind is blowing. It's quite a sight—men and women walking all around the Capitol grounds with their wet index fingers pointed at the sky. The political leaders are really very good at figuring out the direction of the wind, and are quite used to quickly moving in that direction.

It's not a matter of malice for most of them. I've met quite a few politicians, and in fact many came to Washington because they truly wanted to do the right thing. But after a while, they get entrenched in Washington's ways, and change seems ever more distant. Power and wealth are the real governors here, and people adjust to those realities. Even the ones who still really want to make a difference will tell you they can't without public backing, and they don't often find it.

Many of us believe that by replacing one wet-fingered politician with another, we can change our society. But it never really works, and when it doesn't we get disillusioned. We then get tempted to just grumble, withdraw, or give up altogether on ever changing anything. But that's where we make our mistake.

The great practitioners of real social change, like Martin Luther King Jr. and Mahatma Gandhi, understood something very important. They knew that you don't change a society by merely replacing one wet-fingered politician with another. You change a society by *changing the wind*.

Change the wind, transform the debate, recast the discussion, alter the context in which political decisions are being made, and you will change the outcomes. Move the conversation around a crucial issue to a whole new place, and you will open up possibilities for change never dreamed of before. And you will be surprised at how fast the politicians adjust to the change in the wind.

Then I gave them a historical example.

Dr. Martin Luther King Jr. had just won the Nobel Peace Prize and was ready to come home from Norway. The freedom movement had achieved a great victory in securing the Civil Rights Act of 1964, and King was honored as the newest Nobel laureate. But the civil rights leader decided to stop by Washington, D.C., even before heading back home to Atlanta—because he needed to meet with the president of the United States, Lyndon Baines Johnson.

King told Johnson that the next step on the road to freedom was a voting rights act, without which black Americans in the South would never be able to really change their communities. But the nation's master of realpolitik told America's moral leader that he couldn't deliver a voting rights act. Johnson said he had cashed in all his "chits" with the southern senators to get the civil rights law passed and that he had no political capital left. It would be five or ten years, the president told King, before a voting rights act would be politically possible. But we can't wait that long, said King. Without voting rights, civil rights couldn't be fully realized. I'm sorry, Johnson reportedly told King, but a voting rights law just wasn't politically realistic. They would have to wait.

But Martin Luther King Jr. was not one to simply complain, withdraw, or give up. Instead, King and the Southern Christian Leadership Conference (SCLC) began organizing—in a sleepy little town nobody had ever heard of, called Selma, Alabama.

On one fateful day, SCLC leaders marched right across the Edmond Pettus Bridge, alongside the people of Selma, to face the notorious Sheriff Jim Clark and his virtual army of angry white police. On what would be called Bloody Sunday, a young man (and now congressman from Atlanta) named John Lewis was beaten almost to death, and many others were injured or jailed.

Two weeks later, in response to that brutal event, hundreds of clergy from all across the nation and from every denomination came to Selma and joined in the Selma to Montgomery march. Rabbi Abraham Joshua Heschel came down from New York to march beside the black Baptist minister Martin Luther King Jr., as did more ministers from around the country than had ever before come out to support the civil rights struggle—which, for them, had also become a religious one.

The whole nation was watching. The eyes of America were focused on Selma, as they had been on Birmingham before the civil rights law was passed. And after the historic Selma to Montgomery march for freedom, it took only *five months*, not five years or ten, to pass a new voting rights act: the Voting Rights Act of 1965. King had changed the wind.

I remember a panel discussion, many years later, where a famed presidential historian proclaimed, "And Lyndon Johnson, in a dramatic act of presidential courage, went to a joint session of Congress to call for a voting rights act!" I said it was a great thing that Johnson had responded to the challenge as he did (other presidents might not have), but that it was King, not Johnson, who had painted a vivid picture for the world to see that changed the winds of public opinion and made a voting rights act now possible. The Selma campaign had transfixed the nation, dramatically shifted the public debate, and fundamentally altered the political context to make a new voting rights law politically realistic.

I think that's what people of faith and conscience are supposed to be: "wind changers." People motivated by spiritual values that give them a real vision for change are not like those with their fingers up in the air. They already know the direction to head in, and they lead by example. Their commitments, skills, sacrifices, creativity, and, ultimately, moral authority are what make all the difference and change the wind. They also have the capacity to gather people around the vision, and in terms that places like Washington can understand, they demonstrate that they have a critical constituency. But what they really demonstrate is a fundamental principle—that *history is most changed by social movements with a spiritual foundation.* Look at the social movements that have made the most difference for social justice, and that is what you find.

I told the low-income parents that only by uniting together and enlisting the moral authority of the churches could they and we change the wind in the American poverty debate. I told the evangelical and Pentecostal Christian leaders from Latin America, Africa, and Asia that they had both the faith and the critical constituencies to alter the discussions of global trade and economic policy—not by mere lobbying, but by changing the wind. I told the Harvard students that they could either run the systems of power, as expected of the best and the brightest, or they could really test themselves by offering their best energy and gifts to the causes that others deem hopeless but that could become successful initiatives. And I tell audiences around the nation and the world, virtually every week, that a real "faith-based initiative" is much more than providing social services: it is, rather, becoming the prophetic voice and force that can and has moved whole nations to turn in a different direction.

But to change the wind, you have to know what direction you want it to move in. And that is about *vision.* Today, both religion and politics have a vision problem. And changing that will be the substance of this book. So what is the vision?

Write the Vision

An ancient biblical proverb well sums up our political life at the beginning of the twenty-first century, "Without a vision, the people perish." The writer of this Scripture understood that when our public life falters or fails, it is usually for a lack of vision. In our own contemporary society, the "vision problem" is identified in public opinion polls, year after year, as the central reason why American politics seems to lack both principle and purpose and why our political discourse produces lots of blame but not many solutions.

Without real solutions to pressing and increasingly urgent social problems, the people are indeed "perishing," both literally and spiritually. And the painfully obvious reality is that those who perish first are usually the youngest and the most vulnerable. Powerful interests argue, point fingers at one another, and vie for greater position and influence, while far less powerful people suffer and are forgotten. This political pattern has become deeply entrenched, and those left out and behind are always the ones whom Jesus referred to as "the least of these."

But really, we all suffer when there is no vision, no guiding moral compass that steers our public life. We become bereft of meaning and purpose in our social relationships, we lose all sense of the common good or our shared humanity, and the bonds of society themselves become so frayed that each individual feels forced to just fend for themselves. The poorest and weakest among us who are the least able to do that pay the greatest price; but we all are diminished when our social life is reduced to the survival of the fittest.

Another reading of the Proverbs text makes an even deeper point. The New Revised Standard Version of the Bible translates this famous text as, "When there is no prophecy, the people cast off restraint." Perhaps that prophetic insight explains why a lack of vision in a society contributes directly to social unrest, lawlessness, violence, and chaos. It may not just be poverty that leads to social breakdown, but also the absence of any compelling and credible vision, articulated by public leaders and accepted by the people, that serves to hold a society together.

We have seen societies where many people are still poor, the new South Africa for example, where between 45 and 55 percent of the population and between 57 percent and 75 percent of children are living in poverty. But a widely shared trust in the vision for the future, articulated by leaders like Nelson Mandela and Desmond Tutu, enables people to make common sacrifices toward clear goals and even to be patient in the sufferings of the moment. We have also witnessed societies with less suffering, but much more discontent, where trust has been broken or expectations raised only to be dashed.

Vision, therefore, is vitally important. And vision leads directly to values. What is politics for? This is the critical question seldom asked. What is the purpose of our public life, its meaning, its shaping and guiding principles? Where do we want to go, and why? What do we want to achieve? And, most important, what is a good society? Those are all questions about values, and *values will be the most important political question of the twenty-first century*.

Again, the "values problem" has emerged as a deeply felt public concern in survey after survey of political attitudes over the last few decades. Special interests, political ideology, and the naked quest for power continue to control our public discourse and political decision making. But the public alienation caused by the absence of core moral values in our political life seems to grow each year. Issues such as the strength of family life, the meaning of work, the health of neighborhoods, the well-being of children and the shame of child poverty, the moral tone of "entertainment," the truth telling of public officials, the quality and moral content of education, the equity of health care, the stewardship of the environment, and the consistency of foreign policy with expressed national values are all deeply held moral concerns at the heart of contemporary political issues. But many politicians still don't speak the language of ethics and values that could strike responsive chords among many people hungry for a moral political discourse. For example, "the sacred value of human life" is deeply shared by many people (even of diverse political persuasions) yet seldom consistently applied across a wide spectrum of political issues.

The lack of vision in public life and the emptying out of values that visionless leadership creates lead to a politics of complaint. In reaction to politics without values people begin to complain—and there is much to complain about. Moral cohesion unravels, social values crumble, public policies lose their connection to the common good, families lose stability, neighborhoods lose community, leadership loses integrity, poor families and children begin losing everything—and complaint becomes our dominant political discourse.

The biblical prophet Habakkuk described the politics of complaint. He surveyed his own society and saw only violence, destruction, injustice, and deceit and began complaining to God. "O Lord, how long shall I cry for help, and you will not listen? Or cry to you 'Violence!' and you will not save? Why do you make me see wrongdoing and look at trouble? Destruction and violence are before me; strife and contention arise. So the law becomes slack and justice never prevails. The wicked surround the righteous—therefore judgment comes forth perverted" (Habakkuk 1:2–4).

Habakkuk had much to complain about. He lived in a time when Israel was caught between the warring armies of Babylon and Egypt, soon to be captured and destroyed and its people taken into exile. His description of his society reads like the Sunday *New York Times*. His critique is as contemporary as the evening news. He spoke of a situation in which people "heap up what is not your own," where "you get evil gain for your houses," while "the law becomes slack and justice never prevails."

This prophet was a dissenter, a critic, a voice of opposition. You might say that Habakkuk was a liberal or a conservative, whichever one happens to be outside of power at the moment. You know the kind of alarmed critic I mean. They're the ones who are always smacking themselves on the forehead, saying things like, "Oh no! Did you see what the president said now! Unbelievable! Did you see what the Congress did yesterday! Oh my God! Did you see what just happened?" And on, and on. You can tell who the real liberal or conservative critics are. They're the ones walking around with the big red marks on their foreheads! Habakkuk was one of these.

Eventually, the disillusioned and disheartened prophet climbed up into a watchtower to "keep watch" on his confused society and to see what God might say to him—to see "what he will answer concerning my complaint." The "answer" the prophet Habakkuk received is as timely as if it were given today in response to our modern complaints.

The text says, "Then the Lord answered me and said, 'Write the vision; make it plain upon tablets, so he may run who reads it. For still the vision awaits its time; it hastens to the end—it will not lie. If it seems slow, wait for it; it will surely come, it will not delay.' "

The vision is there. It's coming. Indeed, it's already here. The vision waits for us to see it and put it into practice. But what is the vision?

I believe the place to look for that new vision is a very old place—the humble messengers of ancient days who talked about the politics of God. Whether you are religious or not (that isn't the point, really), you'll be drawn to the prophets if you think that values should be the center of politics. Do you long for a politics as if values mattered? This book is about a new vision for the common good that could inspire us all to lives of service and to a whole new set of public (read political) priorities.

Habakkuk says the vision must be written and made clear so that "he or she may run who reads it." I am meeting a whole new generation of "runners" who are ready to go. Many of them are young people that I talk with across the country and around the world. I've spoken on at least fifty university campuses in the last two years and to tens of thousands of youths at summer festivals. They long for a vision to run with, to invest their lives in, to guide their energy, gifts, and passion. And I find runners in every age group now, eager to put their best personal values into the practice of public life. Some are religious, some are not, but most of them hunger for a real spirituality that can undergird their quest for social change. We dare not let them down by offering no vision worth running for.

The "vision" we will put forward in this book for our contemporary society is simply the content of what the Old Testament prophets, Jesus, and the New Testament writers had to say—about our public commitments, our common life, and the social bonds we share in community. The ancient biblical texts are given much lip service in America, but their social and political meaning receives little serious adherence or application. But the vision is always there, waiting for us to grasp, embrace, and implement it.

The vision is also in our own American traditions, in which democratic ideals have prevailed over their many obstacles. And when there has been a clear connection between our prophetic religious traditions and our democratic hopes, the vision becomes most clear—as, for example, when Martin Luther King Jr. changed America with a Bible in one hand and the Constitution in the other.

The vision is there and merely awaits us. When we move toward our prophetic and democratic visions, slaveries are ended, civil rights achieved, freedom established, compassion implemented, justice advanced, human rights defended, and peace made. When we neglect the vision, greed triumphs, selfishness erodes common life, our divisions increase, our weapons expand, and our conflicts proliferate.

When we come closer to the vision, our practice of citizenship is always enlivened; when we move away from it, apathy and withdrawal grow like a cancer in the body politic. Perhaps most importantly, when we embrace our best spiritual and political visions, the renewal of hope is the result. When we forget the moral visions that nourish our public life, cynicism abounds.

Our faithfulness to the best of our prophetic and democratic visions seems to ebb and flow. Today, our fidelity to those visions is at one of its lowest ebbs. But the vision still awaits, and it's time to embrace it again for the sake of our personal and political renewal. Citizenship will not be revitalized merely by improved democratic techniques; it requires a return to the visions with the capacity to reform our public life. That is now our task.

Today we face two related vision problems. One is the lack of vision in public life, as we have already described. But the other is when political leaders have a clear vision—but the wrong one. When politics is being shaped by visions that defend wealth and power, rather than opening up more opportunity; that are more exclusionary than inclusive; that pursue policies that destabilize families and communities; that exalt private interests over the common good; that simply leave too many people behind; that seek national or corporate self-interest over international peace and justice; or that increase conflict rather than reducing it—then such political vision can be as destructive as having no vision at all. It seems that we are afflicted by no vision, on the one hand, and narrow vision on the other. Neither will suffice, neither has the capacity to meet the challenges of our time. And neither is faithful to the compelling public visions contained in both our best religious and democratic traditions.

This book takes up the most important public questions of our time and applies to them the best wisdom of both our prophetic and democratic visions. From family values to foreign policy, from the housing we dwell in to the social values that dwell within us, from health care to healing of our national fears and divisions, from the distribution of our resources to determining the things we value most, from the things that make for peace on a global level to the community level, from our definitions of justice to our practice of it, from what we'd like to change to what gives us hope for ever changing it. Whenever we deal with social and economic decisions and policies, we will always ask what I call the "God question," which is, "How are the kids doing?" What happens to the children, our own and everybody else's, is always a question that illuminates all the others.

Such vision could "change the wind." It could change how we think and feel about our public lives, at a deeper level than the shallow spectator sport that our politics has become. The direction of the wind is a metaphor both for public opinion and the spirit of the age. We'll look at both, because these are spiritual questions at their root. *Ruah* is the word for the wind of God in the Old Testament. Where is that wind blowing today? And how do we read the signs of our times? Why do we seem to be stuck between failed ideological options and unable to find solutions to our most pressing social problems? And how can we move the body politic in better directions? As we consider each major issue in the book, we will ask what the key "wind-changing" shift would be in the ways we approach and deal with the question.

I am convinced that we can make key "wind changing" or paradigm shifts on a wide range of issues. We can create new ways of looking at and talking about crucial questions that could significantly alter the framework and spirit of the current debates, which have deadlocked the public discussion and blocked solutions to some of our most serious problems. In many cases, it means a commitment to stop making false choices and inviting the critical insights from diverse political traditions—many of which are necessary to move us forward beyond mere blaming and posturing. And new ways of thinking lead to new ways of acting.

We are now politically adrift, apart from those best values, and our public life is a bankrupt battlefield of competing special interests without the framework of moral discourse. We will seek to contribute to that discourse by raising the public implications of the spiritual values we often claim to believe but then proceed to ignore when it comes to politics. That practice can and must change, for the sake of our life together, for the sake of our children, and for the sake of our own integrity.

CHAPTER 3

Is There a Politics of God?

God Is Personal, but Never Private

God is personal, but never private. And the Bible reveals a very public God. But in an age of private spiritualities, the voice of a public God can scarcely be heard. Private religion avoids the public consequences of faith. In particular, affluent countries and churches breed private disciples, perhaps because the applications of faith to public life could become quickly challenging and troubling. Can the devotees of private religion even understand the politics of God?

So is there a politics of God? Many of our politicians give a lot of lip service to God these days, but do they really understand the public implications of belief in God? We don't typically hear much about the politics of God, even from our pulpits. Powerful forces would keep God private, or under control, or as an endorser of ideological agendas, or of the political status quo, or (worse yet) of fund-raising activities. Today, religion usually serves more to silence the politics of God than to announce it to the nations.

In the face of suicide bombers who utter the name of God with easy blasphemy, television evangelists who claim God has made them (and could make you) prosperous, and those who use religious institutions to hide hideous crimes, many search today, almost desperately, for true religion and good faith.

Dare we search for the politics of God? It's much easier to just use God to justify our politics. Yet, if we look, really look into our biblical and other holy texts, we find a God who speaks about "politics" all the time, about what believing in God means in this world (not just the next one), about faith and "public life" (not just private piety), about our responsibilities for the common good (not just for our own religious experience). And here's the big news: the politics of God calls all the rest of our politics into question.

The place to begin to understand the politics of God is with the prophets, the ancient moral articulators in the Scriptures who claimed to speak in "the name of the Lord." What were their subjects? Quite secular topics really—land, labor, capital, wages, debt, taxes, equity, fairness, courts, prisons, immigrants, other races and peoples, economic divisions, social justice, war, and peace—the stuff of politics.

Whom were the prophets often speaking to? Usually to rulers, kings, judges, employers, landlords, owners of property and wealth, and even religious leaders. They spoke to "the nations," and it was the powerful who were most often the prophets' target audience; those in charge of things were the ones called to greatest accountability. And whom were the prophets usually speaking for? Most often, the dispossessed, widows and orphans (read: poor single moms), the hungry, the homeless, the helpless, the least, last, and lost. Is God into class warfare? No, God wants the "common good," but speaking for the common good can get one accused of calling for class warfare—usually by the elites who control the political discussion and do not want too much conversation about what God thinks of our political priorities.

But don't take my word for it. Just look at what the prophets say about the substance (or lack thereof) of our politics today. That, quite simply, is the subject of this book. Clearly the politics of God is different than ours—from the Republicans and the Democrats, the liberals and the conservatives, the Left and the Right. The politics of God makes them all look pretty bad and points the way to some very different directions—but some very hopeful ones. Again, the famous biblical proverb says, "Without a vision, the people perish," and this is exactly what's happening to us right now. It's pretty obvious to see that it's time for new vision, and people across the political spectrum seem to agree with that. For many years, poll after poll has reported that many think we are headed in the wrong direction.

The conflict over a private versus public God came early for me. I was born into an evangelical Christian family, raised in the American Midwest, and reared on Republican patriotism. We supported Eisenhower and not Stevenson for president, and I remember taking an early interest in history and politics. My father married his high school sweetheart on the same day he graduated from college and was commissioned as an officer in the U.S. Navy, which was eager to send him and his classmates into the Pacific theater of World War II. My Dad believed the atomic bomb might have saved his life as his destroyer/minesweeper was slated for the invasion of the Japanese islands.

When he returned home to Detroit, my mother and father started their family, and I was the first of five children. We soon moved into a new home, financed by the newly established Federal Housing Administration (FHA) for returning servicemen and their new families—just like us. Our first home was on Riverpark, in Redford Township, and it seemed like all our neighbors were indeed just like us. That was the primary social reality that eventually broke apart my early view of the world.

The thing that made us different than some was a strong Christian commitment that led my parents to help found a new church in the neighborhood. My father was an engineer for Detroit Edison but also a pastor in the Plymouth Brethren Church, which had no clergy but all lay leaders. I remember him getting up most mornings as early as 5:00 a.m. to study his Bible until the family got up at 7:00. My mother was also a gifted leader with natural pastoral skills, but the conservative church allowed no public role for women, who faithfully kept things together behind the scenes.

It was my mother's admonitions that became the first building blocks for my emerging social conscience. Phyllis Wallis told her children to always do two things. First, if there is a child that nobody else will play with, you play with him or her, she ordered. It was like a rule of play for us—nobody gets left out. Second, if there is a bully who is picking on other people, you stand up for them, she courageously commanded. Virtually all my fights in school were with bullies. She also seemed to tie these moral injunctions to our faith: It was what Jesus would have us do. Never do I remember either of my parents applying these life lessons to any political issue, and when I later tried to do that, they didn't understand—at least at first. But their values were clear social commitments nonetheless.

An incident in my church left an indelible impression about the private versus public meaning of faith and permanently altered my own life's direction. My church was all white, as was my neighborhood, my school, and my whole world. I began to notice this homogeneity as a young teenager, and for some reason, it bothered me. In other books, I've recounted the story of my early conversion to Christianity at the age of six years and how later teenage questions about the realities of race in Detroit led to a "second conversion" and, indeed, led me out of the church. I was now regularly making the pilgrimage into inner-city Detroit, taking jobs in the city alongside young black workers my own age, and visiting black churches.

The incident came one day back in my home church when I was arguing with one of the elders about racism and the things I was learning in inner-city Detroit. I will never forget what he said to me, "Christianity has nothing to do with racism; that is a political issue, and our faith is personal." I knew that the questions that were tearing at my heart were not going to go away, and if Christianity had nothing to do with them, I wanted nothing to do with Christianity. I left my childhood faith behind (and many in the church were glad to see me go) and found my new home in the civil rights and student movements of my generation. My return to Christian faith after years of political organizing is also described elsewhere, along with the founding of *Sojourners* magazine and community.

But there was a deep lesson learned in my banishment and return to faith. And it would become the foundation for everything I would do thereafter. The lesson is this: *God is personal, but never private.* If God is not personal, there is little meaning to faith. It merely becomes a philosophy or a set of teachings from religious figures who died long ago. Without a personal God, there is no personal dimension to belief. There is no relationship to God, no redemption, salvation, grace, or forgiveness. There is no spiritual transformation without a personal God, and no power that can really change our lives beyond mere self-improvement. In today's world, there is one overriding and key distinction in all of the religion that is growing—a God who desires relationship with each person. Much of liberal religion has lost the experience of a personal God, and that is the primary reason why liberal Christianity is not growing. And without a personal God, liberal faith will never grow.

However, that personal God is never private. Restricting God to private space was the great heresy of twentieth-century American evangelicalism. Denying the public God is a denial of biblical faith itself, a rejection of the prophets, the apostles, and Jesus himself. Exclusively private faith degenerates into a narrow religion, excessively preoccupied with individual and sexual morality while almost oblivious to the biblical demands for public justice. In the end, private faith becomes a merely cultural religion providing the assurance of righteousness for people *just like us*.

Such righteousness can quickly become self-righteousness, as I learned in my own church tradition. One of the things I am proudest of my father for occurred long after I left my home church. A great crisis arose over a sexual transgression (most of the big ethical issues in our church had to do with sex—a consequence of an overly narrow private God). Two teenagers in the church had got themselves into trouble. The girl was pregnant, and both she and her boyfriend were being ostracized by their Christian community. The elders convened an emergency meeting to decide what to do. "Bring them up before the whole church, and we will denounce their behavior!" said most of the church's spiritual leaders. My father, who was the chief elder, objected to such harsh treatment. Why would such a thing be necessary? he asked. People need to know where we stand, the elders declared. Do you think anybody doesn't know where we stand? my father replied. These two people need to know how wrong this was, asserted another elder. Does anyone here doubt that these kids know they've made a terrible mistake and what a mess they're in? answered my Dad.

The argument continued, and all the other elders were insistent on a very public rebuke of the young people. Okay, okay, my father agreed, we'll do it. Really? his surprised colleagues asked. Sure, he said. Let's bring them up in front of the whole church and declare their sins before everyone. Then let's bring everybody else up too, one at a time, and declare all our sins. And since I've done so much counseling with so many of you and the families of the church, I'll make sure nothing gets left out. At that, the plans for confronting the young couple were dropped, and a discussion followed about how to support them in their perilous situation. A purely sexual transgression of boundaries (and an important one if one takes the real problem of out-of-wedlock births seriously) became also a matter of community equity and compassion. The personal sin was opened up to a social context. The private was held accountable to public ethics. The event reminded me of a certain gospel story about a woman about to be punished for adultery and Jesus's suggestion that those without sin throw the first stones.

Most of the biblical prophets (whom we pass over week after week in our congregations) would offer a quite searing indictment of contemporary American society. Specifically, that we have become a nation of endangered souls and that our society and politics are governed by values quite foreign to the heart of our religious traditions, no matter how religious we like to think we are. Whether conservative or liberal Christians, or members of other faith groups, or just spiritual seekers, we are all guilty of succumbing to a diminished religiosity that is characterized by privatized belief systems, devoid of the prophetic and social witness of Jesus and the prophets—ultimately, nothing more than "small-s" spirituality that is really only ad hoc wish fulfillment or a collection of little self-help techniques we use to take the edge off our materialistic rat-race lives.

What is needed is nothing less than a renovation of our souls and the soul of our politics. We are all—Right and Left—pursuing our innocuous spiritualities ("following our bliss," as the saying goes) as if religion meant nothing to our life together. Whether attending our Zen/Christian retreats as "progressives" or our "seeker-sensitive" megachurches as "conservatives," we've abandoned the heart of "capital R" Religion. In essence this book is arguing a very evangelical point: our souls and our society are in great need of transformation.

But this is not a spiritualized critique. It is instead grounded in the politics of the Bible, in what authentic faith/belief creates in terms of actions and lifestyle. Society and politics both shape and reflect our spiritual values, and these values are increasingly empty. How does a nation of endangered souls recover an authentic faith that is true to the gospel, the example of Jesus, the witness of the prophets, and the crushing needs of our times? What would such a recovery mean for evangelicals, mainline Protestants, Catholics, Jews, Muslims, seekers, and everyday people? What if everyday people made the politics of the prophets their litmus test for political candidates, and for fiscal, social, corporate, and foreign policy issues? Church and state can and must be separate in a modern democracy, but spiritual values still undergird everything and are reflected in the society we live in, the social and political directions we choose, and the candidates we select.

People now very commonly say they aren't "religious," just spiritual . . . and never with any challenge from any quarter. But the problem for all us well-meaning "spiritual folks" is that we've lost the social, unifying, and liberating aspects of biblical faith. Quick to scrub off the ossified forms of irrelevant denominationalism, we've thrown Jesus and prophetic biblical faith out with the bath water! We all need, in a very real and dire sense, to get some old-time Religion! Politics can't save itself. Nor can it rely for its salvation on the debased spirituality so prevalent in the culture.

When religion is relegated merely to the private sphere, it becomes vulnerable to the charge of being "soft" and therefore irrelevant to public life. Politicians themselves often suggest that religion is too soft and weak to be useful to political decision making. But such criticisms are often thinly veiled. So when the infamous former wrestler and insurgent Minnesota governor, Jesse Ventura, made some very bold statements about the "weakness" of religion in a *Playboy* magazine interview, the issue of religion's public role became very controversial. I thought Jesse needed a response from the religious community, so I wrote the following open letter to the governor in an MSNBC column.[1]

An Open Letter to Jesse Ventura

Dear Jesse,

You and I have never met, I think we've lived in different worlds. I've never been a biker, bouncer, or boa-feathered wrestler. And you, I'll guess, have never led a prayer meeting. But you've been preaching a lot of sermons lately. Even one on religion, in *Playboy* magazine no less! I just had to go out and get one, to read what you said about us religious types. I wonder how many preachers you got to buy *Playboy*? No, I'm not worried about the sex; I've just always thought it a little weak-minded.

Anyway, I'm an inner-city pastor who sometimes pumps iron. Maybe we do have something in common? So I read the interview and thought I'd write to straighten out some of your misunderstandings, just to be helpful to you in your new job. There's probably a lot you don't know about religious people, but now that you're a governor you'll want to find out. Lots of Lutherans in your neck of the woods.

You said, "Organized religion is a sham and a crutch for weak-minded people who need strength in numbers . . ." Well, there's a point to that if you mean those who just go to Sunday services and treat it like a nice club. But you should know Jesse, that there are quite a few religious folks who try to live their faith between Sundays, 24/7 as we would say. Let me tell you about some.

First, you might want to visit your own inner-city pastors in Minneapolis, St. Paul, or Duluth. I know a lot of them and they're pretty tough minded, even though they've got big hearts. You see, they live and work in urban war zones where one has to demonstrate the love of God and not just talk about it. I've seen you wrestler guys strut around the ring, but I doubt if many of you would make it in a neighborhood like mine. I think you guys are too used to the good life. You know, like politicians. Anyway, I've been with some of your Minnesota pastors at gang peace summits, no place for the weak-minded, and they could teach you some stuff. I'll send you some names and addresses.

I'm sure people have already reminded you about Martin Luther King Jr. and Gandhi, who were pretty tough. Just ask the Southern governors or the British. And I remember watching South African Arch-Bishop Desmond Tutu face down armed security police inside his Cathedral in Capetown, while Nelson Mandela was still in prison. "You have already lost!" he told them, "So why don't you come and join the winning side!" He was smiling when he said it, but not because he was weak-minded. And Bishop Tutu was right, they lost. By the way, did you know Nelson Mandela is also a religious man and might just be the strongest political leader on the planet?

I wish you could meet my friends Daniel and Phillip Berrigan. They're Catholic priests who've been fighting against nuclear weapons for decades, and have spent years in jail for their often lonely protests. I guess they don't need strength in numbers. Ever spent any time in jail Jesse, I mean for doing something right? Lots of us religious folks have. I remember times in the DC jail when we spent days on a gymnasium floor with 50 prisoners sharing a single toilet.

Most of the people I'm thinking of are not famous. I remember two of them who worked the streets of Olongapo in the Philippines—you know, that place you said you had fun with so many young girls. What you probably didn't know was that most of them were poor, rural girls lured into prostitution with the promise of urban jobs. They became virtual sex slaves, forced to live in barracks-like quarters owned by businessmen who made the profits and kept the women drugged to ease the pain. I've walked those streets with a Catholic priest and Mennonite relief worker who were helping the girls overcome their addictions and diseases. You can imagine how tough minded those two were, especially the Mennonite woman who opened a shelter for the girls and took on the pimps. You wouldn't want to cross her, Jesse. (By the way, did you know that it was U.S. Navy guys who introduced AIDS to the Philippines?)

Many of the folks I wish you could meet are close to home, in neighborhoods all over the country, including Minnesota. They take in refugee families, run homeless shelters and soup kitchens, mentor at risk kids, and walk alongside poor families making the transition from welfare to work.

And remember your exciting day at Harvard last week? Well, lots of those bright young minds are also religious. I know because I teach there, and we have hundreds of students coming to classes and forums on religion and public life. Believe me Jesse, those kids are not weak-minded. You should have seen them turn out for Billy Graham the week before you. What a strong, consistent scandal-free moral guide he has been!

I suspect you're the kind of stand-up guy who would want to know when you got it wrong. So, I thought I'd drop you a line. Hope I've been helpful. Maybe we could arm-wrestle some time. And, by the way, I've assigned your interview to my students at Harvard for next week's class.

Jesse never did reply to my letter (or my challenge to an arm-wrestling match), but the letter was the number one MSNBC feature for three days. It seemed to have struck a chord.

We need a transformation of our religion and our politics that acknowledges that the old ways don't work. But we need more than critique. We must ask what's wrong, but also what the answers are. At its heart, we must offer a challenge to hope, which is the only real path to change.

Our private religions have failed, but we must not lose a personal God. Instead of trying to strike an elusive "balance" between private piety and the social gospel, we must go to the heart of prophetic religion itself in which a personal God demands public justice as an act of worship. We meet the personal God in the public arena and are invited to take our relationship to that God right into the struggle for justice. Indeed, without that personal relationship we will lose the political struggle. That shift—bringing the personal God into the public arena—is at the heart of the prophet's message and will transform both our religion and our politics.

PART II

MOVING BEYOND THE POLITICS OF COMPLAINT

CHAPTER 4

Protest Is Good; Alternatives Are Better

What Are We For?

DURING THE RUN UP to the Iraq war, I learned two very valuable lessons that have stayed with me. One was political and the other personal and, yes, spiritual.

On the eve of the war, my wife, Joy, and I found ourselves unexpectedly in the labor and delivery room of the Washington Hospital Center—our son Jack was coming a month early! I had rushed home from an important Call to Renewal board meeting in Florida, where I was when Joy went into labor prematurely. (Call to Renewal is a national network of churches and faith-based organizations working to overcome poverty.)

Sojourners had just launched the Six-Point Plan, offered by U.S. church leaders, as an alternative to war with Iraq. The plan we offered took the threat of Saddam Hussein seriously. It called for his removal from power through an international criminal indictment, the elimination through coercive inspections of any weapons of mass destruction he might have, and the democratic reconstruction of Iraq under international leadership (not U.S. occupation)—all without war. We said there was a better way than war to solve the problem of Iraq and detailed how it might be accomplished. (See the sidebar of the full text of the Six Point Plan on page 50.)

In less than two weeks, the plan spread around the country and the world. Those of us offering the plan had just met with the British prime minister Tony Blair in London, and discussions with his cabinet leadership continued. In the final weeks before the war, what they called "the American church plan" was being actively discussed at the highest levels of the British government. Top religious leaders in the United Kingdom, in the United States, and around the world were pushing the plan to their government leaders. People at the United Nations, including Secretary General Kofi Annan, were studying it. Officials at the U. S. State Department requested a presentation and discussion of the plan, and even some nonadministration "hawks" on Iraq said it should be tried. Democrats in the House and Senate were calling us to ask for meetings—they hoped that an alternative plan to war from the religious community might help them regain their voice. When the *Washington Post* prominently published the plan on their opinion page under the title "A Third Way Is Possible," a contact at the White House told me that "everybody" over there had seen it.

Perhaps most obvious was how the Six-Point Plan served to empower and energize the peace movement itself. Finally, people said, we have something to be *for* and not just something to be *against*. We were hearing from people all over the world—some reported getting the plan from twenty different e-mail sources. Our own *SojoMail* (which was used to launch the plan on the Internet) more than doubled its readership in those two weeks.

When the midnight phone call came from Joy, I was in my hotel room writing the copy for full-page ads in five leading British newspapers, set to appear in two days, coinciding with the critical debate in Parliament. The ads described the alternative to war and asked Tony Blair and the British people to be "a true friend of America in this critical hour" by helping our government not make "a terrible mistake."

Some friends at Sojourners took Joy to the hospital while others came to the house to look after our five-year-old son, Luke. I e-mailed the ads and grabbed the first flight home in the wee hours of the morning. Joy and I were in regular cell phone contact until I arrived at the hospital, exhausted but very excited. And thankfully, heavy labor had yet to begin.

I realized the cell phone was still on when it began ringing in labor and delivery. On the other end were British cabinet ministers and members of Parliament who had seen the plan and were facing a parliamentary debate the next morning. "Jim, is this a good time to talk about the Six-Point Plan?" they asked. My incredible wife, a woman really committed to peace-making, shouted, "Take the calls! Stop the war! I'm not pushing yet!"

Within a few hours, two happy, grateful, and tired parents were holding their new healthy son. And within two days, the war with Iraq began just as we were taking Jack home from the hospital. Our house in inner-city Washington, D.C., was filled both with the reports of war and the sounds of new life. Our sadness at failing to stop a war was mingled with feelings of delight, awe, deep gratitude, and overwhelming blessing.

At first, I felt almost guilty about feeling joyful at such a time. Then I realized we were being taught two lessons, almost through divine intervention.

The first lesson was this: Saying no is good, but having an alternative is better. Protest is not enough; it is necessary to show a better way. Former British cabinet minister Clare Short has since said she believed the Six-Point Plan might have worked—but came too late; and if given a few more weeks could have gained real momentum. But those making decisions in the Bush administration were determined to go to war with Iraq.

In a press release on the eve of war, several American church leaders made a final appeal to President Bush. We said, "If we wage war on Iraq, it will be a failure of political and moral imagination." We went on to say that "while we acknowledge that we are seconds away from the start of a pre-emptive strike on Iraq, we continue to believe that war is not the answer to the threats posed by Saddam Hussein. Our political leaders have not exhausted all diplomatic options. There is still a credible alternative." And there was, despite the decision of our political leaders to ignore it. There were even people inside the White House who hoped President Bush would meet with the church leaders who had just seen Tony Blair. But the Bush administration refused to listen to religious voices offering an alternative to war. In a final call with the White House that last week, I was told, "The president is scheduling no more meetings." But even the pressure on the White House to meet came not just from protest, but from religious leaders who said there was another answer. Taking the "threats" seriously and offering credible alternatives is absolutely key.

The second lesson is this: Our most difficult and darkest moments are precisely the time to embrace the nurturing relationships that remind us how precious and sacred the gift of life really is. It is those relationships (often with children), along with the other renewing practices and spiritual disciplines we keep, that will remind us of what is truly important. The special and divine activity of affirming life (especially new life) helps us to recall at the deepest level why we care about issues of war and peace, justice and injustice.

Many people will engage in protest, but even more are likely to follow an alternative that offers a better way. To offer an alternative is always more challenging than just protest; it requires more work, creativity, and risk. Like many others, I came of age during the 1960s, when the struggle for justice was embodied in the archetype of protest. We learned our lessons about politics in the streets, and the habit of protest is still deep within us. But protest can become static and formulaic. The aim of effective and transformational protest should be to illumine a society to its need for change. In other words, protest must be instructive to succeed, more than destructive. It should, at its best, point the way to an alternative, rather than just register the anger of its demonstrators. Protest must not become just a ritual of resistance, offering a laundry list of grievances.

Protest should not be merely the politics of complaint, which we discussed in chapter 2. It should instead show the way for both personal and social transformation. That's what excites people and invites them to give their lives for something larger than themselves. The power of protest is not in its anger but its invitation. The test of protest is whether it points and opens the way to change or merely denounces what is. When protest is both instructive and constructive in a society, it becomes something that has to be dealt with and not just merely contained.

In fact, those who protest should be making a promise. They are promising their society that a better way is indeed possible. They are saying that the bigotries, the injustices, the indignities, the indifference, and the unnecessary violence we experience today will not have the last word. Instead their protest reveals the things that can and must be changed for the good and health of the entire society and the world. We need people who pledge themselves, not just to object to what is wrong, but to help find and fashion an alternative. In other words, the best protest is not merely countercultural, it is transformational. It gives a society a better vision for itself and for the future. That is the way of the prophets. They began in judgment but ended in hope for change. The biblical prophets were never just complaining; they were imagining a newer world.

For example, I believe in nonviolence, and we will discuss more about the meaning and relevance of nonviolence today later in the book. But the principle of always seeking an alternative applies to nonviolence as well. It is this: If nonviolence is to be credible, it must answer the questions that violence purports to answer, but in a better way.

Those who would seek alternatives to war must not underestimate the problem of evil in the world or the threat of ruthless and dangerous dictators like Saddam Hussein. It must be admitted that the peace movement sometimes does underestimate the problem of evil, and in doing so weakens its authority and its message. In their protest against the war in Iraq, some in the peace movement seemed to underestimate the evil and dangers of Saddam or, at least, failed to respond adequately to his many atrocities. The public perception was that the peace movement was not determined to oppose Saddam Hussein or to remove him from power. So those who did clearly propose to deal with Saddam appeared to be stronger than those who didn't. When a peace movement appears to be "soft" on the problems that war claims to be able to solve, alternative solutions will seem weak. To avoid or prevent war, we must have answers that effectively deal with the real problems and threats but are *better* than war.

In our world, wars are not ceasing. In fact, war is becoming the primary instrument of foreign policy in many places, especially in the United States. There ought to be an assumption, from a Christian point of view, that this is not a good thing. But vigorous voices now tell us that war is not necessarily a bad thing and is often the best solution to the problems of terrorism or the abuse of human rights. To assert that the resolution of conflict and the prevention of war should be a Christian goal requires that the real causes and consequences of those conflicts be dealt with in a *more effective* way. That is the challenge of nonviolence today.

The same principle of always looking for creative alternatives can be applied to the issue of poverty—both domestically and globally. Decrying poverty is a good thing. One in every six American children still falls below the poverty line in America—and one in three children of color. That is appalling in the richest nation in world history. Even worse is the one billion people globally (all God's children) who are forced to subsist on less than a dollar a day; and the three billion who live on less than two dollars per day. And thirty thousand children die *every day* due to hunger and disease related to utterly preventable causes (like the lack of clean drinking water). Those poverty "facts" shouldn't be tolerable to anyone across the political spectrum. But again, just to decry these facts has not solved the scandal of poverty.

We need solutions. And we need them to go beyond the polarized ideological agendas of partisan politics. The conservatives are right when they say that cultural and moral issues of family breakdown, personal responsibility, sexual promiscuity, and substance abuse are prime reasons for entrenched domestic poverty. The liberals are right when they point to the critical need for adequate nutrition, health care, education, housing, and good-paying jobs as keys to overcoming endemic poverty. So why do we continue to make false choices?

Domestic poverty will not be overcome without investing both public and private resources in the lives of poor children and families *and* by strengthening the bonds of family and community. Poverty will only be significantly reduced by a combination of personal and social responsibility, and then only with a moral commitment that makes possible a new political will—transforming the fight against poverty into a bipartisan commitment and a nonpartisan cause.

Global poverty will not be significantly reduced until we forge new international policies and practices regarding aid, trade, and debt while also honestly facing the real issues of governance and transparency in poor countries. Only a new commitment to moral and political responsibility on the part of the wealthy nations *and* a new level of public integrity to root out the corruptions that poverty brings in the poor nations will create the capacity for genuine partnerships.

In the fight against poverty, clear and compelling alternatives are needed in every area: in education, health care, housing, and economic development and in personal, family, and community renewal. Gratefully, those alternatives exist. They must be learned from, further developed, and scaled up into larger solutions in strategic partnership with a whole new set of national and international policy decisions that will effectively and dramatically reduce poverty.

The spiritual component in all this is absolutely crucial. An understanding of how sacred the blessing of life is must undergird all our efforts for justice and for peace. It may be that only theology and spirituality can save the poor and the victims of war. In our new global economy, the poor are really not needed and can safely be forgotten, as can the countless refugees and casualties of war. But each of those forgotten souls was made in the image of God and carries that sacred value. To remember that is a religious duty, and to remember them is part of our obedience to God. In fact, the gospel reminds us that God is actually *particularly* present in those very victims when Jesus says, "As you have done it to the least of these, you have done it to me."

We must never be satisfied with mere protest or complaint about the things we believe are wrong. Rather we must do the harder, more creative, and ultimately more prophetic work of finding and offering alternatives. And we must always seek to remember and remind each other of the deeper reasons and roots for our public visions and commitments. In both instances, we will seek the biblical wisdom to "Choose life; so that you and your children may live."

The clear shift from protest to alternatives could change the political framework of debate. Protest and protesters are too easily written off. While the noble and necessary tradition of citizen protest should, of course, continue, it should always point and concretely connect to viable policy alternatives that could actually solve the issues at hand. Offering better visions is in the best of our traditions of dissent and protest. Saying "yes" and not just "no" gives political leaders something to consider and debate and not merely something to stop. And alternatives give citizens something to advocate and not just something to oppose. A political alternative brings more energy and possibilities to the public debate than political opposition can by itself. Being "for" is simply better than only being "against." And, ultimately, it will be more successful. It is the tradition of the prophets who claimed that injustice will not have the last word and that a newer world is not only possible but, indeed, is the promise of God.

Here is the complete text of the Six-Point Plan, as published in the *Washington Post* on March 14, 2003.

There Is a Third Way

By Jim Wallis and John Bryson Chane

It is the eleventh hour, and the world is poised on the edge of war. Church leaders have warned of the unpredictable and potentially disastrous consequences of war against Iraq—massive civilian casualties, a precedent for preemptive war, further destabilization of the Middle East and the fueling of more terrorism.

Yet the failure to effectively disarm Saddam Hussein and his brutal regime could also have catastrophic consequences. The potential nexus between weapons of mass destruction and terrorism is the leading security issue in the world today.

This is the moral dilemma: a decision between the terrible nature of that threat and the terrible nature of war as a solution.

The world is desperate for a "third way" between war and ineffectual responses—and it must be strong enough to be a serious alternative to war. The threat of military force has been decisive in building an international consensus for the disarming of Iraq, for the return of inspectors and for pressuring Hussein to comply. The "serious consequences" threatened by the Security Council need not mean war. They should mean further and more decisive actions against Hussein and his regime, rather than a devastating attack on the people of Iraq.

On Feb. 18 a group of U.S. church leaders, accompanied by colleagues from the United Kingdom and the Anglican communion, met with Prime Minister Tony Blair and his secretary of state for international development, Clare Short, to discuss alternatives to war. The following elements of a "third way"—an alternative to war—were developed from those discussions and subsequent conversations within our U.S. delegation:

• Remove Hussein and the Baath Party from power. The Bush administration and the antiwar movement are agreed on one thing—Hussein is a brutal and dangerous dictator. Virtually nobody has any sympathy for him, either in the West or in the Arab world, but everybody has great sympathy for the Iraqi people, who have already suffered greatly from war, a decade of sanctions and the corrupt and violent regime of Hussein. So let's separate Hussein from the Iraqi people. Target him, but protect them.

As urged by Human Rights Watch and others, the Security Council should establish an international tribunal to indict Hussein and his top officials for war crimes and crimes against humanity. This would send a clear signal to the world that he has no future. It would set into motion both internal and external forces that might remove him from power. It would make clear that no solution to this conflict will include Hussein or his supporters staying in power. Morton Halperin has pointed out: "As we have seen in Yugoslavia and Rwanda, such tribunals can discredit and even destroy criminal regimes."

• Pursue coercive disarmament. Removing Hussein must be coupled with greatly intensified inspections. This would mean not just more inspectors but inspections conducted more aggressively and on a much broader scale. The existing U.S. military deployment should be restructured as a multinational force with a U.N. mandate to support and enforce inspections. The force would accompany inspectors to conduct extremely intrusive inspections, retaliate against any interference and destroy any weapons of mass destruction it found. There should be unrestricted use of spy planes and expanded no-fly and no-drive zones.

• Foster a democratic Iraq. The United Nations should begin immediately to plan for a post-Hussein Iraq, administered temporarily by the United Nations and backed by an international armed force, rather than a U.S. military occupation. An American viceroy in an occupied Iraq is the wrong solution. An internationally directed post-Hussein administration could assist Iraqis in initiating a constitutional process leading to democratic elections.

• Organize a massive humanitarian effort for the people of Iraq now. Rather than waiting until after a war, U.N. and nongovernmental relief agencies should significantly expand efforts to provide food, medical supplies and other humanitarian assistance to the Iraqi people now. Focusing on the suffering of the Iraqi people, and immediately trying to relieve it, will further help to protect them from being the unintended targets of war. It would also help to further isolate Hussein from the Iraqi public by contrasting the world's humanitarian concern with his indifference to his own people.

Finally, to ensure a lasting peace in that troubled region, two other points are necessary.

First, we should recommit to a "road map" to peace in the Middle East. The United States, Britain and other European Union nations must address a root cause of Mideast conflict with a peace plan resulting in a two-state solution to the conflict between Israel and the Palestinians by 2005, structured to include meaningful deadlines enforced by the international community.

Second, we should refocus the world's energies on the greatest threat it faces—networks of suicidal terrorists. The international campaign against terrorism has succeeded in identifying and apprehending suspects, freezing financial assets and isolating terror networks. But it is in danger of being disrupted, both by acrimony and by lack of attention, as the world focuses on the impending conflict with Iraq.

Unless an alternative to war is found, a military conflagration will soon be unleashed. A morally rooted and pragmatically minded initiative, broadly supported by people of faith and goodwill, might help to achieve a historic breakthrough and set a precedent for effective international action in the many crises we face in the post-Sept. 11 world.

Jim Wallis is editor and executive director of Sojourners and convener of Call to Renewal. John Bryson Chane is Episcopal bishop of Washington, D.C.

This is the complete text of the full-page ad that ran in five major British newspapers on the day of the Parliament debate on going to war with Iraq.

A Letter from Concerned Americans

On February 18, [2003,] five U.S. church leaders met with Prime Minister Tony Blair to discuss alternatives to war on Iraq. The plan in the following letter was developed from that discussion and subsequent conversations in the U.S.

Prime Minister Blair,

The world needs you to find a "third way" between war and inaction. It is two minutes before midnight, and the world's people are desperate for an alternative to war.

Yes, we must disarm Saddam Hussein.

Yes, we must remove him from power.

But, we need a better solution than war to accomplish our goals. The people of Iraq have already suffered greatly, and we must not inflict even more suffering upon them.

As Americans, we have a special relationship with the British people. We need you to be a true friend of America in this critical hour. We need you to help our government not make a terrible mistake.

American church leaders are offering an eleventh-hour initiative that proposes an effective response to the threat of Saddam Hussein and his weapons of mass destruction, but avoids the terrible dangers and suffering of war.

Mr. Prime Minister, we appeal to you as a man of moral and religious convictions to persuade our President not to take our countries into a war that could have terrible, bitter, and divisive consequences. Let us instead:

1. Indict Saddam Hussein for his crimes against humanity and send a clear signal that he has no future in Iraq, setting into motion the internal and external forces that could remove him from power and bring him to trial at the International Court in The Hague. History has shown, as with Slobodan Milosevic, that this can help bring down a criminal regime.

2. Pursue coercive disarmament with greatly intensified inspections backed by a U.N. mandated multinational force.

3. Foster a democratic Iraq through a temporary post-Hussein U.N. administration, rather than a U.S. military occupation.

4. Organize a massive humanitarian effort through the U.N. and nongovernmental relief agencies for the people of Iraq now, rather than only after a war.

5. Commit to implement the "roadmap" to peace in the Middle East, with a clear timetable toward a two-state solution that guarantees a Palestinian state and a secure Israel by 2005.

6. Re-invigorate and sustain international cooperation in the campaign against terrorism, rather than having it disrupted by a divisive war against Iraq that intelligence officials believe will likely lead to further attacks.

Such a morally rooted and pragmatic initiative could help achieve an historic breakthrough and set a precedent for decisive and effective international action instead of war.

We urge you to provide the principled leadership and real friendship that could resolve this international crisis without war.

We are grateful for the recent meeting we had with you. You have our hopes and prayers.

Sincerely,

Jim Wallis, Executive Director and Editor-in-Chief of Sojourners;

John Bryson Chane, Episcopal Bishop of Washington, D.C.;

Clifton Kirkpatrick, Stated Clerk of the Presbyterian Church USA;

Melvin Talbert, Ecumenical Officer of the United Methodist Council of Bishops;

Daniel Weiss, Immediate Past General Secretary of the American Baptist Churches in the USA;
and millions of concerned Americans.
Sponsored by Business Leaders for Sensible Priorities

CHAPTER 5

How Should Your Faith Influence Your Politics?

What's a Religious Voter to Do?

THE POLITICS OF GOD is often not the same as the politics of the people of God. The real question is not *whether* religious faith should influence a society and its politics, but *how.*

Democratic and Republican Religion

As the Democratic candidates for president attended religious services during the 2004 election campaign, their worship was tempered by an uncomfortable fact: Churchgoing Americans tend to vote Republican.

An overwhelming majority of Americans consider themselves to be religious. Yet according to a study released in late 2003 by the Pew Research Center for the People and the Press, people who attend church more than once a week vote Republican by 63 percent to 37 percent; people who seldom or never attend vote Democratic by 62 percent to 38 percent.[1] This was borne out by the 2004 exit polls—people who attend church more than once a week voted 64 percent for President Bush to 35 percent for Senator

Kerry; those who never attend voted 62 percent for Kerry to 36 percent for Bush.[2]

This disparity should concern Democrats—if not as a matter of faith, then as a matter of politics. More important, it should concern anyone who cares about the role of religion in public life. By failing to engage Republicans in this debate, the Democrats impoverish us all.

The Republicans clearly have an advantage with people of faith, as recent election year results have suggested. Republicans are more comfortable talking about religious values and issues, and they are quick to promise that their faith will affect their policies (even if, like their Democratic counterparts, they don't always follow through on their campaign promises).

President Bush is as public and expressive about his faith as any recent occupant of the White House. Among his first acts as president was to establish the Office of Faith-Based and Community Initiatives, which helps religious and community groups get federal financing for some of their work. Although the "faith-based initiative" turned out to be more symbolic than substantial, symbolism matters—in religion as well as politics.

The Democratic candidates in 2004, in contrast, seemed uncomfortable with the subject of religion. (The exception was Joseph Lieberman, though even he seemed less comfortable than he was in 2000.) Democrats stumble over themselves to assure voters that while they may be people of faith, they won't allow their religious beliefs to affect their political views.

For too many Democrats, faith is private and has no implications for political life. But what kind of faith is that? Where would America be if the Reverend Dr. Martin Luther King Jr. had kept his faith to himself?

Howard Dean, the early front-runner who challenged President Bush, illustrated the Democrats' problem. Dr. Dean said he left his church in Vermont over a dispute about a bike path, and he often explained that his faith did not inform his politics. Later he sought to reverse himself and talked more about his religion when he was attacked for being too "secular," but then he made matters worse by showing an embarrassing ignorance about faith matters. Among other things, Dean made the mistake of saying his favorite New Testament book was Job. Job, of course, is in the Old Testament, and talking in such uninformed ways about religion is often a big problem for Democrats. Dean had previously said the presidential race should stay away from the issues of "guns, God, and gays" and focus on jobs, health care, and foreign policy. But by framing the issue in this way—declining to discuss overtly "religious" topics—Democrats allow Republicans to define the terms of the debate. The "religious issues" in an election get reduced to the Ten Commandments in public courthouses, gay-marriage amendments, prayer in schools, and, of course, abortion.

These issues are important. But faith informs policy in other areas as well. What about the biblical imperatives for social justice, the God who lifts up the poor, the Jesus who said, "Blessed are the peacemakers"?

How a candidate deals with poverty is a religious issue, and the Bush administration's failure to support poor working families should be named as a religious failure. Neglect of the environment is a religious issue. Fighting preemptive and unilateral wars based on false claims is a religious issue (a fact not changed by the defeat and capture of Saddam Hussein).

Such issues could have posed problems for the Bush administration among religious and nonreligious people alike—if someone were to define them in moral terms. The failure of the Democrats to consistently do so is not just a political miscalculation. It shows they do not appreciate the contributions of religion to American life.

As the campaign progressed, Democratic nominee John Kerry did begin to make these connections. In a speech to the National Baptist Convention, he quoted from the letter of James: "It is not enough, my brother, to say you have faith, when there are no deeds. . . . Faith without works is dead." Kerry went on to say, "As you know, my friends, we are taught to walk by faith not by sight. And when we look around us—when we look around neighborhoods and towns and cities all across this country, we see faith to be lived out, and so many deeds to be done. . . . We see jobs to be created. We see families to house. We see violence to stop. We see children to teach—and children to care for. We see too many people without health care and too many people of color suffering and dying from diseases like AIDS and cancer and diabetes."[3] But it was too often only in speeches to religious bodies.

The United States has a long history of religious faith supporting and literally driving progressive causes and movements. From the abolition of slavery to women's suffrage to civil rights, religion has led the way for social change.

The separation of church and state does not require banishing moral and religious values from the public square. In fact, America's social fabric depends on such values and vision to shape our politics—a dependence the founders recognized.

It is indeed possible (and necessary) to express one's faith and convictions about public policy while still respecting the pluralism of American democracy. Rather than suggesting that we not talk about "God," Democrats should be arguing—on moral and even religious grounds—that all Americans should have economic security, health care, and educational opportunity and that true faith results in a compassionate concern for those on the margins.

Democrats should be saying that a just foreign and military policy will not only work better, but also be more consistent with both our democratic and spiritual values. And they must offer a moral alternative to a national security policy based primarily on fear, and say what most Americans intuitively know: that defeating terrorism is both practically and spiritually connected to the deeper work of addressing global poverty and resolving the conflicts that sow the bitter seeds of despair and violence.

Many of these policy choices can be informed and shaped by the faith of candidates and citizens—without transgressing the important boundaries of church and state.

In June of 2004, I made the long trip to Santa Fe, New Mexico, and testified before the Democratic Platform Drafting Committee, which wanted testimony from some faith-based organizations, especially those working on poverty. Democrats haven't always asked for religious input, and by not doing so have too often conceded the issue of religion to the Republicans. I decided to accept the committee's invitation to testify, on behalf of Call to Renewal and Sojourners, because it was a good opportunity to link the issues of faith and justice. I had met with President George W. Bush and the White House policy staff to discuss the "faith-based initiative" and how "compassionate conservatism" ought to produce a domestic policy that effectively assists low-income families in escaping poverty. So I was happy to speak also to the Democrats. I would have willingly testified before the Republican Platform Committee as well, if they had asked.

I quoted Isaiah to the Democrats and urged them not to avoid moral and religious language in expressing their concern for economic justice. One member of the committee responded to my testimony by saying that Democrats did have "moral issues" at the heart of their agenda when they stood up for poor people. I said I thought that was true, but I suggested that they have often hesitated to use moral and religious language when they spoke to those questions, and it was time to do so. Heads were nodding all around. The committee was made up of elected officials, civil rights and labor leaders, academic experts, and grassroots organizers. I could tell that some were clearly religious people, and a few were struggling not to keep saying "Amen" as I went along. But all were very attentive and seemed quite taken by the discussion of faith and politics, even at the end of a long day of hearings. At the end of my testimony, questions, and discussion, the acting chair, Los Angeles city councilman Antonio Villaraigosa, thanked me by saying, "Reverend, you have given us a spanking that we needed."

I also spoke to the first-ever People of Faith Luncheon at the Democratic National Convention in Boston, where I encouraged the delegates to let their moral and religious values shine through.

The Democrats are trying to take religion seriously, more than they have in recent years. Some Democrats remember when the party was allied with the civil rights movement in the 1960s—which, of course, was led by black churches. And no one in American history ever linked religion and politics better (or more prophetically, democratically, and inclusively) than Dr. Martin Luther King Jr.

Two weeks before Santa Fe, the newly formed Center for American Progress held a significant and well-attended conference on "progressive religion" in Washington, D.C. Led by John Podesta, the Center is challenging Democrats to remember that religion has fueled most of the progressive social change movements in American history. The conference served to "validate" the importance of religion in an election year and beyond. The message was that "religion is progressive, and progressives are religious."

All these are encouraging signs. Religion should not be the exclusive possession of the Republican or Democratic Party, the Right or the Left, but must be able to critique and challenge both. And clearly, in the election of 2004, Christians voted both ways, because of their faith.

Again, God is always personal, but never private. The Democrats are wrong to restrict religion to the private sphere—just as the Republicans are wrong to define it solely in terms of individual moral choices and sexual ethics. Allowing the right to decide what is a religious issue has become both a moral and political tragedy.

Not everyone in America has the same religious values, of course. And many moral lessons are open to interpretation. But by withdrawing into secularism, the Democrats deprive Americans of an important debate.

A Tale of Two Movements

So how best do we "put God in politics"? We never seem to learn: Power does corrupt, and both political and religious leaders have been regularly seduced by it. Does that mean Christians should stay out of politics?

Everybody loves to quote the famous dictum by Lord Acton, "Power corrupts, and absolute power corrupts absolutely." Heads nod all around, and then everyone ignores what the wise old Englishman said. Power does indeed corrupt human beings: It compromises principles, it quiets conscience, and it mellows morality. Power tells us to just get along, instead of getting upset; it encourages us toward smooth sailing, and it discourages us from rocking the boat.

But it's bad theology to say that power, per se, is always bad. The Bible speaks of the power of God, the power of the gospel, and the power of truth. The New Testament Greek word *dunamis* means "spiritual power." Even political power (which is what Acton was really talking about) isn't always evil. Look at the moral power and authority that Nelson Mandela exercised to free South Africa or that Dr. Martin Luther King Jr. and the civil rights movement used to change the landscape of American life. Yet power, and especially political power, is very dangerous. It's often riddled with the hubris and illusion to which we all are so susceptible.

Human beings seem not to handle power very well. Of all people, religious leaders ought to know that best. Instead, religious leaders are often among the most easily corrupted by power, especially when they get close to political power. Doug Coe, the principal leader of the annual Presidential Prayer Breakfast in Washington, D.C., once told me that the best way to get religious leaders together was to invite them to a meeting with a powerful political leader—hence the sold-out successes of each year's prayer breakfast. He said most church leaders generally ignored Jesus's suggestion to take the humbler places at a banquet and wait until they are invited to "come up higher." Instead they jostle for the best positions and places at the events where the powerful gather. It regularly amazes me how good religious folk get so excited about sharing an intimate breakfast with the president and three thousand other people.

I believe it is very instructive to compare the two major faith-inspired movements of the last fifty years that have tried to influence national politics: the black-church-led civil rights movement of the 1950s and 1960s and the religious Right movement of the 1980s and 1990s, exemplified by Jerry Falwell's Moral Majority and Pat Robertson's Christian Coalition. Let's call it a tale of two movements. Both were rooted in the churches, both advanced an agenda of "moral issues," and both sought to influence the direction of American life. But there were important differences.

When I first learned about a book called *Blinded by Might*,[4] written by two former leaders of the religious Right, I was fascinated. Authors Cal Thomas and Ed Dobson (no relation to James Dobson) told the story of how the Moral Majority and their movement had been seduced by power.

It recounts the exhilaration many conservative evangelicals felt when Ronald Reagan was elected president in the 1980 landslide, and how many of them (the authors included) bristled with pride when the media gave the newly organized religious Right a substantial part of the credit for the victory. Thomas and Dobson were as excited as everyone else at the post-election celebration at Jerry Falwell's Liberty Baptist College in Lynchburg, Virginia. The ecstatic crowd, so proud of their pastor, who had now become a major national figure, leaped to their feet when Falwell strode into the packed auditorium as a band struck up "Hail to the Chief." "Hail to the Chief"? The authors said it was almost as if Jerry Falwell had been elected president. All of a sudden, conservative evangelicals who felt ignored and ridiculed for so long in the cultural backwaters of American life, almost since the infamous Scopes trial in the 1920s, were now in the national spotlight and getting their pictures taken in the Oval Office with the president.

President Reagan and Reverend Falwell apparently spoke often, sometimes several times a week, as the fundamentalist minister became an insider to power. Revealing stories in the book demonstrate the cost of becoming a political insider. At one point, Reagan was about to appoint Sandra Day O'Connor to the Supreme Court—not the kind of unequivocal pro-life justice the religious Right was hoping for. But Reagan called Falwell, saying, "Jerry. . . . I want you to trust my judgment on this one." Perhaps anxious to be a player, a winner, an insider, Falwell went along, and a series of compromises began. Direct mail strategy and fund-raising came to dominate the religious Right's political agenda over its previous moral concerns. Political success, defined as keeping political power, eventually became more important than the issues that initiated the formation of the religious Right in the first place. It's an old story.

Of course, religious conservatives are not the only ones to be mesmerized by political power. If the Moral Majority and the Christian Coalition have loved the attention from the Republican Party, too many liberal religious leaders have often been satisfied with photo-ops at a Democratic White House. After twelve years in exile during the Reagan-Bush administrations, many mainline Protestant leaders were eager to be back in the loop with the Clinton-Gore victory. And many of them were reduced to defending Clinton's indefensible moral behavior in a sexual and political scandal (or at least maintaining an awkward silence). Access clearly has its price.

There is no doubt that the religious Right's leaders and constituency have helped to elect Republican presidents. But Thomas and Dobson, lamenting how little of their agenda has actually been accomplished, say the religious Right has failed. On most of the issues the religious Right has identified as their priorities, the culture has moved in opposite directions, despite Republican victories. Abortion rates were declining, but this is apparently due more to economic progress for poor women during the Clinton administration than to any initiatives by religious Right–supported Republican presidents. And, during the first term of the Bush administration, some evidence indicates that the number of abortions again rose, due to the declining status of low-income women. Pornography has exploded on the Internet and cable television. Cultural acceptance for gay civil rights is expanding, despite successful state referendas against gay marriage. Critics of Thomas's and Dobson's book say the two authors are suggesting a shift away from political engagement to personal piety and church building. But that is a misreading of their message. I hear Thomas and Dobson asking not *whether* Christians should be involved in the world, or even in "politics," but *how*.

In comparison, the civil rights struggle succeeded because it first built a movement that was morally based and politically independent. It certainly tried to change political structures and policies, but mostly operated outside of them. The movement's strength and its base were not primarily inside of politics, but rather at the grassroots. Indeed, it was the movement's outside strength and moral argument that proved key to its ultimate successes inside the political system: the Civil Rights and Voting Rights Acts of 1964 and 1965. Because the movement changed the way the American people *thought* about race and sought to affect the very cultural values of the country, it opened up the possibility of political change.

The religious Right, on the other hand, went for political power right away. Their strategy was less movement building and values changing than immediate electoral organizing. In fact, their hope was to take over the Republican Party (and they were successful in several states and local areas) and then implement their legislative agenda. But the critical step of persuading by moral argument and building a constituency for change was neglected. Ironically, the group that opted for an insider political strategy was ultimately relatively ignored by political power, in part because it failed to build the independent popular base and to fashion the moral argument that is so critical for real social transformation. The religious Right is still an important voter base for the Republican Party, but the political discussion is now more often about how the party can keep its conservative evangelical base by appeasing it with symbols and rhetorical commitments, rather than offering real substance. We will see if that changes in a second Bush term.

There have been countless books, documentaries, plays, television shows, and motion pictures telling the story of the civil rights movement. Will we be seeing anything similar about the religious Right? The religious Right went wrong by forgetting its religious and moral roots and going for political power; the civil rights movement was proven right in operating out of its spiritual strength and letting its political influence flow from its moral influence. Other great social causes led by religious communities—abolition of slavery, child labor reform, women's suffrage, and so on—all followed the same strategy.

Religious and Secular Fundamentalists

Some people just want candidates and citizens to keep their religion to themselves—at least in the public arena. They don't want anybody to speak out of their own faith tradition or to affirm the role of religion in shaping values for American public life. They believe in an absolute separation of church and state, that faith and politics just don't mix—or shouldn't. But most of the nation is willing to engage in the important discussion about the proper roles of religion, values, and public policy, and it is a conversation that will be with us far beyond any election. The Pew Forum on Religion and Public Life, which has done extensive polling on these issues, concludes that "people of faith, and the institutions they build, play a critical role in our nation, and these contributions are not a purely private matter. . . . Religious beliefs play an important role in where people stand on important issues of the day."[5]

It is, of course, not a new conversation, but one that goes back to the nation's founding. The founders saw the wisdom of separating church and state, making sure there would be no state-sponsored religion in the new nation or state interference in the devout diversity of the citizen's religious practice. Religion has always been a part of American politics, with frequent reference to God and faith on the part of many presidents, from George Washington, Thomas Jefferson, and Abraham Lincoln to Dwight Eisenhower, Jimmy Carter, Ronald Reagan, Bill Clinton, and George W. Bush. A myriad of activist preachers and social reformers, from William Jennings Bryan to Martin Luther King Jr., from Jesse Jackson and Al Sharpton to Pat Robertson, Jerry Falwell, and James Dobson, have repeatedly invoked the language of faith for a variety of social and political purposes.

At its best, faith in God has been used to hold the nation to divine accountability, as in Lincoln's expressions of collective penitence and the need for national forgiveness after the Civil War, and when King called his country to its best religious and political ideals in his *Letter from the Birmingham Jail.* But at its worst, biblical prooftexting to support ideological causes has made both religion and politics look bad. "Are we on God's side?" has always been a better question than "Is God on ours?" Religion has and will always be a part of American politics. Again, the real question is not *whether* religious values should help shape our national political discussion, but *how.*

Being raised evangelical in the Midwest gave me a personal experience of the phenomenon called "religious fundamentalism." A story illustrates. When I was a boy in high school, I was interested in a girl from our church. It was an evangelical church, although some might have called it a bit fundamentalist—taking a hard line on cultural issues. But I took a chance and invited her to a movie, which was certainly frowned upon back then in our church culture (though my own parents snuck us out to Walt Disney movies at the drive-in, where we were unlikely to be spotted). I chose *The Sound of Music,* thinking it was "safe." Who could object to Julie Andrews, I confidently thought? I was wrong. As we left the house, my girlfriend's father stood in the doorway, blocking our exit, and said to his daughter, "If you go to this film, you'll be trampling on everything that we've taught you to believe." She fled downstairs to her bedroom in tears. We missed the movie, and the evening was a disaster. A year later, the fundamentalist father watched *The Sound of Music* on his television—and liked it.

Fundamentalism is essentially a revolt against modernity. It is a reaction usually based on profound fear and defensiveness against "losing the faith." My girlfriend's father instinctively knew that his religion should make him different than the world. That is a fair religious point, and to be honest, there is much about modernity that deserves some revolting against. But I wish he had chosen to break with America at the point of its materialism, racism, poverty, or violence. Instead, he chose Julie Andrews.

But most of the fundamentalism in Christianity, Judaism, or Islam (and it exists in all three) does not result in what we witnessed on September 11. That takes a shift, or a turn—to theocracy, to violence, and to a reach for power.

Conventional wisdom suggests that the antidote to religious fundamentalism is more secularism. But that is a very big mistake. The best response to bad religion is better religion, not secularism. The three great monotheistic traditions are all religions of "the book." The key question is how do we interpret the book? For example, in Christian faith, we have the interpretation of Martin Luther King Jr., but also that of the Ku Klux Klan. Better interpretations of the book are a much more effective response to fundamentalism than throwing the book away.

Southern slave masters gave their captives the Bible to keep their eyes trained on heaven, instead of their plight on earth. But in the Bible those same slaves found Moses and Jesus, who became the foundations for their liberation struggle. We must always acknowledge that our religious traditions can be both a cause for oppression and an inspiration for liberation. Religious arguments have fostered terrible sectarian division, hatred, and violence, but faith has also helped to set people free. We must be honest about both. In the very same traditions that have been used to sanction injustice are found the seeds of justice, peace, and freedom. Those of us from religious communities must be the first to be critical of our own traditions when they are used to foster more conflict and violence while, at the same time, holding out the prophetic possibilities in every one of our religious faiths.

It is also often said that fundamentalism comes from taking religion too seriously. The answer, then, is to take religion less seriously. Wrong again. The best response to fundamentalism is to take faith *more seriously* than fundamentalism usually does. The best critique of fundamentalism comes from faith itself, which challenges the accommodations of fundamentalism to theocracy, power, and violence. It is faith that leads us to assert the vital religious commitments that fundamentalists often leave out, namely compassion, social justice, peacemaking, humility, tolerance, and even democracy as a religious commitment.

At heart, I am a nineteenth-century evangelical; I was just born in the wrong century. Before the movement was humiliated as a result of the famous Scopes trial in 1925, fundamentalism was often socially allied with the Left in supporting the kind of economic reform that would benefit its mostly working-class constituency. As we have said, evangelical and fundamentalist reformers led battles for the abolition of slavery, for child labor laws—even for women's suffrage. But after being banished to the cultural backwaters of American life, fundamentalism became an increasingly conservative and isolated enclave of defensive faith, "separate from the world." It was, in part, the assaults of secularism that helped turn fundamentalism into the right-wing force it has become today in the United States.

Of particular concern is how modern fundamentalism has made the move to theocracy—in Christianity, Judaism, and Islam. That move is really a betrayal of the biblical faith that regards political power much more suspiciously. Al Qaeda, the Taliban, and American fundamentalists like my old debate partners Jerry Falwell and Pat Robertson are, indeed, all theocrats who desire their religious agenda to be enforced through the power of the state. And that is, primarily, a religious mistake.

With the move to theocracy, modern fundamentalism too easily justifies violence as a tool for implementing its agenda. The fatal attraction of fundamentalism to violence is certainly at stake in the crisis of terrorism. It is evident in Islamic calls for holy war against the West, but also in the Christian Right's denunciations of all Islam and its religious affirmation of the war against terrorism with the language of "crusade" much too close to the political surface.

The fundamentalist arguments for violence quickly become more political than religious. I remember my debates with the Reverend Jerry Falwell about nuclear weapons. We would be talking about the potential annihilation of hundreds of millions of people. And yet his arguments were always more political than religious. The little bracelet on the wrists of so many teenagers today that reads "WWJD"—"What would Jesus do?"—gets quickly to the heart of the violence debate. It's always striking to me that when I listen to the Christian fundamentalist justifications for violence I don't hear them asking that question, "What would Jesus do?" From a fundamentalist Christian point of view, shouldn't that be the key question to ask? What is more "fundamental" to Christianity than Jesus? Perhaps the teachings of Jesus most unpopular with Christian fundamentalists (and other Christians too) are his statements about loving our enemies and not just seeing the "specks" in your adversary's eye, but also the "log" in your own.

Genuine faith, in all our traditions, either forbids violence as a methodology or, at most, says it must always be limited and lamented, never glorified or celebrated. True religion always seeks alternatives to violence that try to break its deadly cycle. Modern fundamentalism, instead, offers what Walter Wink calls "the myth of redemptive violence"—that somehow violence can save us after all.

The lesson of September 11 must not be to relegate religion to the private sphere. Rather, the future of politics must become a discourse about values, which includes moral and religious ones. We should talk less about the ideological categories of Left and Right, and more about what kind of people we want to be, what kind of community, what kind of world. Even those who don't trust organized religion often do want to talk about spiritual or moral values. Applying spiritual values to politics will be the key.

The Anglican archbishop Desmond Tutu is a good example of that endeavor. As chair of South Africa's Truth and Reconciliation Commission, Tutu provided a clear and compelling example of how to deal with the societal conflicts we must resolve if we're going to move forward, and how the cycle of revenge and violence can be broken and even healed. Tutu continually speaks of the need for a "spirituality of transformation," one that can even change politics.

In a large living room just blocks from Harvard Square, I gave a talk on the topic of religion and public life to a group of Boston's best and brightest Left/liberal intelligentsia. I thought I had given a reasonable talk, but after I had finished, the first questioner made me pause. He asked, "But Jim, what about the Inquisition?" I said, "Well, I was against it at the time. And I'm still opposed to it!" Then I challenged him, "Unless you want me to raise the specter of the Communist butcher Pol Pot and his brutal Khmer Rouge regime in Cambodia every time you talk about the need for a comprehensive national health plan, why don't we move on to a better discussion?" "Touché!" replied my liberal questioner, acknowledging the point about the unfairness of straw men and agreeing to have a better conversation on faith and politics.

Today there are new fundamentalists in the land. These are the "secular fundamentalists," many of whom attack all political figures who dare to speak from their religious convictions. From the Anti-Defamation League, to Americans United for the Separation of Church and State, to the ACLU and some of the political Left's most religion-fearing publications, a cry of alarm has gone up in response to anyone who has the audacity to be religious in public. These secular skeptics often display an amazing lapse of historical memory when they suggest that religious language in politics is contrary to the "American ideal." The truth is just the opposite. As we have discussed, many of the most progressive social movements in American history—anti-slavery, women's suffrage, the fight for child labor laws, and the civil rights movement—had overt religious roots and motivations.

Why do so many liberals seem supportive of religious language when it is invoked by black civil rights leaders like Martin Luther King Jr., but recoil when such language is employed by white political leaders? Is there a subtle kind of racism going on here, where religion is okay for liberals as long as it comes from black or poor people? Are black people supposed to be culturally religious (love those black choirs), while white believers are intellectually suspect?

Secular fundamentalists make a fundamental mistake. They believe that the separation of church and state ought to mean the separation of faith from public life. While it is true that some conservative religionists do want to blur the boundaries between church and state, most advocates of religious and moral values in the public square do not. Most of us do not support state-sanctioned prayer in public schools, or school-backed prayers at high school football games in the South, or huge granite blocks inscribed with the Ten Commandments in every courthouse.

A good and fair discussion of how a candidate's faith shapes his or her political values should be viewed as a positive thing—it's as relevant and appropriate as many other facts about a politician's background, convictions, and experience for public office. The more talk about values, the better in political campaigns, and religion is a primary source of values for many Americans. Clearly, minority religions and nonreligious people must always be respected and protected in our nation. But the core commitments of religious liberty need not be compromised by an open discussion of faith and public life. Indeed, the right kind of talk about religion and politics represents, according to astute political observer and columnist E. J. Dionne, "not a threat to religious liberty but its triumph."[6]

The secular fundamentalists tell us that religion should be restricted to one's church or family. No talk of faith, they seem to be saying, ought to be allowed to seep into the public arena for fear of violating the First Amendment or alienating the nonreligious. Perhaps we should make sure all our church and living room windows are shut tight so that no words of faith are overheard in the wider society. When vice presidential candidate and devout Jew Joe Lieberman said, "The Constitution guarantees freedom of religion not freedom from religion,"[7] one of his most virulent critics was prompted to admit that she really did want a society free from religion. Fortunately, the Constitution protects the free speech of believing *and* unbelieving citizens.

The secular fundamentalists want to hide faith "under a bushel," but the gospel specifically instructs us not to do that. The purpose of biblical faith is not simply to comfort the believers but to transform the world. But that must always be done in ways that both respect religious liberty and enhance democracy. Catholic teaching says it well—religious faith should serve the common good. And with practitioners like Dr. Martin Luther King Jr., religious faith should help "bend the world toward justice."

Religion is not inherently undemocratic, as the secular fundamentalists want to make it out to be. But religiously motivated citizens must learn that bringing faith into public life doesn't best happen by the takeover of the mechanisms of the state—the school boards in Orange County, for example. They must learn the dynamics and disciplines of prophetic religion. And prophetic faith is the best counterpoint to fundamentalist religion. We bring faith into the public square when our moral convictions demand it. But to influence a democratic society, you must win the public debate about why the policies you advocate are better for the *common good.* That's the democratic discipline religion has to be under when it brings its faith to the public square. And some religious fundamentalists haven't learned that yet. But religious people shouldn't be told just to be quiet, they should be invited to participate as *citizens* who have the right and the obligation to bring their deepest moral convictions to the public square for the democratic discourse on the most important values and directions that will shape our society.

CHAPTER 6

Prophetic Politics

A New Option

PROPHECY IS NOT FUTURE TELLING, but articulating moral truth. The prophets diagnose the present and point the way to a just solution. The "prophetic tradition," in all of the world's great religions, is just what we need to open up our contemporary political options, which are, honestly, grossly failing to solve our most pressing social problems. The competing ideological options, from which we are forced to choose, are perhaps at their lowest ebb in compelling the involvement of ordinary citizens in public life. It is not that people just don't care, but that they feel unrepresented and unable to vote for anything that expresses their best values. That is a serious political crisis, and we need better options.

What would it mean to evaluate the leading current political options by the values of the prophets? What would happen if we asserted that *values* are the most important subject for the future of politics? What if we proposed a "prophetic politics?"

After the 2002 midterm elections, I attended a private dinner for Harvard Fellows in Cambridge. Our speaker was a Republican political strategist who had just won all the major senatorial and gubernatorial election campaigns in which he was involved. Needless to say, he was full of his success and eager to tell us about it. This very smart political operative said that

Republicans won middle-class and even working-class people on the "social" issues, those moral and cultural issues that Democrats don't seem to understand or appreciate. He even suggested that passion on the social issues can cause people to vote against their economic self-interest. Since the rich are already with us, he said, we win elections.

I raised my hand and asked the following question: "What would you do if you faced a candidate who took a traditional moral stance on the social and cultural issues? They would not be mean-spirited and, for example, blame gay people for the breakdown of the family, nor would they criminalize the choices of desperate women backed into difficult and dangerous corners. But the candidate would decidedly be pro-family, pro-life (meaning really want to lower the abortion rate), strong on personal responsibility and moral values, and outspoken against the moral pollution throughout popular culture that makes raising children in America a countercultural activity. And what if that candidate was also an economic populist, pro-poor in social policy, tough on corporate corruption and power, clear in supporting middle- and working-class families in health care and education, an environmentalist, and committed to a foreign policy that emphasized international law and multilateral cooperation over preemptive and unilateral war? What would you do?" I asked. He paused for a long time and then said, "We would panic!"

Virtually every time I'm out speaking on "prophetic politics" during any election year campaign, somebody asks the following question: "How can I vote for what I've just heard?" Some very interesting polling in the last few years shows how increasingly important voters' perceptions of "values" are to their electoral behavior. And most voters feel they can't really vote for their values, at least not *all* their values. In the polling, the values question now goes beyond traditional family and sexual matters and now also includes matters such as "caring for the poor." The problem is that politics is still run by ideological polarities that leave many people feeling left out.

There are now three major political options in our public life. The first political option in America today is conservative on everything—from cultural, moral, and family concerns to economic, environmental, and foreign policy issues. Differences emerge between aggressive nationalists and cautious isolationists, corporate apologists and principled fiscal conservatives, but this is the political option clearly on the ascendancy in America, with most of the dominant ideas in the public square coming from the political Right.

The second political option in contemporary America is liberal on everything—both family/sexual/cultural questions and economic, environmental, and foreign policy matters. There are certainly differences among the liberals (from pragmatic centrists to green leftists), but the intellectual and ideological roots come from the Left side of the cultural and political spectrum—and today most from the liberal/Left find themselves on the defensive.

The third option in American politics is libertarian, meaning liberal on cultural/moral issues and conservative on fiscal/economic and foreign policy. The "just leave me alone and don't spend my money option" is growing quickly in American life.

I believe there is a "fourth option" for American politics, which follows from the prophetic religious tradition we have described. It is traditional or conservative on issues of family values, sexual integrity and personal responsibility, while being very progressive, populist, or even radical on issues like poverty and racial justice. It affirms good stewardship of the earth and its resources, supports gender equality, and is more internationally minded than nationalist—looking first to peacemaking and conflict resolution when it come to foreign policy questions. The people it appeals to (many religious, but others not) are very strong on issues like marriage, raising kids, and individual ethics, but without being right-wing, reactionary, or mean-spirited or scapegoating against any group of people, such as homosexuals. They can be pro-life, pro-family, and pro-feminist, all at the same time. They think issues of "moral character" are very important, both in a politician's personal life and in his or her policy choices. Yet they are decidedly pro-poor, for racial reconciliation, critical of purely military solutions, and defenders of the environment.

At the heart of the fourth option is the integral link between personal ethics and social justice. And it appeals to people who refuse to make the choice between the two.

Who are these people? Many are religious: Catholics, black and Hispanic Christians, evangelicals who don't identify with the religious Right, and members of all our denominational churches who want to put their faith into practice. They are Jews and Muslims who are guided by an active faith and not just a personal background. They are people who do not consider themselves religious, but rather spiritual. And they are people—religious, spiritual, or not—who consider themselves shaped by a strong sense of *moral* values and who long for a political commitment that reflects them.

As I travel the country, I find many people who share this perspective. Still, it is not yet a political option. It should be. As one who has called for a new moral politics that transcends the old categories of both the secular Left and the religious Right, I believe it is time to assert a clear fourth political option. In a recent conversation I had with E. J. Dionne of the *Washington Post,* the columnist said there was a huge constituency of "non-right-wing Christians" and other morally concerned people in the country who need to get organized. Like E. J., they are moderate to conservative on personal moral questions and very progressive on social justice.

As detailed in the beginning of chapter 5, recent polling shows that the more religious voters are, the more likely they are to vote for the conservatives. Given how negatively much of the political Left seems to regard religion and spirituality, this is not surprising. But what if a new political option regarded personal ethics to be as important as social justice and saw faith as a positive force in society—for progressive social change? I think the fourth option could be a real winning vision, and I believe many are very hungry for it. While the political elites and many special-interest groups resist the personal ethics/social justice combination (perhaps because it threatens many special interests), many ordinary people would welcome it.

What we need is nothing less than prophetic politics. We must find a new moral and political language that transcends old divisions and seeks the common good. Prophetic politics finds its center in fundamental moral issues like children, diversity, family, community, citizenship, and ethics (others could be added, like nonviolence, tolerance, and fairness) and tries to construct national directions that many people across the political spectrum could agree to. It would speak directly to the proverb quoted earlier, "Without a vision, the people perish," and would offer genuine political vision that arises out of biblical passages from prophetic texts. Our own ancient prophetic religious traditions could offer a way forward beyond our polarized and paralyzed national politics and could be the foundation for a fourth political option to provide the new ideas politics always needs.

The political class is at war, while the media focus more on the *process* of politics than its content. And while there are certainly very committed partisans on both sides, they seem to be fighting more for their careers than their principles. In the meantime, most Americans are not terribly passionate about their political choices. During election seasons, many voters are undecided until the very end, speak in "lesser of evils" language about their decisions, and wonder if this is really the best America has to offer.

Most simply put, the two traditional options in America (Democrat and Republican, liberal and conservative) have failed to capture the imagination, commitment, and trust of a clear majority of people in this country. Neither has found ways to solve our deepest and most entrenched social problems. Record prosperity hasn't cured child poverty. Family breakdown is occurring across all class and racial lines. Public education remains a disaster for millions of families. Millions more still don't have health insurance or can't find affordable housing. The environment suffers from unresolved debates, while our popular culture becomes more and more polluted by violent and sex-saturated "entertainment." In local communities, people are more and more isolated, busy, and disconnected. Our foreign policy has become an aggressive assertion of military superiority in a defensive and reactive mode, seeking to protect us against growing and invisible threats instead of addressing the root causes of those threats. The political Right and Left continue at war with each other, but the truth is that these false ideological choices themselves have run their course and become dysfunctional.

Prophetic politics would not be an endless argument between personal and social responsibility, but a weaving of the two together in search of the common good. The current options are deadlocked. Prophetic politics wouldn't assign all the answers to the government, the market, or the churches and charities, but rather would patiently and creatively forge new civic partnerships in which everyone does their share and everybody does what they do best. Prophetic politics wouldn't debate whether our strategies should be cultural, political, or economic but would show how they must be all three, led by a moral compass.

Perhaps most importantly, prophetic politics won't be led just by elected officials, lawyers, and their financial backers. Look for community organizers, social entrepreneurs, nonprofit organizations, faith-based communities, and parents to help show the way forward now. Pay particular attention to a whole generation of young people forged in community service. They may be cynical about politics, but they are vitally concerned with public life. The politics we need now will arise more from building social and spiritual movements than from merely lobbying at party conventions. And ultimately, it will influence the party conventions, as successful movements always do.

The prophetic role churches are now undertaking is illustrative of the larger public vocation that may now be required. That role has become more clear in the wake of the election. Without a clear and compelling vision, Democrats had nothing to offer the American people as an alternative to the Bush administration.

With the Republicans offering war overseas and corporate dominance at home, and the Democrats failing to offer any real alternatives, who will raise a prophetic voice for social and economic justice and for peace? Never has there been a clearer role for the churches and religious community. We can push both parties toward moral consistency and their best-stated values and away from the unprincipled pragmatism and negative campaigning that both sides too often engaged in during the recent election.

The courage many church leaders showed in opposing the war with Iraq is an early sign of that prophetic role. So is the growing unity across the spectrum of the churches on the issue of poverty. The truth is that there are more churches committed to justice and peace than belong to the religious Right. It's time the voice of those congregations be heard and their activism be mobilized to become the conscience of American politics in a time of crisis.

We have seen many moments in recent history when the churches emerged as the leading voice of political conscience. Certainly there were key times in the South African struggle against apartheid when the churches there became the critical public voice both for political challenge and for change. Leaders such as Desmond Tutu, Alan Boesak, and Frank Chikane served as both church and public figures at the same time. Who can forget the role of Archbishop Oscar Romero in El Salvador during the 1980s, church leaders in the Philippines during the revolution that ousted dictator Ferdinand Marcos, or the critical opposition to communist rule in Poland? In many other oppressive circumstances, churches and church leaders have risen to the call to prophetic public leadership.

Even in democracies, churches have responded to that same prophetic vocation. In New Zealand during the 1990s, when conservative forces ripped that society's long-standing social safety net to pieces, it was the churches in partnership with the indigenous Maori people who led marches, ignited public protest, and emboldened a wobbly Labour party to recapture the government and restore key programs in health care, housing, and social services. And, of course, it was in the United States that black churches, under the leadership of Baptist ministers such as Reverend Martin Luther King Jr., provided the moral foundation and social infrastructure for a powerful civil rights movement that reminded the nation of its expressed ideals and changed us forever.

In a bitterly divided nation, we face historic challenges. But the political "tie" that the nation is caught in might be a moment of opportunity. It shows that the old options and debates have created a deadlock. This very crisis could open the way for some new and creative thinking and organizing. And that could be very good news indeed. Our political leaders must learn the wisdom that the way to reach common ground is to move to higher ground. And we citizens should start by showing the way.

Voting All Your Values

With religious fundamentalists on one side and secular fundamentalists on the other, how should religious people express their vision of faith and politics? How do we do more than conform to existing options and instead help create better alternatives for solving the pressing problems of our society? And, most particularly, how are we to vote?

First, we should ask what the religious profession of candidates actually means for their policy making. If religious affirmation is not tied to real political values and directions, what good is the expression of such public piety? And how could genuine discussion of moral, spiritual, or religious values in public life contribute to the formation of needed new political options?

When George Bush said in the 2000 presidential campaign that Jesus was "his favorite philosopher," it was fair to ask what that meant for his political philosophy. Would the poor get prioritized on his policy agenda as they did with Jesus? And the same with Al Gore's claim that he often asked himself "What would Jesus do?" Which candidate's political commitments were more likely to focus on the bottom 20 percent of American families, including thirteen million children who are living in poverty and shut out of our national well-being? Bush should have been asked if child poverty would get more attention in his administration than oil and gas interests; Gore should have been asked how he would implement his new populist rhetoric when it inevitably collided with the interests of the corporations that now finance both political parties. It's an appropriate moral concern to ask candidates about their plans to extend health care and education to those without them. And protecting the environment is as much a religious issue as caring for the poor. What if religious voters asked which candidates see human rights and workers' protections to be as important as global trade?

How does the religious principle of the sacredness of human life challenge all the candidates, for example, on abortion, capital punishment, military spending, missile defense, or gun control? Religious people may disagree on the answers to those questions (for example, Joe Lieberman supports both capital punishment and legal abortion while the Catholic bishops do not), but shouldn't the highest rates of both abortion and death-row executions in the Western world be a concern to all those who profess moral values? Couldn't both pro-life and pro-choice political leaders agree to common ground actions that would actually reduce the abortion rate, rather than continue to use abortion mostly as a political symbol? Instead of imposing rigid pro-choice and pro-life political litmus tests, why not work together on teen pregnancy, adoption reform, and real alternatives for women backed into dangerous and lonely corners? Do we really want to dramatically reduce abortion and make it "rare," as Bill Clinton once suggested, or have both sides just continued to treat this deeply important issue as a political football? And here is another tough values question for both parties: How does allegiance to the Christian Prince of Peace or the God of the Hebrew prophets square with a national security policy that still relies on threatening the use of nuclear weapons—something all of our religious traditions abhor? What does the same "sacredness of human life" principle mean for foreign and military policy? Chicago's late Catholic cardinal Joseph Bernardin often spoke of a "seamless garment" of life, which applied to any issue where human lives are at stake, from abortion and euthanasia, to capital punishment and nuclear weapons, to poverty and racism. Applying that consistent life ethic to politics would upset the ideological apple carts of both Republicans and Democrats, which would be a very good role for the religious community.

Speaking of moral concerns, whose policies will better strengthen family life and values (something both liberal and conservative households with children could support), combat the spread of violence both in the popular culture and on our streets, and advance racial justice and reconciliation? Moral character and leadership are important. The attempted separation of personal and public behavior by occupants of the White House doesn't square well with biblical religion or make most of us very happy about the message our kids are getting.

How does one distinguish between symbolic choices and responsible ones? Is voting for candidates who are far from perfect—"the lesser of evils"—a moral compromise or an ethical decision to seek incremental change? The answers to these questions are far from easy. But sorting out the meaning of faith in the world seldom is. One thing is clear: True faith cannot be kept inside the narrow boundaries of the "sacred," as some would suggest, but is intended to be "salt and light" in the midst of what is often called the secular world. Indeed, to change the world is a vocation of faith. Elections are not the most important part of that, but they are indeed one significant piece of faithful citizenship in a political democracy. So think, pray—and then vote.

It is important to remember that the particular religiosity of a candidate, or even how devout they might be, is less important than how their religious and/or moral commitments and values shape their political vision and their policy commitments. If one's religious and ethical convictions don't shape a candidate's (or a citizen's) public life—what kind of commitments are they? Yet in a democratic and pluralistic society, we don't want to evaluate candidates by which denomination or faith tradition they belong to (and only vote for the candidate in our group) or how often they attended church or synagogue (like a tally of votes missed by a member of Congress), but rather to understand the moral compass they bring to their public life and how their convictions shape their political priorities.

There were some positive signs among the Democrats in the 2004 campaign. While Howard Dean's initial forays into religion were clumsy at best (surely someone on his staff must have known that his "favorite New Testament book" of Job was, in fact, in the Old Testament), his concern about losing our "sense of community" in America was a deeply moral and religious one. Perhaps knowing what is contained in the books of the Bible is ultimately more important than knowing where they all are! Dick Gephardt talked about health care as a "moral issue," and John Edwards, who became the Democratic candidate for vice president, sounded like a preacher when he spoke of two Americas and declared that poverty is not only an economic concern, but is "about right and wrong," and that "poverty reduction is a moral responsibility." Joe Lieberman seemed to be regaining his religious voice when he spoke about the poor, and John Kerry, who became candidate for president, was talking about a "broken value system" and not just his war record. Wesley Clark seemed to be comfortable relating his faith journey to social justice, Dennis Kucinich spoke of his moral values all along the campaign trail, and of course, Reverend Al Sharpton spoke like the Pentecostal preacher he is. As we've noted earlier, Kerry and Edwards spoke more strongly about their faith as the campaign wore on. George Bush continued to speak of his personal faith as a motivator and was pushed to be more explicit about how his personal faith applied to social issues like HIV/AIDS.

Perhaps the most mistaken media perception of our time is that religious influence in politics only equates to the politics of the religious Right. The biggest story that the mainstream media has yet to discover is how much that reality is changing. My prediction is that moderate and progressive religious voices will ultimately shape politics in the coming decades far more significantly than the religious Right will.

First, it is a mistake to regard all the conservative Christians who are sympathetic to the religious Right as people who want to impose their religious values on fellow citizens. To be conservative is not necessarily to be fundamentalist. In my experience, most are motivated more by *defensive* feelings than *offensive* intentions. Many people of faith (of all different political stripes) are concerned about the coarsening of American life, the unraveling of traditional values, and threats to religious precepts like the sacredness of human life. What our children are being subjected to on television, in school yards, and now over the Internet has been a greater motivating force for political involvement than keeping the Ten Commandments in state courthouses or any number of other issues on the agenda of the political right wing.

It is true that some of the religious Right's leaders are indeed theocrats—those who would impose their versions of morality on the nation if they ever had the chance. As we have discussed, their mistake was the attempt to take political power in and through the Republican Party to impose a "moral" agenda from the top down. But even more disconcerting has been what is missing from the agenda of the religious Right—such as concern for the poor and racial justice—and the appeals of the movement's leaders to affluent self-interest over biblical imperatives. How did tax cuts for the rich become a religious imperative? Those blatant hypocrisies of the religious Right are becoming more and more evident, especially to a new generation of evangelical Christians.

History does teach us that the most effective social movements are also spiritual ones, which change people's thinking and attitudes by appealing to moral and religious values. Those movements change the cultural and political climate, which then makes policy changes more possible, palatable, and, yes, democratic. The best example of doing it right, as we have said, is the American civil rights movement, which was led by ministers who appealed directly to biblical faith. I believe that will be the more likely pattern for future movements that combine faith and politics, replacing the more politically conformist model of the religious Right.

It is another mistake to be always fighting against the religious Right, as many frightened liberals continue to do. The electoral strength of the Moral Majority or the Christian Coalition was always exaggerated by both themselves and the media, but now their ability to "deliver" decisive blocs of votes is greatly diminished. The Republican Party is now careful at party conventions to hide its religious fundamentalists, as mainstream voters have soured on both their message and style.

There was indeed a period in the 1980s and 1990s when the perception abounded that "Christian" involvement in politics meant Christian Right. But that hasn't been true for some time now. The U.S. Catholic Conference of Bishops has provided real social policy leadership in the last decade and has become a clear alternative to the dominance of groups like the Moral Majority and the Christian Coalition. The Catholic bishops are opposed to abortion, but also to capital punishment, increased military spending, and notions of welfare reform that neglect poor working families.

Influential organizations like World Vision and World Relief are playing leadership roles in sensitizing and mobilizing evangelicals for disaster relief, comprehensive economic development, and global campaigns to combat HIV/AIDS. Influential groups like Evangelicals for Social Action led by Ron Sider, the Christian Community Development Association led by John Perkins, and the Evangelical Association for the Promotion of Education led by Tony and Bart Campolo, have defined poverty reduction as a compelling moral, biblical, and decidedly *evangelical* issue. Several conservative evangelical denominations, like the Evangelical Covenant Church, have now made the critical links between evangelism, compassion, and social justice. Evangelical Christian colleges have shown a deepening social conscience in their curriculums over issues of poverty and race, often putting them at odds with the Evangelical right wing.

New leadership in the mainline Protestant churches also promises greater ecumenical collaboration on social issues with both evangelicals and Catholics. Many of those denominations have played critical leadership roles in faith-inspired movements like Jubilee 2000, a broad coalition for debt cancellation for the world's poorest countries, which has demonstrated a real impact on international governmental policy and has shown the power and potential for crossing old dividing lines in the religious community. We will discuss the Jubilee campaign in more detail in chapter 17, and a new anti-poverty mobilization led by the National Council of Churches that promises to revive the old social gospel.

Overtly Christian organizations like Bread for the World (a nationwide Christian citizens' movement seeking justice for the world's hungry people by lobbying our nation's decision makers) and Habitat for Humanity (a non-profit, ecumenical Christian housing ministry) have demonstrated a strong advocacy for faith-based justice in the public square and have established a clear place for moderate and progressive religious perspectives. Call to Renewal has been successful in pulling the churches together on the issue of poverty across the wide spectrum of the church's life, thus offering another alternative to the Religious Right's silence on that issue. More and more

Christians are openly advocating the kind of biblically based social justice agenda that has long characterized the readers of magazines like *Sojourners* and *The Other Side*. Thus the religious Right is now only one of many voices on issues of political ethics, as it should be.

The public conversation about religious and moral values now has the potential to be a serious and thoughtful discussion in America. The good news for religion and public life in America is that the word *religious* will no longer always be followed by the word *Right*.

To move away from the bifurcating politics of liberal and conservative, left and right, would be an enormously positive change and would open up a new "politics of solutions." Right now, Washington responds to a problem or crisis in two ways. First, politicians try to make us afraid of the problem, and second, they look for somebody to blame for it. Then they watch to see whose political spin succeeded, either in the next poll or the next election. But they seldom get around to actually solving the problem. The media make everything worse by assuming that every political issue has only two sides, instead of multiple angles for viewing and solving the problem. Addicted to conflict as their media methodology, they want always to pitch a fight between polarized views instead of convening a public discussion to find serious answers. The answer is to put values at the center of political discourse and, in every public debate, to ask what kind of country and people we really want to be. We would find new agreements across old political boundaries and new common ground among people who agree on values and are ready to challenge the special interests on all sides who are obstructing the solutions most Americans would support. Ideologies have failed us; values can unite us, especially around our most common democratic visions.

PART III

Spiritual Values and International Relations

When Did Jesus Become Pro-War?

CHAPTER 7

Be Not Afraid

A Moral Response to Terrorism

My travel and speaking schedule is still pretty busy, but fatherhood has changed all my routines. I try not to be away for more than one or two nights running, even coming home in the middle of the night to be there in the morning when Luke and Jack wake up. Being away from the boys and Joy for longer than a day or two is too much for me. And now that my son Luke is six years old, we can have great phone conversations whenever I am away, often several times a day.

When Luke was just four, I was speaking at a conference in Florida and had already talked with him a couple of times during the day. But when I got back to my hotel room that night, there was a voice mail waiting for me—from Luke. After all, several hours had passed since we had spoken, and there were lots of things to tell me about! His little voice always brings a smile to my face, as he enthusiastically gives me a blow-by-blow account of his day's activities. Luke's phone sign-off has become especially wonderful and always warms the hearts of his mom and dad: "Daddy, I love you, I like you, and you're incredible!" It's just the kind of unconditional affirmation we all need but find hard to accept. (I suspect it's close to what God wants to tell us sometimes.) But it's probably easier to receive it from a child. But then

Luke finished with something he had never said before. Completely out of the blue, my four-year-old son said, "Daddy, don't be afraid."

I could hear his Mom catch her breath in the background. Where did a little child find those words? They hadn't been talking about anything that could have prompted it. "Be not afraid" were the most frequent words of Jesus to his disciples in the New Testament. He kept repeating the instruction over and over, as if he knew how much his followers needed to hear it.

Today, in the United States, we have a foreign policy based primarily on fear. We move back and forth between new national color codes indicating the level of danger from terrorist attack. The Office of Homeland Security regularly moves the nation to Orange Alert—the second highest state of risk from terrorist violence.

September 11, 2001, changed our lives, and since then, we have become a nation always living in fear. We were terrified by the murderous attacks on the World Trade Towers and the Pentagon, and even after dramatic war victories in Afghanistan and Iraq, America is still afraid. Indeed, the war in Iraq was argued and justified, almost entirely, on the basis of fear. It was Trappist monk Thomas Merton who said many years ago, "The root of war is fear."[1]

There are indeed real dangers prowling about our world. Prudence and strategic action are called for, as is much deeper reflection on the causes of those dangers. But fear can cause us to give up important things, to accept other things that violate our own best values, and even to do terrible things to other people. Fear has led us into a new foreign policy based on preemptive and potentially endless wars—which are not likely to remove our fears and could likely make the dangers we face even worse.

September 11 shattered the American sense of invulnerability. But instead of accepting the vulnerability that most of the rest of the world already lives with, and even learning from it, we seem to want something nobody can give us—to erase our vulnerability. We want it to just go away. If the government says more wars can do that, many people will say fine. If they say suspending civil liberties can do that, many will say fine. If they claim spending more and more of our tax dollars on the military and homeland security will do it—at the expense of everything else—many will say fine. But we simply can't erase our vulnerability, not in this world and not with the human condition being as it is. To be prudent and vigilant in the face of danger is good. But when a government offers to take away our vulnerability, it borders on idolatry.

I am convinced that mere political action to counter a foreign policy based primarily on fear will not be enough. We must go deeper, to the roots of the fear. Courage is not the absence of fear, but the resistance to it. For people of faith, it means trusting God in the face of our vulnerabilities. We need nothing less than the healing of the nation, beginning with our own fears. That healing will be essential to make peacemaking possible. I'm not sure what such healing will mean, but the prophetic words of a four-year-old, "Daddy, don't be afraid," keep me asking the question.

Several weeks after the attacks of September 11, I spoke at an interfaith prayer service at St. Aloysius Church in Washington, D.C., introducing a ceremony of lighting candles as a sign of commitment and community. These were my remarks:

> Lighting a candle at an interfaith service is something many of us have done more times than we can remember. Speaking the language of darkness and light on religious occasions and in liturgical seasons has also become a matter of habit. But our darkness feels very real and powerful in this moment—almost impenetrable and threatening to close in on us. And the light we need is feeling almost urgent.
>
> Old familiar spiritual words must take on a new reality for us now—and a new sense of mission: words like "Let there be light!" and "The light has come into the darkness, and the darkness has not overcome it."
>
> So tonight, we don't just light candles; we make a commitment. More than we knew before September 11, there are many dark places in the world where unspeakable violence against large numbers of innocent people is being planned. Let those places be exposed to the light of day and the violence be thwarted.
>
> There are many other dark places in the world, where grinding, miserable, and life-destroying poverty carries out daily violence to other innocents—out of the view of a fast-moving, affluent world. Let those hidden places and people be exposed to the light of justice.
>
> There are also dark places within us and in our nation that might lash out from our deep woundedness, grief, and anger, carelessly inflicting more pain on innocent people. Let the light of compassion, and reason, prevent us from spreading our pain.

There are dark places within and among us that might retaliate from fear and revenge against even our fellow citizens who happen to be Muslim or Arab American—many of whom tonight feel the darkness of their own fear. Let the light of tolerance and solidarity bind us together and not let us be torn apart.

And tonight, most of us feel the darkness of our own confusion about how this happened, why this happened, and how to protect ourselves, our families, and the world. Let us be illumined by the light of understanding.

Let us not just cling to old ways of thinking, but go deeper than we ever have before to seek new answers. We must learn to carefully comprehend the connections between the violence of the world, while never allowing ourselves to tolerate any of it—ever again.

And let the light of courage equip us to face the darkness that lies so thick and heavy before us. Courage to heal the darkness in ourselves. Courage to reveal the darkness in the very structure of our world. Courage to confront the darkness in the face of evil we saw on September 11. And let us remember that courage is not the absence of fear, but resistance to it.

So now, let us light our candles as an act of commitment so that the darkness will not overcome the light.

September 11, 2001, is a day we will always remember. Each of us has a story about where we were when the terrorist attacks struck. Being in Washington, D.C., on that fateful day was especially harrowing. At our house, we were all getting ready for the day's activities, with Luke going off to preschool and me about to head for the office. National Public Radio was on, as usual, and I vaguely heard the report of a plane hitting one of the towers at the World Trade Center. It appeared to all just to be an accident. But minutes later, a second plane that had been turned into a lethal weapon exploded into the other tower, and Americans began to realize with fright that we were under attack.

Then the Pentagon was struck by another hijacked airliner, and we could actually see the smoke rising over downtown Washington. Immediately, there were rumors of more suicidal planes in the air, some rumored to be on the way to Washington, D.C. Joy, Luke, and I huddled to hold on to each other and to pray. After it appeared the danger of more attacks had passed, I rushed to the office to gather with the rest of our staff. We all joined together in the meeting room, keeping vigil with both the television news and our corporate prayers. Calls were coming in from around the country from family, friends, and magazine subscribers. After people were told that we thought we were okay, the questions turned to how Christians ought to respond to these horrific events. What I most remember from that extraordinary day were the two questions on our minds. The first was, "Are we safe?" And the second, which soon followed, was "What should be our response?"

The next day, I sat down and tried to craft a first response. The result was the statement called "Deny Them Their Victory: A Religious Response to Terrorism." We released the statement, along with other leaders from the faith community who helped shape it, and it traveled quickly via the Internet. Within two weeks, nearly five thousand religious leaders had signed on to it. "Deny Them Their Victory" became an early sign of the faith community's quest to find a moral response to terrorism. This was our statement:

Deny Them Their Victory: A Religious Response to Terrorism

We, American religious leaders, share the broken hearts of our fellow citizens. The worst terrorist attack in history that assaulted New York City, Washington, D.C., and Pennsylvania has been felt in every American community. Each life lost was of unique and sacred value in the eyes of God, and the connections Americans feel to those lives run very deep. In the face of such a cruel catastrophe, it is a time to look to God and to each other for the strength we need and the response we will make. We must dig deep to the roots of our faith for sustenance, solace, and wisdom.

First, we must find a word of consolation for the untold pain and suffering of our people. Our congregations will offer their practical and pastoral resources to bind up the wounds of the nation. We can become safe places to weep and secure places to begin rebuilding our shattered lives and communities. Our houses of worship should become public arenas for common prayer, community discussion, eventual healing and forgiveness.

Second, we offer a word of sober restraint as our nation discerns what its response will be. We share the deep anger toward those who so callously and massively destroy innocent lives, no matter what the grievances or injustices invoked. In the name of God, we too demand that those responsible for these utterly evil acts be found and brought to justice. Those culpable must not escape accountability. But we must not, out of anger and vengeance, indiscriminately retaliate in ways that bring on even more loss of innocent life. We pray that President Bush and members of Congress will seek the wisdom of God as they decide upon the appropriate response.

Third, we face deep and profound questions of what this attack on America will do to us as a nation. The terrorists have offered us a stark view of the world they would create, where the remedy to every human grievance and injustice is a resort to the random and cowardly violence of revenge—even against the most innocent. Having taken thousands of our lives, attacked our national symbols, forced our political leaders to flee their chambers of governance, disrupted our work and families, and struck fear into the hearts of our children, the terrorists must feel victorious.

But we can deny them their victory by refusing to submit to a world created in their image. Terrorism inflicts not only death and destruction but also emotional oppression to further its aims. We must not allow this terror to drive us away from being the people God has called us to be. We assert the vision of community, tolerance, compassion, justice, and the sacredness of human life, which lies at the heart of all our religious traditions. America must be a safe place for all our citizens in all their diversity. It is especially important that our citizens who share national origins, ethnicity, or religion with whoever attacked us are, themselves, protected among us.

Our American illusion of invulnerability has been shattered. From now on, we will look at the world in a different way, and this attack on our life as a nation will become a test of our national character. Let us make the right choices in this crisis—to pray, act, and unite against the bitter fruits of division, hatred, and violence. Let us rededicate ourselves to global peace, human dignity, and the eradication of the injustice that breeds rage and vengeance. As we gather in our houses of worship, let us begin a process of seeking the healing and grace of God.

"Let us make the right choices" in response to terrorist violence remains, even today, the key issue. Just a few weeks after the attacks, I went to Ground Zero in New York City. Awestruck was the only word that came to mind after seeing the scope of the devastation at what was once the World Trade Center on the Lower West Side of Manhattan. Thanks to the good graces of the local clergy and Red Cross, I was able to get right onto the site. All the pictures I'd seen couldn't fully capture the enormity of the destruction that stood before me. Standing on the pile, I prayed and wondered what might come out of this incredible and atrocious event.

I met people doing extraordinary things that day, but I was especially struck with several groups of firefighters from cities around the country and Canada. They had come to attend memorial services and "just to be here," as one young fireman told me. After speaking to several of these men and women, I realized what was going on. They were pilgrims visiting a holy site. That's what their faces and voices revealed to me. I've been to other holy sites and seen other pilgrims and now saw the same things at a place that would forever be remembered.

Much has been said about the heroism in response to September 11 on the part of the firefighters, police officers, and rescue workers who literally risked and even gave their lives for other people. I visited St. Paul's Church, close to Ground Zero, which had been set up as a support center and spiritual retreat for exhausted and overwhelmed emergency personnel. Everything from new boots, clean socks, eye drops, sandwiches, shoulder rubs, personal counseling, and the Holy Eucharist was being lovingly offered. The sanctuary was filled with notes of thanks, photos of the victims, and teddy bears for the lost children in this place that was "being the church" for believers and non-believers alike.

We don't visit holy sites just for the ritual; we go to be changed. How will the events of September 11 change us, I wondered? Will they change us? I began to believe that September 11 could either become a doorway to transformation—or set us back for years. America had been attacked more massively and viciously than at any other time in our history. How would we respond? What were the choices before us?

That Sunday, I preached at New York City's Cathedral of St. John the Divine. I could see the pain and fear on the parishioner's faces when I said,

> Our gospel text from Luke 21 today sounds eerily contemporary. "Not one stone will be left upon the other. . . . You will hear of wars, insurrections, but do not be terrified. . . . Nations will rise up against nations, kingdoms against kingdoms. . . . There will be earthquakes [or perhaps earthshakes in New York and Washington]. . . . There will be famines in various places. . . . There will be dreadful portents and great signs, arrests, persecutions, disaster, calamity, war and suffering. . . ."

> The focal point of this text is verse 13, "This will give you an opportunity to testify." I believe that people of faith, in this critical moment, have an opportunity "to testify." Verse 15 says, "I will give you words and wisdom that none of your opponents will be able to withstand or contradict, and by your endurance, you will gain your souls."

September 11 was a *teachable moment*—a time to understand and critically examine ourselves and our relationship to the world. The loss of so many lives, which were sacred, and the heroism of sacrifice offered there made Ground Zero in New York City a sacred place. In a way, we've all become pilgrims now. My hope remains that this pilgrimage could still change us.

Something extraordinary had just happened to us as a nation. We experienced both our vulnerability and our unity as never before. Out of horrible death and destruction came powerful acts of compassion, courage, and heroism. Thousands of us were killed, and the whole nation felt deeply connected to the lives lost. Our reaction to these momentous events would truly be "a test of our national character," as the religious leaders' statement said.

Most every American wanted a strong response to the attacks of September 11. But two paths emerged in reaction to the terror that had been visited upon us. One spoke the language and spirit of justice and invoked the rule of law in promising to bring the perpetrators of terrorist violence to accountability. Those who so violated the standards of civilized life, and the human values we hold most dear, should never be allowed to escape judgment and punishment—and the danger of even more terror must be urgently prevented.

The other path used the language of war and invoked a spirit of retribution and even vengeance, emotions we can all understand. A "war on terrorism" summons up the strength and resolve to stop these horrific acts and prevent their cancerous spread. But the war language fails to provide moral and practical boundaries for that response. It could lead to indiscriminate retaliation and the "collateral damage" of even more loss of innocent life.

Americans have seldom seen up close or felt the pain that comes from the deliberate destruction of innocent life on such a scale. Until September 11, it has only been in foreign lands where we have observed the horrible loss that accompanies the massive and violent rending of families and relationships in unspeakable events. Now we understood what many people who inhabit this planet with us have been forced to live with. I even noticed a difference in the response to September 11 between the youth of inner-city neighborhoods and their counterparts in the suburbs. The city kids were certainly horrified by September 11, but they were less shocked. Random violence that suddenly removes a loved one from your life is an experience that many young people who live in drug- and violence-torn urban war zones already knew. But for most Americans, it was a terrifyingly new and utterly shocking experience.

But it is just that collective experience of terrible pain that could shape our response. In a radio interview shortly after September 11, I heard one woman speak against the danger of simple retaliation: "Mr. President, don't spread our pain." Even from some of the family members of those lost on September 11, we began to hear pleas for a response born of our best selves, and not a reaction from our worst impulses. Some of them formed a new organization called "Families for a Peaceful Tomorrow," which sought to bring healing out of suffering. I have been very impressed by those family members whom I have met and the courageous role they continue to play. From them and others, we have heard voices warning that we must not become the evil we loathe in our response to it and that we should respond out of *our* deepest values, not the terrorists'.

It is, indeed, the victory of the terrorists that must be denied. They and what they represent must be soundly defeated, but the question we faced was how to do that. The religious leaders said, "We can deny them their victory by refusing to submit to a world created in their image. . . . We must not allow this terror to drive us away from being the people God has called us to be." While they demanded that "those responsible for these utterly evil acts be found and brought to justice," they insisted that "we must not, out of anger and vengeance, indiscriminately retaliate in ways that bring on even more loss of innocent life."

Shortly after September 11, a Gallup poll reported that 73 percent of the American people believed our response should be targeted "only against those responsible" for the terrorist attacks on America.[2] I believe such public opinion was motivated not only by moral considerations, but also by pragmatic concerns. Bombing the children of Kabul and Baghdad created utter glee among the Osama bin Ladens of the world, who finally are able to raise the armies of terror they've always dreamed of. It also deprived us of the moral high ground that the United States held in world opinion immediately after September 11—for the first time in many years.

Many of us said a more courageous response would be required. Discipline, patience, and perseverance in vanquishing the networks, assets, and capabilities of violent terrorists was the path most likely to be effective rather than merely cathartic. An even more courageous national commitment would be to face honestly the grievances and injustices that breed rage and vengeance and are continually exploited by terrorists to recruit the angry and desperate.

The debate about which path to follow began to take place in conversations across America, including at the highest levels of political power. Despite American anger at the attacks, there was significant public opinion opposing a massive military response. At home, President Bush's admirable call to respect and protect Arab Americans and Muslims helped to defend them against reprisals in all our communities.

But in addition to the vocation of protecting innocent lives against military retaliation and defending our Arab or Muslim fellow citizens, some American religious leaders began to take on the prophetic role of answering why this happened or, as many have put the question, "Why are so many people angry at us?" The first two tasks, while major undertakings, were easier to define. It was the third challenge that required our best discernment and genuine soul-searching.

It is indeed impossible to comprehend adequately the terrorist attacks of September 11 without a deeper understanding of the grievances and injustices felt by millions of people around the world. That is a painful subject that the U.S. government mostly refuses to engage, the mainstream media avoid, and many Americans are unable to hear when they are feeling such mourning, grief, and anger. Indeed, the discussion has the potential to further divide, hurt, and blame ordinary people who already feel very vulnerable and under attack.

In bringing to justice the few thousand estimated to be involved in murderous terrorism, our response must not inflame and infuriate the tens of millions more in the Arab world (and elsewhere) who, to use even former

Secretary of State Colin Powell's words, "hate our presence in parts of the world that they think we should not be in."[3] That sentence must be explained for the American people. It means that many people in the world may not be mad at us for "our values," as the Bush White House continues to say, but for "our policies." But the U.S. policies that most anger people around the world are generally unknown to most Americans—and therein lies the problem of talking about this issue.

Perhaps the religious community can play a crucial role here because it is itself an international community and not just an American one. We also should have the capacity for self-criticism and even repentance, while national governments are seldom good at either. The truth that most of the world knows is that the U.S. government has far too often supported military dictators in Latin and Central America, Asia, and Africa who have murdered as many or more innocent people as Saddam Hussein. The truth is that the United States has not been an honest broker for Middle East peace and has not sought the proper balance between Israeli security and Palestinian human rights. The truth is that American and Western appetites for oil have led to a corrupt and corrupting relationship with despicable Arab regimes. The truth is that the United States sits atop and is the leader of a global economy in which half of God's children still live on less than two dollars a day, and the United States will be blamed around the world for the structures of injustice that such a global economy daily enforces. To speak these truths is very hard, sometimes especially in American middle-class congregations, but speaking hard truths is part of the prophetic religious vocation.

Yet such a hard conversation could illuminate the confusion many Americans feel and could actually help in the necessary process of national healing while offering practical guidance for preventing such atrocities in the future. It takes courage to face difficult questions. President Abraham Lincoln, unlike most American presidents, pushed the nation to look at its own sins in a time of crisis, to dig deep into our spiritual traditions and ask whether we are on God's side, rather than the other way around. We desperately need a Lincolnesque quality of self-examination in this historical moment, but President Bush's speeches since September 11 have lacked it. So far, the White House has missed that opportunity in the path against terrorism that it has chosen.

But despite the initial path the nation has taken, we still have the task of going to the roots of global terrorism, and the religious community must help the nation do that. But talk of national self-examination is complicated in this situation. Some critics on the political Left have sounded almost like another version of the Reverends Jerry Falwell and Pat Robertson, blaming America for these attacks as if we deserved it for our many national sins—though they would substitute U.S. foreign policy, militarism, and global economic domination for the religious Right's diatribes against abortion, gays, feminists, liberals, and the ACLU as being the moral causes of the terrorist attacks.

For the crucial work of self-examination, at least three things are important. First, in the necessary prophetic ministry of telling the truth about American global dominance and its consequences, we should be clear not to imply that America—including the victims of the September 11 attacks and their families—deserved that great day of evil as some kind of judgment, as Falwell, Robertson, and some left-wing critics have suggested. In a powerful statement released September 17, 2001, by Palestinian poets, writers, intellectuals, and political leaders—who all have deep grievances with American foreign policy in the Middle East—the line was drawn: "No cause, not even a just cause, can make legitimate the killing of innocent civilians, no matter how long the list of accusations and the register of grievances. Terror never paves the way to justice, but leads down a short path to hell."[4] Their statement was called, "But Then, Nothing, Nothing Justifies Terrorism," a sentence that should serve as a fitting final word in any discussion about all the injustice that lies behind terrorist acts. We must draw that same line.

Second, we must not make the mistake of thinking that these terrorists are somehow freedom fighters who went too far. On the contrary, the people responsible for the September 11 attacks were not out to redress the injustices of the world. Osama bin Laden's network of terror simply creates great new oppressions, as was evidenced in the Taliban, the regime that represented their vision for the future. Their terror is not about correcting the great global gulf between haves and have-nots, about the lack of even-handed Middle East policies, or about the absence of democratic freedoms in corrupt Gulf States.

The terrorists don't want Saudi Arabia to respect human rights but to be more like the Taliban regime—under which girls couldn't go to school, acid was thrown on the faces of women without head covers, and any religion or lifestyle different than their fanatical extremism was exterminated. For these terrorists, the only "just" solution for the Middle East and the whole Arab world would be to expel all Jews and Christians. And their willingness, even eagerness, to inflict weapons of mass destruction on whole populations is beyond dispute.

The root of the terror attacks is not a yearning for economic justice for the poor and oppressed of the world. It is motivated rather by the ambition of a perverted religious fundamentalism for regional and global power; one that rejects the values of liberty, equality, democracy, and human rights. However much the United States has fallen short of those professed values and has often contradicted them, this terrorism is also an attack on the values themselves; it is not violence in their name or on their behalf.

And if we are to tell the truth about America, let us also tell the truth about the terrorists. We are accustomed to thinking in a political and economic framework. This time, we need to shift and understand motivations that are more ideological and theological. The evil of bin Laden and his network of terror may have been foolishly strengthened by the support of the CIA during the Cold War, but this evil is not a creation of American power. Indeed, to suggest, as some on the Left have done, that this terrorism is an "understandable consequence of U.S. imperialism" is a grave mistake of both moral and political analysis. The terror of bin Laden's al Qaeda network is not just a reaction to "American Empire," but the radical assertion of an ambitious new religious and political empire of their own.

Third, we must carefully distinguish between seeing global injustice as the cause of terrorism and understanding such dramatic inequalities as the breeding and recruiting ground for terrorism. Grinding and dehumanizing poverty, hopelessness and desperation, clearly fuel the armies of terror, but a more ideological and fanatical agenda is its driving force. Therefore, the call for global justice, as a necessary part of any response to terrorism, should never be seen as an accommodation, surrender, or even negotiation with the perpetrators of horrific evil. A serious agenda of global poverty reduction, for example, would be an attack on the terrorists' ability to recruit and subvert the wounded and angry for their hideous purposes, as well as being the right thing to do. Evidence shows that at times when the prospects for peace appeared more hopeful in the Middle East, the ability of terrorist groups operating in the region was greatly diminished. We must drain the swamps of injustice in which the mosquitoes of terror breed, but without seeming to justify or excuse the utterly inexcusable acts we witnessed in New York and Washington. We must not shirk from rooting out the terror we have experienced, but we must also commit ourselves to the prophetic task of examining its deeper roots. Above all, we must find a way to make this a teachable moment rather than merely a blame game.

Despite the famous arrogance of too many American travelers overseas, many people around the world have a real affection for the American people while feeling a real antipathy toward the policies of the U.S. government. If ordinary Americans are to find a deeper understanding of why so many people are angry at us, we will need to overcome our appalling ignorance of world geography and international events and develop a much deeper comprehension of what the American government is doing in our name.

And a real international effort against terrorism must demonstrate a new compassion, generosity of spirit, and commitment to relieve the suffering of precisely those people who have been abandoned and abused. Yes, let us stop terrorist plans to hurt more people, but then let us undertake a massive and collective effort that clearly demonstrates the relationship between halting terrorism and removing injustice. Suffering people everywhere would see the clear signal, and the recruiters of pain would be dealt a death blow.

It is indeed time for justice—for the perpetrators of terror and for the people our global order has, for so long, left out and behind. How we respond to these murderous events will shape our future even more than the terrorists can. As the religious leaders' statement pleaded, "Let us make the right choices."

But since September 11, our government has made many bad choices. Military spending and national security needs have demanded so much of our attention and resources that there is little left over for anything else. We've seen the loss of crucial civil liberties in the name of national security. The healthy desire for energy independence has been used to justify drilling in sensitive environmental habitats. Primarily military solutions to terrorism threaten to dominate the nation's foreign policy for years to come, leading to wider wars, recruiting even more terrorists, and fueling an unending cycle of violence. And the critical questions of *how* to defeat terrorism and resolve conflicts in less violent ways have been given far less attention than they deserve.

We saw evil on September 11, and the world was horrified. But to describe our own nation as only "good," as President Bush has done, is to miss the critical moment for self-reflection. Instead of facing the many roots of this evil, we have turned away from our own responsibilities. Yes, September 11 has set us back. Instead of looking to ourselves or to American foreign policy to see how we have contributed to the grievances and injustice that breed terrorism, we now face the spiritual danger of defining the struggle as simply a battle between good and evil.

Poverty has fallen off the national agenda. Food banks, soup kitchens, shelters, and many other social service providers are suffering from diminishing donations. Faith-based organizations around the country are reporting shortfalls, and fearful people have not been challenged to be generous people.

But I still believe there are opportunities to respond to the horrible events of September 11 in better ways—invitations, even, for national transformation. In the days following the attacks, it was difficult to find parking spaces outside our churches, synagogues, and mosques. The deeply felt need for spiritual resources and insight in a confusing and fearful time can drive us to deeper places, not just provide a religious affirmation for American patriotism and civil religion. Faith communities can still help provide moral and political guidance in this crisis and even offer prophetic words on what kind of people we ought now to be. It still depends on the choices we make.

All those who were not included in our national unity before September 11 could now be invited to the table—including the thirteen million children in America who are still poor. To abandon our poorest and most vulnerable citizens mocks the God of compassion and undermines any notion of national strength and security. If we ever hope to experience a new and deeper understanding of unity after September 11, it would be a good time to make sure everybody was included in it.

During the crisis, my son Luke really connected to "the helpers." We all did. But did we just admire the heroism after September 11, or did we learn from it? As I watched the horrific images of the World Trade Center towers collapsing over and over again, a clear mantra emerged for me: We all suffered together; now we must all heal together. It has become a strong realization and almost a calling. CEOs and janitorial workers, senior vice presidents of brokerage firms and waiters and cooks, law partners and firefighters and police officers—all were lost on September 11. Their families and friends felt a deeply shared sense of pain and loss. The enormous tragedy of New York and Washington, and the extraordinary heroism in its aftermath, knew no boundaries of race, class, country, or religion. The evil that assaulted us created equality in our vulnerability, and a commonality in our response, which I believe was partly responsible for all the talk of "national unity" afterward.

Yet the same day in October 2001 that the U.S. Postal Service issued a new stamp declaring "United We Stand," the House of Representatives passed its "economic stimulus" package loaded with tax breaks and handouts for our richest corporations and wealthiest citizens. The majority of unemployed workers hit hardest by the economic fallout of the terrorist attacks got little or nothing. Somebody wasn't getting the message of unity.

More than five hundred thousand workers were laid off in the first months after September 11—many already on the bottom end of the employment ladder who were only one paycheck away from poverty. The layoffs occurred in more than half the country, and most were in the transportation and hospitality/tourism sectors. An economic stimulus package should have had as its goal to immediately "stimulate" the economy by providing emergency assistance for those families in greatest jeopardy—who, in fact, are most likely to spend the money. Instead, 80 percent of the House package benefited corporations and the wealthiest taxpayers, who are in least jeopardy and least likely to spend the money. That failed the tests of both common sense and unity.

The stimulus included $70 billion in corporate tax cuts, with $24.5 billion going to cover the repeal of the alternative minimum tax (designed to make sure that corporations do not evade all their taxes through a variety of deductions and loopholes)—with refunds retroactive back to 1986! Seven corporations alone received $3.3 billion. An additional $28 billion in individual tax cuts was also a part of the bill, heavily skewed toward high-income earners but, gratefully, including rebates for some lower-income workers who did not receive any tax refund because they made too little. Republicans said that giving corporations money would keep them from cutting jobs. But a $15 billion bailout of airline companies did not include a plan to sustain the livelihood of approximately 140,000 airline workers who were laid off.

Conservative commentator Kevin Phillips said the House stimulus package should cause a "public outcry" and urged Americans to point their fingers at the politicians who voted for it, saying "Shame, shame, shame!" At a time when the country was being urged to make sacrifices for the common good, the spectacle of the most well-to-do Americans lining up for huge tax breaks was, indeed, appalling.

A more fair stimulus package would have included targeted help to those displaced by the crisis, an end to taxing unemployment benefits, a fund to allow laid-off workers to maintain their families' health care coverage, retraining and reemployment programs, and home energy and child care assistance. Equitable help for those in the most need should be a bipartisan effort. We need a "Republican" initiative that sends money to the state and local governments who are also reeling from the crisis, and a "Democratic" initiative that invests in the nation's infrastructure, from vital nutrition and transportation needs, to health-care and education reform, to affordable housing and sustainable energy programs. These would all be part of an overall effort to bolster true national security.

There was a better way—and a great opportunity in our response to the horrible events we suffered. I had some fictitious newspaper headlines in my head for several days that fall—headlines from the future. One read, "After September 11, Americans Became Less Tolerant of Their Own Gaping Inequalities." Another one says, "Attack on America's Democracy from Without Strengthens Democracy Within." Sadly, many of our citizens were not really a part of American unity before September 11, including thirteen million of our poorest children. Will our unity include them now?

In the big celebrity telethon to raise money for the victim's families after the attacks, Julia Roberts made a statement that has endeared her to me as my favorite "theologian" of this crisis. The popular sweetheart of America went right to the heart of things when she said, "We're learning that in a crisis, we don't just save ourselves, we save each other." Everything we do now should reflect that moral wisdom.

On September 11, Americans joined the world. Our sense of national invulnerability was shattered. Instead of just bombing every country where terrorists live and telling the world, "You're either with us or against us," we might actually reach out to a world that, until now, too few Americans have really understood. There have been some positive signs. In public schools, students have begun to study Arabic. In churches where I have preached, local Imams have been invited to speak about Islam and attracted far more people than expected.

Now could be the perfect time to finally demand a Middle East peace settlement that is just for both Palestinians and Israelis, thereby addressing the most grievous source of anger among Arabs and Muslims. It could also be the right time to support democracy in the Arab world, where official stupidity, greed, and oppression have helped to create the dangerous grievances that lead to terrorism. It could also be the right time to undertake major economic relief and reconstruction in the entire region of the Middle East—which could win more hearts and minds away from terrorism than will endless bombing campaigns.

The debate about which path to take—about what our response to terrorism should be—is far from over. It is still taking place in conversations across America, in the religious community, and, I hope, at the highest levels of political power. The outcome of that debate will shape our future even more than the terrorists can. We can still make the right choices in this crisis and, in so doing, point the way to a safer and better world. September 11 will ultimately become either a doorway to transformation or the excuse for our worst instincts and old habits. It depends on the choices we still have to make.

On the first anniversary of the September 11 attacks, I wrote the following column entitled, "Ten Lessons to Defeat Terrorism."

1. *Treat the threat of terrorism as very real.* Don't underestimate it or politicize it. Cells of terrorists around the world are trained and ready to strike again. To prevent further terrorist violence is a worthy cause. The question is not whether, but how. I live with my wife and four-year-old son on a terrorist target, only 20 blocks from the White House. I want to stop potential terrorist threats against my family and other innocents with all my being—but not in ways that risk and kill other people's four-year-olds.
2. *Avoid bad theology.* The American Bush theology sees a struggle between good and evil—we are good, they are evil. And everyone else is either with us or against us. If we can't see the face of evil in the events of Sept. 11, we have been corrupted by the post-modern world of moral relativism. On the other hand, we are not the good. That's bad theology. Jesus teaches us to see the beam in our own eye, and not just the mote in our adversary's eye. George Bush is a Methodist, but he sees no beams in the American eye. But there is also a bad anti-American theology that suggests that evil resides only in Washington, D.C. Bin Laden is not a freedom fighter. He cares nothing for the have-nots of the world. He's only recently become interested in the Palestinians. His is a twisted ideology and pathology of hate, vengeance, and lust for power. And he would turn Islam into a religion of violence against innocents. We must act so that the world will not be remade in the image of the terrorists; and we deny the terrorists their victory when we refuse to be changed into people God has not called us to be.
3. *Listen to the different perceptions of Sept. 11 around the world.* Random, senseless violence, which can take loved ones at a moment's notice, is not a new experience for most of the world's people in

places like Sarajevo, San Salvador, Johannesburg, or Jerusalem. Our illusions of invulnerability must be shattered—so we can join the rest of the world.

4. *Let's define terrorism the right way, and allow no double standards.* Terrorism is the deliberate taking of innocent lives. It applies to individuals, groups, and nations alike—all of which can and have supported and committed acts of terrorism. Those who turn airplanes into missiles to attack skyscrapers full of people, those who become suicide bombers, and those who order military strikes against apartment buildings full of civilians and children are all terrorists, not religious devotees, martyrs, or defenders of national security.
5. *Attack not only the symptoms, but also the root causes of terrorism.* Poverty is not the cause of terrorism, but impoverishment and hopelessness are among terrorism's best recruiters. We must drain the swamp of injustice in which the mosquitoes of terrorism breed. Justice really is the best path to peace, and there is no security but common security.
6. *The solutions to terrorism are not primarily military.* Drying up the financial resources of terrorism, coordinating international intelligence, and multi-national policing are much more effective weapons against terrorism than bombing Iraq. Dealing with root causes is the best strategy of all.
7. *It's time to move beyond the old debates of pacifism vs. just war,* and focus on the promising common ground of conflict resolution. We must ask what are the transforming initiatives and practices that will actually prevent, reduce, contain, and, ultimately overcome the inevitable eruptions of violence in our world.
8. *It is time to end the era of unilateral action by any nation,* even the world's last remaining superpower—no matter how strong it seems to be. Nobody can go it alone. No victory over terrorism is possible without a whole new level of international judicial, political, and financial collaboration. Only a real world court to weigh facts and make judgments, with effective multinational law enforcement, will be able to protect us.

9. *This is not a time for peace-loving, but rather for peacemaking,* which is much more demanding. And peacemaking is, finally, less a position than a path—the path Jesus has clearly instructed us to take. That path cost him dearly, and no doubt will us too. But the alternatives are both impractical and frightening.
10. *Finally, the fight against terrorism is a spiritual struggle, not just a political one.* It causes us to ask what is really important, what our closest relationships really mean to us, and what we are really doing with our lives and the gifts God has given us. Like firefighters who make pilgrimages to Ground Zero, we are all pilgrims now.

CHAPTER 8

Not a Just War

The Mistake of Iraq

SADDAM HUSSEIN WAS an evil ruler, no doubt about it. Add to that description—extraordinarily brutal and unbelievably cruel. But that was not enough for a war. Other heads of state have also been horrifically evil, including some who have been allies of the United States (like Saddam during Iraq's war with Iran). Iraq had used chemical weapons against the Iranian people and the Kurds, and the world had every reason to believe that he hoped to acquire more of them and would use them again if it suited his purposes. But without solid evidence of an imminent threat from his chemical and biological weapons or irrefutable intelligence that he was close to possessing nuclear weapons, that was not enough for a war either. Many other nations have weapons of mass destruction too, including the United States and many of its allies—Israel, in particular. The question was what Saddam's evil meant for the world, whether there was an urgent threat from his weapons, and of course, what response would be both effective and consistent with Christian ethics.

The only consistent commitment Saddam Hussein had ever shown was to the preservation of his own power. Those who minimized his evil were morally irresponsible, and those who underestimated his willingness to commit mass murder (again) were making a serious mistake. But what was the best response?

The tradition of Christian nonviolence and pacifism rules out war as a way to resolve conflicts, and the just war doctrine, which many more churches accept, demands that a decision for war be subject to rigorous criteria and conditions. Those are the only two Christian traditions regarding war—unless we want to bring back the understanding of war as crusade, an idea which seems to be gaining in popularity with radical elements of Islam but not anywhere in the churches that I am aware of.

Whatever Christians decide about wars like the one in Iraq, they must do it on the basis of Christian theology. Liking and trusting President George W. Bush, as many conservatives do, or hating him, as many liberals do, is just not relevant here. Patriotism means loving your country and its best ideals, enough even to oppose it when it is grievously wrong. And Christian faithfulness always supercedes patriotism. American Christians often need to be reminded that we are a worldwide church—the body of Christ—and what other Christians around the world think about what the United States does ought to be at least as important to us as the views of our fellow citizens. Judging from all the letters, statements, and articles that came across my desk, most churches around the world (and across the theological spectrum) did not support the U.S. arguments for going to war with Iraq. Which gets us back to theology.

Christian peacemaking always calls churches to seek alternatives to war in resolving conflicts. Both the just war and pacifist traditions agree with that. And there were alternative ways to deal with Iraq's potential development of weapons of mass destruction. What was needed was a "carrot and stick" diplomacy. United Nations Security Council resolutions had called for the "destruction, removal, or rendering harmless" of nuclear, chemical, and biological weapons along with ballistic missiles in Iraq. And the United States and the international community were right to force the return of U.N. weapons inspectors to Iraq (it should have happened earlier). The inspections had begun to accomplish their purpose in enforcing U.N. resolutions and should have been aggressively completed. The containment of Saddam Hussein was possible without the risks and costs of military attack.

Saddam and his Iraqi regime should have been disarmed of any and all weapons of mass destruction, and it could have been done without war.

The international community could have united in an effective strategy to isolate, contain, disarm, and ultimately undermine and remove the brutal and dangerous regime of Saddam Hussein. Instead it was pressured to support the war agenda of the world's last remaining superpower. As for the reasonable goal of regime change, the Iraqi people themselves could have been supported internationally to create civil resistance within their own country to achieve that goal. And something like the Six-Point Plan, discussed in chapter 4, could have assisted them.

When All You Have Is Hammers, Everything Looks Like a Nail

During the summer before the war with Iraq, I was in Britain, on a panel with the newly selected archbishop of Canterbury, Rowan Williams, along with Jewish and Muslim leaders from the United Kingdom. We were reflecting on the meaning of September 11. I suggested that the attacks on America could become a teachable moment for the world and even a doorway to necessary transformation. Or those horrible events could be used in America and the West as an excuse for our very worst instincts and habits.

Archbishop Williams offered an observation that became for me the most insightful statement of the year-long run-up to the war with Iraq. He said (quoting psychologist Abraham Maslow), "When all you have is hammers, everything looks like a nail."

The United States has the biggest and best hammers in the world. But they are the only "tools" we seem to know how to use. And all we seem able to do is look for more nails to pound. Iraq is the nail the U.S. government chose to pound after Afghanistan in its "war against terrorism." We now know that Bush administration officials were planning a war with Iraq even before September 11. By pounding the nail of Saddam Hussein, the Bush administration repeatedly said a blow would be struck against terrorism.

Saddam Hussein and his government had cruelly repressed the Iraqi people and were a real threat to other countries in the region and potentially to the world. He had used chemical weapons and had stockpiled biological weapons, and he was trying hard (though unsuccessfully, we have learned) to acquire nuclear weapons. The United Nations had repeatedly demanded that Iraq stop its violations of human rights, stop threatening peace in the region, stop its attempts to develop additional weapons of mass destruction, and respect the role of the world community through the United Nations in accomplishing these important goals. But a military attack was simply not the only way to pursue these legitimate goals. Instead, international law, political wisdom, and collective action around moral principles could have guided our actions.

Neither international law nor Christian just war doctrine allow preemptive military action by one state against another. The U.N. Charter and international law only allow for individual states to go to war in situations of self-defense or after an armed attack. And in both international law and just-war doctrine, there are scrupulous conditions even for self-defense. For the United States to unilaterally initiate military action was a dangerous precedent; it severely undermined the system of international security established since World War II.

Iraq had not attacked the United States and had not been credibly implicated in the September 11 terrorist attacks—despite the continual rhetoric from the White House implying Saddam *was* linked to them. (With the constant Bush administration drumbeat against Iraq, combined with an acquiescent media, an unbelievable 69 percent of the American people were led to believe that Iraq was involved in September 11.[1]) Nor was a clear and convincing case ever made that Iraq's weapons of mass destruction were an immediate or imminent danger to other countries in the surrounding region or the rest of the world. While Saddam Hussein would obviously have liked to have had nuclear weapons, there was no evidence that his finger was resting on the button that would trigger mass murder. The manipulation of intelligence to support a preemptive war with Iraq became one of the worst distortions of fact in modern history.

In an interview with *Sojourners,* former CIA analyst Ray McGovern said, "It's the first time that I've seen such a long-term, orchestrated plan of deception by which one branch of our government deliberately misled the other on a matter of war and peace." And another former CIA analyst, David MacMichael, added, "The use of deception to frighten Congress and secure its consent for the October resolution reflects the way our government has been functioning in the area of war and peace for more than half a century. Congress has effectively resigned its power in these areas to the executive."[2]

And when the victorious Americans arrived in Iraq, no weapons of mass destruction could be found. Instead of being embarrassed by the revelation that the chief justification for preemptive war was false, the Bush administration acted as if it didn't matter.

A war against Iraq, many warned beforehand, could have serious consequences for future American foreign policy if the United States and the United Kingdom acted alone. Many of our allies—both in Europe and in the Arab world—were opposed to war. It would create explosive and long-term consequences in a region that was already facing serious unrest. And, as later events demonstrated, very little attention had been given to what would follow the war to remove Saddam Hussein. America simply didn't have an exit strategy or a plan for the years of rebuilding, international economic aid, and development of new political institutions that would be required to avoid postwar chaos.

The ongoing international intelligence gathering and law enforcement efforts against terrorism were producing real results, with significant arrests of terrorist leaders and the exposure of secret cells. The cooperation of countries around the world—particularly Arab and Muslim nations—in this effort was undercut by war with Iraq. Hatred against the United States has grown, leading to new volunteers for further terrorist attacks against the West in new countries like Spain and even in Saudi Arabia. Prophetic religious voices said the United States would certainly win a war against Iraq, but could lose the struggle against international terrorism.

Most important, a war with Iraq raised serious moral questions. Untold thousands of Iraqi civilians would be killed, as well as many young Americans. The impact on families and loved ones in our own society would be enormous.

Despite White House self-confidence, the United States has yet to recognize how the real threats of terrorism are very different from what we have known as "war," to which we simply respond with our habitual military solutions. Instead of seeing the events of September 11 as a new call for international cooperation and collaboration in addressing a host of global problems, the Bush administration quickly retrenched into Pax Americana, "going it alone" as the world's only superpower. Our choices included the rule of law or the habit of war, collective action or unilateral decisions, effective containment or unpredictable escalation. And the Bush administration made the wrong choices.

The Church's Unity on Iraq

In a demonstration of unity never seen before, virtually every world church body in the United States and the world that spoke out on the prospect of war with Iraq (with the notable exception of the American Southern Baptists) concluded this was not a "just war." And churches everywhere spoke out against it—including Catholic, Protestant, Evangelical, and Pentecostal bodies from around the world. Many churches didn't speak on the subject, and of course, many American congregations were in sympathy with the president on Iraq. But it was highly significant that whenever a church body considered the theological merits of a U.S. war with Iraq and decided to speak, they spoke against the war. They claimed a war with Iraq would be illegal, unwise, and immoral. The worldwide churches' opposition to the war in Iraq has been shown by subsequent events to be quite prophetic. One of those church statements, which we drafted, was the only joint declaration from U.S. and U.K. church leaders. I include it here to show how, before the war began, church leaders were raising all the right questions.

Disarm Iraq Without War

A Statement from Religious Leaders in the United States and United Kingdom

"Nation shall not lift up sword against nation, neither shall they learn war any more." Isaiah 2:4

As the calls for military action against Iraq continue from our two governments, despite the new opening for U.N weapons inspections, we are compelled by the prophetic vision of peace to speak a word of caution to our governments and our people. We represent a diversity of Christian communities—from the just war traditions to the pacifist tradition. As leaders of these communities in the United States and the United Kingdom, it is our considered judgment that a preemptive war against Iraq, particularly in the current situation would not be justified. Yet we believe Iraq must be disarmed of weapons of mass destruction; and that alternative courses to war should be diligently pursued.

Let there be no mistake: We regard Saddam Hussein and his regime in Iraq as a real threat to his own people, neighboring countries, and to the world. His previous use and continued development of weapons of mass destruction is of great concern to us. The question is how to respond to that threat.

We believe the Iraqi government has a duty to stop its internal repression, to end its threats to peace, to abandon its efforts to develop weapons of mass destruction, and to respect the legitimate role of the United Nations in ensuring that it does so. But our nations and the international community must pursue these goals in a manner consistent with moral principles, political wisdom, and international law. As Christians, we seek to be guided by the vision of a world in which nations do not attempt to resolve international problems by making war on other nations. It is a long-held Christian principle that all governments and citizens are obliged to work for the avoidance of war.

We therefore urge our governments, especially President Bush and Prime Minister Blair, to pursue alternative means to disarm Iraq of its most destructive weapons. Diplomatic cooperation with the United Nations in renewing rigorously effective and thoroughly comprehensive weapons inspections, linked to the gradual lifting of sanctions, could achieve the disarmament of Iraq without the risks and costs of military attack.

We do not believe that preemptive war with Iraq: is a last resort, could effectively guard against massive civilian casualties, would be waged with adequate international authority, and could predictably create a result proportionate to the cost. And it is not clear that the threat of Saddam Hussein cannot be contained in other, less costly ways. An attack on Iraq could set a precedent for preemptive war, further destabilize the Middle East, and fuel more terrorism. We, therefore, do not believe that war with Iraq can be justified under the principle of a "just war," but would be illegal, unwise, and immoral.

Illegal

Whether we oppose all war, or reluctantly accept it only as a last resort, in this case the U.S. government has not presented an adequate justification for war. Iraq has not attacked or directly threatened the United States, nor is it clear that its weapons of mass destruction pose an immediate and urgent threat to neighboring countries or the world. It has not been credibly implicated in the attacks of September 11. Under international law, including the U.N. Charter, the only circumstance under which individual states may invoke the authority to go to war is in self-defense following an armed attack. In Christian just war doctrine, there are rigorous conditions even for an act of self-defense. Preemptive war by one state against another is not permitted by either law or doctrine. For the United States to initiate military action against Iraq without authorization by the United Nations Security Council would set a dangerous precedent that would threaten the foundations of international security. And under our domestic governance, the U.S. Congress and the U.K. Parliament must also play a key role in authorizing any contemplated military action.

Unwise

The potential social and diplomatic consequences of a war against Iraq make it politically unwise. The U.S. and the U.K. could be acting almost entirely alone. Many nations, including our European allies and most of the Arab world, strongly oppose such a war. To initiate a major war in an area of the world already in great turmoil could destabilize governments and increase political extremism throughout the Middle East and beyond. It would add fuel to the fires of violence that are already consuming the region. It would exacerbate anti-American hatred and produce new recruits for terror attacks against the United States and Israel. A unilateral war would also undermine the continued political cooperation needed for the international campaign to isolate terrorist networks. The U.S. could very well win a battle against Iraq and lose the campaign against terrorism. The potentially dangerous and highly chaotic aftermath of a war with Iraq would require years of occupation, investment, and a high level of international cooperation—none of which have yet to be adequately planned or even considered. And the Iraqi people themselves have an important role in creating non-violent resistance within their own country with international support.

Immoral

We are particularly concerned by the potential human costs of war. If the military strategy includes massive air attacks and urban warfare in the streets of Baghdad, tens of thousands of innocent civilians could lose their lives. This alone makes such a military attack morally unacceptable. In addition, the people of Iraq continue to suffer severely from the effects of the Gulf War, the resulting decade of sanctions, and the neglect and oppression of a brutal dictator. Rather than inflicting further suffering on them through a costly war, we should assist in rebuilding their country and alleviating their suffering. We also recognize that in any conflict, the casualties among attacking forces could be very high. This potential suffering in our own societies should also lead to prudent caution.

We reaffirm our religious hope for a world in which "nation shall not lift up sword against nation." We pray that our governments will be guided by moral principles, political wisdom, and legal standards, and will step back from their calls for war.

In hindsight, the concerns and warnings of the religious leaders have proven profoundly prophetic. There seems to have been no official monitoring of Iraqi casualties by the United States and their coalition forces, but international human rights groups and other credible analysts put the number of Iraqis killed in the war at tens of thousands, and more die each day of the occupation. And there is no doubt in my mind that the importance the churches and the peace movement placed during the prewar debate on protecting the innocents was very influential in keeping them from being indiscriminately targeted by military planners during the initial invasion.

All the other concerns have also been borne out since the war. What did it mean when the leaders of the international body of Christ were united in opposition to a war? When huge majorities of the populations of European countries were opposed to war with Iraq? When the Middle Eastern countries most threatened by Saddam Hussein opposed war as the solution to that threat (except for Israel and Kuwait)? When several former U.S. defense secretaries, many former military officers, and former office holders from previous Republican administrations (including many of those who served with George H. W. Bush) were also against the war? Didn't it at least mean we should have had a serious national debate before going to war?

But we did not. And the Bush administration seemed to equate dissent and even debate with a lack of patriotism. In a sermon at Harvard's Memorial Church entitled "God, Country, and the Duty of Dissent," Reverend Peter Gomes preached, "How can we have an intelligent conversation on the most dangerous policy topic of the day without being branded traitors, self-loathing Americans, anti-patriotic, or soft on democracy? . . . This is a frightening time, and if one cannot speak out of Christian conscience and conviction now, come what may, then we are forever consigned to moral silence. . . . What is and has always been lovely about our country is our right and our duty to criticize those in power, to dissent from their policies if we think them to be wrong, and to hold our alternative vision to be as fully valid as theirs."[3]

The Lessons of War

The national media in the United States were for this war. Depending on how the question was asked, up to 50 percent of the American people questioned the war or the way it might be fought. But let's say only 40 percent of the American people were against this war. Did we have anything like 40 percent of the media coverage of the war issue giving voice to those who opposed it? How's that for democracy? Why were the political and media elites so afraid of an honest debate about such a momentous decision?

Now imagine this. Imagine what would have happened if we had been able to really stop this war before it started. Let the U.N. weapons inspectors (who didn't find any weapons of mass destruction) do their jobs and disarm Iraq of whatever weapons or programs it had (as they had successfully done four years earlier). Then when the inspectors' job was done, President Bush could have declared victory, claimed credit for his tough stance, and brought the troops home. We would have avoided war. And it would have been a major victory for democracy. Imagine the empowerment that all the people who spoke and acted for peace would have felt. We would actually have stopped a war before it started—for the first time in memory. People would look back for years to remind each other of the difference we all made before the potential war with Iraq. Citizens would believe they could actually make a difference. Maybe that's part of what the people who control national politics and the media were so afraid of.

Nobody ever said the United States wouldn't win the war. No one knew how easy or hard it might be, but the outcome was never in doubt. Those of us who opposed the military action simply said that a mostly American pre-emptive war was the wrong answer to the threats of Saddam Hussein. Many of us also supported regime change in Iraq, citing the tremendous human cost of Saddam's brutality and the danger posed by his love for weapons of mass destruction. Admittedly, not everybody in the anti-war movement was as clear about the need not only to disarm Saddam, but also to remove him from power—and they all should have been. Many of us didn't apologize for saying we were glad to see Saddam gone. But we didn't think this war was the right or best way to accomplish that, as mentioned before. Just before the war began, American church leaders offered a Six-Point Plan that many thought had a real chance for success. Since then, present and former government and military officials from several countries, including American general Wesley Clark, have stated that such an approach could have worked and should have been tried.

The end of the U.S. invasion turned out not to be the end of the war and has left Iraq and the world with horrible problems. The Bush administration chose an American military occupation of Iraq instead of an internationally supported U.N. initiative in humanitarian aid and reconstruction. Terrible violence has been the result ever since, more than anyone expected or predicted. The U.S. military still hasn't found any weapons of mass destruction in Iraq, and U.S. officials are not confident about ever doing so. The central justification for the war turned out not to be there. But what is there are huge numbers of American soldiers, perhaps for a very long time. The cost in lives, both American and Iraqi, in the postwar period has been utterly tragic, and the sacrifice of American military families has gone on longer than anybody contemplated. The specter of endless chaos and conflict, and even eventual civil war between Iraq's contentious factions, looms ever before us. Three large questions remain.

First, American leaders feel vindicated in prosecuting a preemptive and mostly unilateral war and are justifying that policy for the future. The media are already speculating on who will be next. Iran and Syria seem to be prime candidates. Is this the kind of world we want, where the wealthiest and most powerful nation in the world makes all the big decisions?

Second, the "humanitarian" war fought to "liberate" Iraq raises the question of why only this gross violator of human rights was chosen. What about the many others? Will the United States now choose to liberate people in countries where American geopolitical interests and the economic stakes are not as high, say in the Congo or Sudan?

Third, what about the emerging doctrine of Pax Americana, where the neoconservatives who now virtually run the U.S. government are openly speaking the language of dominance and empire as the moral foundation for American foreign policy. The Pentagon has yet to renounce plans for permanent military bases in Iraq.

Perhaps of special concern for Christians is the president's increasingly religious language to justify American war and domination. Do we really believe that America and George W. Bush have been divinely appointed to root out evil in the world? That's bad theology and a very dangerous one, which is the subject of the next chapter.

All this has resulted from the war with Iraq. Yes, I am very glad that Saddam is gone. But there were better ways to accomplish that worthy goal and set into motion precedents that would have left us in a much more healthy and hopeful place than we now find ourselves. This war isn't yet over, and its consequences have just begun.

When the war finally did break out, I offered ten lessons for understanding and surviving war, especially for people of faith. They were applicable, not only for Iraq, but for what all modern warfare has become. Originally written in the early days of the Iraq conflict, they appear even more telling at this writing—more than a year after the American "victory." These are the "Lessons of War," which have proven to be painfully true.

1. Nobody should be surprised that a vastly superior American fighting force will vanquish a vastly inferior Iraqi army. But one of America's worst characteristics is hoping that success wipes away all the moral questions. In the long run, it won't. War is always ugly, and this one was too.
2. There are many more civilian casualties in modern warfare than military casualties. Smart bombs are never as perfect as boasted, and not all Iraqis may want to be "liberated" by an American occupation. Above all, we must remember that "collateral damage" is never collateral to the families and loved ones of those killed in war. Don't accept the first reports on casualties from governments (on either side) or "embedded" journalists—many of whom now sound more like cheerleaders than reporters. Be sure that technology does not ultimately usurp theology or morality. Find alternative sources for information. Watch and wait for the real story.
3. Humanitarian aid must never be co-opted by the military. Instead it is the painful task to be taken on after the destruction caused by war. Many predict that the aftermath of this war could be far more dangerous and costly, in human terms, than the military campaign. Listen to the nongovernmental relief organizations as we move forward.
4. If an evil, dangerous, and unpopular regime does collapse quickly, that is not an endorsement of war as the answer but a sign that a better way to resolve the threat might well have been possible. The best wisdom of most church leaders, Nobel Peace Prize laureates, and a majority of international political figures and diplomats around the world was that al-

ternatives to a full-scale military assault on Iraq were not adequately tested. This was not a war of last resort.

5. A preemptive war of choice, rather than of necessity, fought against overwhelming world opinion and without approval by the United Nations, will not create an atmosphere of cooperation for postwar reconstruction or, most significantly, for the crucial international collaboration needed to defeat the real threats of terrorism.
6. A new world order based on unilateral rather than multilateral action, military power rather than international law, and the sole decisions of the world's last remaining superpower over the deliberations of the community of nations will not create a framework the world can or should trust for peace.
7. Unresolved injustice—such as the Israeli-Palestinian conflict, feudal Arab regimes protected by oil, and globalization policies that systematically give advantage to wealthy nations over poor countries and people—remains a root cause of violence and will not be overcome by the imposition of American military superiority.
8. Dissent in a time of war is not only Christian, it is also patriotic. A long and honorable record of opposition to war in church tradition and American history puts dissent in the mainstream of Christian life and American citizenship. Rather than acquiesce to war, prayerful and thoughtful dissent is more important than ever.
9. The churches have demonstrated the most remarkable unity in our history in opposition to war, even before the war with Iraq started. The American churches didn't just say no to war, but offered compelling and credible alternatives. These alternatives were seriously considered by many political leaders around the world, but not by our own government. An American president who increasingly uses the language of Christian faith refused even to meet with American church leaders for discussion and prayer as he made momentous decisions to go to war. The American churches are now in deep solidarity with the worldwide body of Christ and may have to choose between their Christian alliances and the demands and policies of their own government. We must learn to be Christians first and Americans second.

10. The onset of war with Iraq does not demonstrate the failure of the peace movement, but rather the failure of democracy. Tens of millions of people around the world have become engaged in active citizenship against the policies of preemptive war for resolving the greatest threats to peace and security. It is time to build on that movement, rather than withdraw from collective action. We must learn the differences between grief and despair, between lamenting and languishing, between hope and hostility. We are stronger now, not weaker. Our action has just begun.

Mission Accomplished?

That was in May 2003. It now seems like a very long time since George W. Bush donned an Air Force flight jacket and made a dramatic helicopter landing on the aircraft carrier USS *Abraham Lincoln* that same month to declare "Mission accomplished." As everyone can now see, our mission is far from accomplished.

As of this writing, 1,254 American and 146 coalition soldiers have now died in Iraq—most after the invasion and during the occupation. More than nine thousand have been wounded or maimed. Reliable accounts say as many as one hundred thousand Iraqis lost their lives.[4] No weapons of mass destruction have been found in Iraq. U.S. weapons inspector David Kay reported that they probably weren't there and that U.S. government officials should honestly admit that they were mistaken. They haven't. President Bush and his administration have repeatedly said the fact that the principal argument for going to war with Iraq has turned out to be false doesn't matter. There was no "imminent" or "urgent" threat from chemical or biological weapons, and Iraq wasn't developing a nuclear threat, as was clearly claimed before the war.

The best explanation is that intelligence was manipulated and selectively reported to justify a worst-case scenario that had previously been arrived at on political grounds. The worst explanation is that the case was fabricated. Either way, the president of the United States misled the American people into going to war. Former treasury secretary Paul O'Neill's book, *The Price of*

Loyalty, makes clear that the Bush administration decided to go to war with Iraq even before September 11, and the "facts" were never the decisive factor. Bob Woodward's book, *Plan of Attack,* confirms that alarming truth. Former CIA Director George Tenet has virtually said his agency's efforts to prevent the Bush administration from "overstatement" on Iraq were a failure.

Iraq remains chaotic, unstable, and highly lethal. Divided factions threaten any political solution. The United States made much of its plans to "turn over sovereignty" to a new Iraqi government, the timing of which was clearly tied to the American election. The real issue of how successful the transfer to Iraqi sovereignty will be remains to be seen.

It is indeed a good thing that the brutal regime of Saddam Hussein is over. But that worthy goal should and could have been accomplished, over time, in much better ways than a preemptive and largely unilateral war that has proved to be both unnecessary and unjust. Iraq is now a huge mess with no clear U.S. exit strategy in sight and is likely to remain so for a very long time.

The Bush administration's argument that the war with Iraq was a critical battle in the war on terrorism is also not compelling. In his enormously influential book, *Against All Enemies: Inside America's War on Terror,* and his riveting testimony before the 9/11 Commission, Richard Clarke, former counterterrorism czar in both the Clinton and Bush administrations, shocked the nation by charging that the U.S. government ignored clear warning signs of the impending al Qaeda attack and then pursued an unnecessary war against Iraq that has distracted America's attention from the real terrorist threats and made us less safe.

The Iraq war may ultimately make the defeat of terrorism more difficult because of the division it caused among key allies, the deeper resentment it has triggered in Muslim countries, and the failure of the war policy to produce the promises of democracy in Iraq or the beginning of a Middle East peace agreement—all of which seem further away than ever. The war in Iraq has, indeed, proved to be a great distraction and diversion from the fight against terrorism, rather than a necessary component. Already, one-third of Afghanistan is back under Taliban control. Osama bin Laden has yet to be found, and the networks of terror are more dispersed throughout the world. And al Qaeda has carried out more successful and massive terrorist attacks in Spain and elsewhere, now even in the Arab world.

In Spain, 191 people died from bombs on commuter trains in the spring of 2004, with more than two thousand injured. Right-wing American columnists had the audacity to attack the Spanish people for caving in to terrorists because they defeated the pro-American Spanish party in subsequent elections. What they don't say is that the vast majority of the Spanish people were against their government's decision to support the U.S. war with Iraq, which was therefore an undemocratic decision. Or that the ruling party lied to the Spanish people in an attempt to blame the commuter train attacks on a local group and distract attention from Al Qaeda. What the U.S. government and its media allies seemed to be saying was that the proper response of the Spanish people after being bombed should have been to vote for the continuing policies of George W. Bush. They didn't because, like most of the world's people, they had seen enough.

Who Will Sacrifice?

Although President Bush and his top advisers have yet to admit it, continuing events in Iraq reveal fundamental miscalculations and multiple policy failures on the part of the administration. Those failures have left the White House looking out of control of the situation in postwar Iraq because, despite the postwar bravado of Donald Rumsfeld, Dick Cheney, and the president's "bring 'em on" statesmanship, they are losing their grip. Instead of at least acknowledging miscalculations on the issue of weapons of mass destruction, the reception of the Iraqi people to their American "liberators," the unexpected level of resistance to American occupation, the cost and scope of reconstruction, the American unilateralism that has made needed international help so difficult to obtain, and the abysmal lack of a postconflict plan (to name just a few), the president just calls for more "sacrifice."

But who will do the sacrificing? President Bush has so far, in mid-2004, requested nearly $200 billion to pay for the American war, occupation, and reconstruction of Iraq. News reports already reveal the comparative costs and "sacrifices" of this enormous expenditure: The entire proposed fiscal year discretionary budget for the Department of Health and Human Services is $66 billion; for the Department of Education, $53 billion. The total amount for all fifty states to meet their projected budget shortfalls this year is $78 billion. Meanwhile, the administration now projects the federal deficit to rise to as much as $521 billion in 2004.

Who else was being forced to sacrifice was made clearer in a Department of Defense announcement the day after one of the president's speeches. Thousands of National Guard and Reserves, who had been expecting their one year call-up to end this fall, had their duty extended for up to another year. The ongoing occupation of Iraq has changed everything for many American families. And then there are the continuing casualties. Deadly attacks occur every day, while American families nervously await news from Iraq.

Truth telling has also been sacrificed in the president's speeches about the war with Iraq. There is still no mention of embarrassing facts like no weapons of mass destruction yet found—the once primary reason for the war. And there are still frequent references to September 11, linked to the war in Iraq. The U.S. victory in Iraq has now become rolling back the terrorist threat to civilization. The Bush administration has continued its unsubstantiated insinuations of connections between Iraq and al Qaeda, without any proof from U.S. or any other intelligence sources. And the result is that a majority of the American people still believe that Saddam Hussein was involved in the attacks of September 11—despite the lack of evidence. Ironically, the American occupation of Iraq has now resulted, as some are already noting, in a circular self-fulfilling prophecy: By claiming Iraq harbored terrorists and by going to war, the United States has now created a new breeding ground and gathering point for all manner of terrorists in Iraq, united by a common American enemy.

Who will bear *no* sacrifices is also clear: the beneficiaries of the Bush tax cuts and the recipients of the lucrative contracts for Iraqi reconstruction, which are going to carefully selected American corporations like Halliburton, where Dick Cheney was the former chairman of the board. Those who will not sacrifice, in other words, are the wealthy and powerful allies of the Bush administration and its core constituency. It is not hyperbole to say that those beneficiaries of wartime tax cuts and contract deals should be called war profiteers. The record deficits under this Republican White House and Congress have been caused by three factors: first the economic downturn, then the huge increases in military spending, and finally the enormous tax cuts that the Bush administration wants to make permanent. Steven Mufson, who covered economic policy for the *Washington Post,* has called the billions for the occupation of Iraq "chump change"[5] compared to what has been given away in the Bush tax cuts to the wealthy—only 5 percent of the costs of the president's tax cuts over ten years. If politics is the allocation of power, resources, and sacrifices, the politics of the sacrifices for the war on terrorism are becoming increasingly clear.

It is high time for the White House to admit its miscalculations and policy failures. And those responsible for the failures should be the first to sacrifice. Therefore, the chief architects of the failed Iraqi policy—Secretary of Defense Donald Rumsfeld and his deputy, Paul Wolfowitz—should both be asked to resign. These chief unilateralists have presided over the policy failures and, if a better direction of international cooperation is to be restored in Iraq, Rumsfeld and Wolfowitz must step aside.

Removing these hard-liners would open the way for greater involvement by the United Nations. It is clear that the only way other countries will contribute peacekeeping troops and reconstruction aid is if there is genuine U.N. control of the political and military situation. Economist Jeffrey Sachs of Columbia University has controversially claimed that a U.N. reconstruction of Iraq would be far less costly, bringing political stability much sooner and allowing Iraq's economy and oil industry to get up and running more quickly. He predicts only about $10 billion. That would leave many more resources to fight AIDS, tuberculosis, and malaria in Africa and for critical economic development in poor countries across the globe (a critical task that British chancellor of the exchequer Gordon Brown repeatedly pleads is essential for defeating terrorism). Sachs claims, "The US occupying army is therefore delaying rather than accelerating Iraq's reconstruction and recovery." It's the commitment to U.S. political control and hegemony in the region that will make rebuilding Iraq so difficult and costly, says Sachs. "The United States will continue to destabilize Iraq as long as the occupation continues, and the American people will end up paying a high price for the fantasy of hegemony."[6]

And if the White House's calls for sacrifice are to have any moral credibility, the administration's tax cuts to the wealthiest Americans must be immediately rescinded, and the no-bid contracts to U.S. corporations favored by the Bush administration, like Halliburton and Bechtel, must be suspended. Neither the poor nor our children and their children should be forced to pay for the war in Iraq while those with the greatest ability to sacrifice are reaping a world of benefit. That is morally unconscionable, and the only responsible course of action now is to repeal the egregious tax cuts and honestly seek international assistance in rebuilding Iraq.

Whether either the resignations of Rumsfeld and Wolfowitz or the repeal of the tax cuts are politically likely at this moment (they aren't) is not the point. There are fundamental issues of moral accountability here that go beyond political calculation.

Stop the Occupation, Start the Rebuilding

Events in Iraq have dramatically revealed that the U.S. occupation is out of control. In April 2004, U.S. Marines began a siege of the city of Falluja following the death and mutilation of four American private contractors. Intense battles ensued, including street-to-street fighting between Marines and Iraqi insurgents during the day, followed by attacks from AC-130 gunships and jets using 500- and 1,000-pound bombs at night. The few reports from the city told of total chaos, with people fleeing for safety. During the lulls in fighting, casualties were collected and buried in the soccer stadium.

The month-long heavy fighting and intense bombing killed nearly 140 American troops and ten times that many Iraqi civilians, with unknown hundreds more wounded.

One of the first journalists into Falluja after the lifting of the siege, *London Observer* reporter Patrick Graham, quoted Dr. Mohammed Samarae's descriptions of the casualties treated at his hospital. "Ninety percent of the injured were civilians—children, old people, women." The characteristics of the wounds show they were American inflicted, said the doctor. "We have had a lot of experience of U.S. weapons." Mustafa, a twenty-two-year-old student said, "All these people were killed because of four dead Americans. . . . The Americans are killing people who had nothing to do with the death of those four Americans."[7] In November 2004, ten thousand U.S. and two thousand Iraqi troops, with heavy air support, began an all-out assault on Falluja. After two weeks of heavy fighting, with more than fifty U.S. deaths and an unknown number of civilians dead, an estimated two thousand insurgents had died and the rest had fled the city.

I was in London the first week of May 2004. For days, the lead story in all the British media (as well as the U.S. press) was the abuse of Iraqi prisoners by American soldiers. An official military investigation by Major General Antonio Taguba was soon leaked to the press. It reported evidence of "war crimes" against the inmates and found that military police and intelligence officials had committed "sadistic, blatant and wanton criminal abuses" and that personnel were directed to "set physical and mental conditions for favorable interrogation of witnesses."[8] The official investigation reportedly began in January when a soldier turned over evidence of the abuse, including photographs. But it was only admitted by the U.S. military and government when the pictures were obtained by the media.

Both President Bush and Prime Minister Blair rightly expressed their condemnation of the behavior but insisted it was limited to a few individuals. But Amnesty International reported "patterns of torture," with "scores" of allegations dating back to July 2003, and said the media reports were just "the tip of the iceberg."[9] Brigadier General Janis Karpinski, the former head of U.S. military prisons in Iraq (who was relieved of her command), also described "patterns of abuse." She and others say that military intelligence officers and "private contractors" were directing the intimidation and mistreatment of prisoners. Karpinski claims she warned her superiors about problems at the prison, but they ignored her because "they wanted it to go away."[10] The International Committee of the Red Cross said it had repeatedly asked American authorities to take "corrective measures" at the prison.[11]

Since then, there have been more horrifying revelations, including reports of as many as twenty-five deaths of prisoners, and many more photographs. President Bush appeared on Arab television to describe the abuse as "abhorrent" and pledged that "we will find the truth, we will fully investigate."[12]

The debate then centered on "bad apples" versus the great majority of American servicemen and women who wouldn't do such things (which is undoubtedly true), whether the punishment was going high enough in the chain of command, and whether the so-called "private contractors" (let's just call them paid mercenaries) are accountable enough. Some called for an independent investigation and others for the resignation or firing of Defense

Secretary Donald Rumsfeld, who had led the complete mismanagement of postwar Iraq and had personally relaxed the rules of how prisoners may be treated. Both are still excellent ideas.

By the fall of 2004, a Pentagon commission headed by former Defense Secretary James Schlesinger issued its report. Among its conclusions was that at least partial responsibility for the abuse went to the top—to Secretary Rumsfeld, the Joint Chiefs of Staff, and other top military commanders. Yet only seven low-ranking soldiers had been indicted.

But as important as all those questions are, they miss the heart of the matter. Such abuse and atrocities are the direct consequence of military occupation. They always have been, and they will continue to be. In Vietnam, a brutal American war and occupation created bloody insurrection. Viet Cong fighters did terrible things to American soldiers, and in turn, the soldiers did terrible things to Vietnamese civilians. It is simply the cycle of violence.

The administration's plans to increase deployment with nearly 140,000 troops through the end of 2005 are not the solution. It will, if anything, make the situation worse. Here is the real issue: The Americans and the British cannot and should not run Iraq. The American-led occupation is leading to more suffering on all sides, and it will just get worse. The unilateral American occupation must be stopped and the rebuilding of Iraq begun, but under international authority and control. The United Nations was never given the full political authority to appoint a transitional Iraqi government and lead the process to clear elections and real Iraqi sovereignty. Security is, indeed, the immediate question, but a unilateral American military presence will never be able to provide it. We are the targets now and the biggest cause of the security problem. The international community must not simply be brought in to help the U.S. agenda succeed; it must be given the authority to repair Iraq. American occupation is not the solution; it is the problem. And it must end.

Putting Down the Mighty from Their Thrones

The news of Saddam Hussein's capture came in December 2003, as I was finishing preparations for a Christmas sermon. The text at the heart of my message was Mary's song of praise from the second chapter of Luke's gospel.

He has shown strength with his arm;
He has scattered the proud in the imaginations of their hearts.
He has brought down the powerful from their thrones,
And lifted up the lowly;
He has filled the hungry with good things,
And sent the rich away empty.

Mary's Magnificat proclaims a great reversal of the power relations of this world, brought about by the birth of Christ. That social revolution is absolutely central to the Christmas narratives, including a murderous rampage by King Herod against innocent children in vain search for the one child whom the Roman vassal rightly feared might spell the end of his rule and others. None of that, however, gets into the commercial and cultural narrative of Christmas in America; no challenge to the political and economic powers will interrupt our mall shopping, and the only breaks in the constant advertising juggernaut were the holiday tributes to our military forces in Iraq, fighting "the war against terrorism."

My first thought on hearing the breaking news was that Mary had predicted the downfall of the brutal and tyrannical Saddam, just as the hearers of this gospel in her time would have understood it to mean the collapse of their oppressive Roman rulers. Rulers everywhere have reason to be concerned about the new Kingdom brought about by the birth of Christ, and the more unjust their rule, the more they ought to worry.

Mary's words are, of course, a thanksgiving for the blessing she has received from God, but they are not the pious and sentimental expressions we so often hear around Christmas. She simply doesn't sound like a "social service provider" inviting us all to be more generous during this special season toward those less fortunate than we are. This new king, says Mary, is more likely to turn the world upside down.

Mary's stunning announcement about the high and mighty being brought low and the lowly exalted is at the heart of the Christmas story—this is how the Scriptures portray the social meaning of the Son of God born in an animal stall. Mary is herself a poor young woman, part of an oppressed race and living in an occupied country. Her prayer is the hope of the downtrodden everywhere, a prophecy that those who rule by wealth and domination, rather than by serving the common good, will be overturned because of what has just happened in the little town of Bethlehem. Mary's proclamation can be appropriately applied to any rulers or regimes that prevail through sheer power, instead of by doing justice.

But the leaders of the world's last remaining superpower, who now claim credit for Saddam's downfall, will likely miss the point of Mary's song and certainly show no understanding of how her words might also apply to them. It is theologically accurate to say (and was proved historically true) that Mary was actually prophesying the end of Pax Romana (the "peace" of Roman rule) in her great Magnificat—but not only of Rome. And if those who would enforce a new Pax Americana (a term that they themselves now like to use) continue their vision of success through unilateral dominance, they too could suffer the same fate as Rome, or even Saddam. That is part of the meaning of Christmas that we didn't hear in the media's messages of good cheer.

America finally found Saddam. But have we found safety or security? More and more say we are less and less secure, both in Iraq and at home, as long as the American occupation continues.

Saddam should be brought to justice in the kind of fair trial he never gave others. The Iraqis should be the primary ones to hold him accountable for his horrendous crimes, but the international community should have an important part to play as well. The Bush administration should finally move to genuinely internationalize the peacemaking and decision-making strategy in Iraq and allow the United Nations to oversee the process of writing constitutions and having elections—they are simply better at that than we are.

The capture of Saddam should have marked the beginning of genuine nation building, if the Americans would have given up their control and allowed other nations to help shape and secure a new Iraq. But the United States seems to savor its more imperial style. Yet if we take Mary seriously, Pax Americana not only won't work, it won't ultimately prevail. That's both a Christmas promise and a Christmas hope.

Stop . . . in the Name of Humanity

In the months before the start of the war in Iraq, Pope John Paul II and the Vatican were among the strongest voices in the religious community opposing war. Archbishop Renato Martino, president of the Pontifical Council for Justice and Peace, issued a series of stronger and stronger statements.

On March 11, 2003, contrasting the looming war with the first Gulf War that followed Iraq's invasion of Kuwait, he said that in this case, "there is no aggression and so this preventive war is, in itself, a war of aggression."[13] A week later, two days before the war began, he upped the ante: "It is a crime against peace that cries out with vengeance before God. Let us pray so that the Pharaoh's heart will not be hardened and the biblical plagues of a terrible war will not fall on humanity."[14]

His (and our) prayers were not answered. Pharaoh's heart was hardened, war began, and more than a year later the plague continues. More than twelve hundred U.S. servicemen and women have died, along with tens of thousands of Iraqis, the reports of prison abuse by the U.S. military continue to unfold, and the administration now intends to ask Congress for additional funding to increase and maintain the U.S. military through the end of next year while dramatically cutting vitally needed programs here at home.

Before a meeting in Rome between President Bush and the pope in June 2004, Cardinal Pio Langhi, the former U.S. papal nuncio who visited Bush in the spring of 2003 to personally deliver the pope's message against war with Iraq, again spoke to an Italian newspaper.

" 'Stop' is the cry expressed by the Church in the name of abused humanity," the Vatican's Zenit News Service reported Cardinal Langhi saying. "The United States must also stop and I think it has the strength to do so. It must re-establish respect for human beings and return to the family of nations, overcoming the temptation to act on its own. If it does not stop, the whirlwind of horror will involve other peoples and will lead us ever more to the abyss."

Echoing growing calls for an end to the U.S. military occupation, the cardinal concluded: "The forces present in Iraq not only must not be in fact under the command of the United States, but they must not even give the impression that they are. There should be a multilateral presence, which is not under those who organized and wanted the war."[15] This was not a just war. And the recognition of that is the beginning of both moral and political wisdom—a necessary prerequisite to a better future.

Meeting with Prime Minister Tony Blair

On Tuesday, February 18, 2003, the prime minister of Great Britain, Tony Blair, met with five American church leaders about the decision to go to war with Iraq. President George W. Bush never agreed to meet with American religious leaders to hear their concerns about the U.S. rush to war.

The meeting at number 10 Downing Street lasted longer than the usual fifteen to twenty minutes for such encounters. Tony Blair met with us for a full fifty minutes and was very engaged in the discussion about the moral and even theological issues at stake in this momentous choice.

Sojourners organized and led the delegation, which included Bishop John Bryson Chane, Episcopal Diocese of Washington; Bishop Melvin Talbert, ecumenical officer, Council of United Methodist Bishops; Reverend Clifton Kirkpatrick, stated clerk, Presbyterian Church USA; and Reverend Dan Weiss, immediate past general secretary, American Baptist Churches in the USA. I represented Sojourners.

We were joined by international church leaders Archbishop Njongonkulu Ndungane, Anglican archbishop of Cape Town; Bishop Clive Handford, Anglican bishop of Cyprus and the Gulf; Bishop Riah Abu El-Assal, Anglican bishop of Jerusalem, Jordan, Lebanon, and Syria; Reverend Dr. Keith Clements, Conference of European Churches; and our United Kingdom church counterparts: Bishop Peter Price, Anglican bishop of Bath and Wells; Bishop John Gladwin, bishop of Guildford; Reverend David Coffey, general secretary,

Baptist Union of Great Britain; Reverend John Waller, moderator, United Reformed Church; Reverend David Goodbourn, general secretary, Churches Together in Britain and Ireland (CTBI); and Paul Renshaw, coordinating secretary for international affairs, CTBI.

The trip was made in partnership with similar delegations to Berlin, Paris, Rome, and Moscow, coordinated by the National Council of Churches. In London, the organization Churches Together in Britain and Ireland graciously hosted us.

We affirmed that Tony Blair, a practicing Christian, was bringing "moral concerns" into the debate over Iraq. And we agreed with the prime minister that the issues of terrorism and the threat of weapons of mass destruction were deeply moral and theological issues. We also agreed, unequivocally, that Saddam Hussein was a real threat to his own people and to the entire world.

But we shared with Tony Blair how American church bodies have never before in our history been more united in their opposition to a war. While American and British leaders point out how terrible the regime of Saddam Hussein is (and rightly so), the churches want also to remind the world (and our political leaders) how terrible war is. In moving personal statements, the church leaders testified to our conviction that war is not the answer to the real threats posed by Saddam Hussein. The unintended and unpredictable consequences of war make it far too dangerous and destructive an option. We told the prime minister that the answer to a brutal, threatening dictator must not be the bombing of Baghdad's children.

It was neither hyperbole nor high drama to recognize, we told Tony Blair, that the British people and their prime minister were in a position to influence the decision about a war with Iraq more than any other people or leader in the world. We said that must be a terrible burden to bear and offered our genuine prayers and support to Mr. Blair as he charted the course his leadership would take in the coming critical weeks.

As Americans, we told the British leader that it would be a dangerous thing for the world, and for America, if an issue of such importance were to be decided solely or mostly by American power. We strongly affirmed that the issue of Iraq, with all its possible consequences, must be decided by the world community, in the Security Council of the United Nations, and not by the unilateral decision making of the world's last remaining superpower. We said that the United States was becoming a "new Rome" in claiming a singular and preemptive moral authority to act in the world today, and that this was both bad theology and bad policy.

We respected the "convictional core" of the British prime minister around the legitimate concerns regarding the juxtaposition of terrorism and weapons of mass destruction, but we urged him to persevere in finding another way to resolve the problem with Iraq apart from an American-led war. In fact, we suggested he, more than any other world leader, might help forge or even broker a better way, even a "third way," beyond doing nothing about Iraq or submitting to the inevitability of an American war, which could lead to a postwar regime in Iraq ruled by an American general. We talked of other directions, especially with a strong role for the United Nations—even a U.N. mandate or protectorate in Iraq—with rigorous inspections and continual monitoring of Saddam Hussein, backed by international force.

The critical need for a resolution to the Israeli-Palestinian conflict also figured prominently in our discussions. The bishop of Jerusalem, Bishop Riah, spoke with great authority and clarity and told Prime Minister Blair, "The road to Baghdad leads through Jerusalem." The British government is making the critical connection between Middle East peace and the problem of terrorism and even Iraq, much more than the U.S. government has. We committed ourselves to helping change that. British Secretary of State Clare Short also met with our delegation for an hour and a half and joined us in the meeting with Mr. Blair. Secretary Short had become an important adviser to church efforts to find a solution to Iraq that was both effective and humanitarian.

I was impressed by how Prime Minister Blair entered into a real dialogue with us, shared our concerns for the people of Iraq for a genuinely international and U.N. solution, and recognized how crucial a Middle East peace was to this moment. I also saw a Christian political leader seriously wrestling with crucial matters of theology and moral discernment as we all approached the hour that was, in Martin Luther King Jr.'s words, "five minutes before midnight."

CHAPTER 9

Dangerous Religion

The Theology of Empire

> *Religion is the most dangerous energy source known to humankind. The moment a person (or government or religion or organization) is convinced that God is either ordering or sanctioning a cause or project, anything goes. The history, worldwide, of religion-fueled hate, killing, and oppression is staggering. The biblical prophets are in the front line of those doing something about it.*
>
> *The biblical prophets continue to be the most powerful and effective voices ever heard on this earth for keeping religion honest, humble, and compassionate. Prophets sniff out injustice, especially injustice that is dressed up in religious garb. They sniff it out a mile away. Prophets see through hypocrisy, especially hypocrisy that assumes a religious pose.*
>
> —Eugene H. Peterson, *The Message,* NavPress, 2002, p. 1641.

"THE MILITARY VICTORY IN IRAQ seems to have confirmed a new world order," Joseph Nye, dean of Harvard's Kennedy School of Government, wrote in the *Washington Post.* "Not since Rome has one nation loomed so large above the others. Indeed, the word 'empire' has come out of the closet."[1]

The use of the word *empire* in relation to American power in the world was once controversial, often restricted to left-wing critiques of U.S. hegemony. But now, on op-ed pages and in the nation's political discourse, the concepts of empire, and even the phrase "Pax Americana," are increasingly mentioned in unapologetic ways.

William Kristol, editor of the influential *Weekly Standard,* admits the aspiration to empire: "If people want to say we're an imperial power, fine."[2] Kristol is also chair of the Project for the New American Century, a group of conservative political figures who began in 1997 to chart a much more aggressive American foreign policy. In their own words, the Project's papers lay out the vision of an "American peace" based on "unquestioned U.S. military preeminence." These imperial visionaries write that "America's grand strategy should aim to preserve and extend this advantageous position as far into the future as possible." It is imperative, in their view, for the United States to "accept responsibility for America's unique role in preserving and extending an international order friendly to our security, our prosperity and our principles." And, they warn, "The failure to prepare for tomorrow's challenges will ensure that the current Pax Americana comes to an end." That, indeed, is empire.[3]

There is nothing secret about all this; on the contrary, the views and plans of these powerful men have been quite open. These are far-right American political leaders and commentators who ascended to governing power and, after the trauma of September 11, 2001, have been emboldened to carry out their agenda.

In the run-up to the war with Iraq, Kristol told me that Europe was now unfit to lead, because it was "corrupted by secularism," as was the developing world, which was "corrupted by poverty." Only the United States could provide the "moral framework" to govern a new world order, according to Kristol, who recently and candidly wrote, "Well, what's wrong with dominance, in the service of sound principles and high ideals?"[4] But whose principles and ideals? The American right wing's definition of American principles and ideals?

Bush Adds God

To this aggressive extension of American power in the world, President George W. Bush adds God. And that changes the picture dramatically. It's one thing for a nation to assert its raw dominance in the world; it's quite another to suggest, as this president does, that the success of American military and foreign policy is connected to a religiously inspired "mission" and even that his presidency may be a divine appointment for a time such as this.

Many of the president's critics make the mistake of charging that his faith is insincere at best, a hypocrisy at worst, and mostly a political cover for his right-wing agenda. I don't doubt that George W. Bush's faith is sincere and deeply held. The real question is the content and meaning of that faith and how it impacts his administration's domestic and foreign policies.

George Bush reports a life-changing conversion, around the age of forty, from being a nominal Christian to a born-again believer—a personal transformation that ended his drinking problems, solidified his family life, and gave him a sense of direction. He changed his denominational affiliation from his parents' Episcopal faith to his wife's Methodism. Bush's personal faith helped prompt his interest in promoting his "compassionate conservatism" and the faith-based initiative as part of his administration.

The real theological question about George W. Bush was whether he would make a pilgrimage from being essentially a self-help Methodist to a social-reform Methodist. God had changed his life in real ways, but would his faith deepen to embrace the social activism of John Wesley, the great English preacher and founder of Methodism, who said poverty was not only a matter of personal choices but also of social oppression and injustice? Would Bush's God of the twelve-step program also become the God who required social justice and challenged the status quo of the wealthy and powerful, the God of whom the biblical prophets spoke?

Then came September 11, 2001. Bush's compassionate conservatism and faith-based initiative rapidly gave way to his newfound vocation as the commander in chief of the "war against terrorism." Close friends say that after 9/11, Bush found "his mission in life." The self-help Methodist slowly became a messianic Calvinist, promoting America's mission to "rid the world of evil." The Bush theology was undergoing a critical transformation.

In an October 2000 presidential campaign debate, candidate Bush warned against an overactive American foreign policy and the negative reception it would receive around the world. Bush cautioned restraint. "If we are an arrogant nation, they will resent us," he said. "If we're a humble nation, but strong, they'll welcome us."[5]

The president has come a long way since then. His administration has launched a new doctrine of preemptive war, has fought two wars, and now issues regular demands and threats to other potential enemies. After September 11, nations around the world responded to America's pain—even the French newspaper *LeMonde* carried the headline, "We are all Americans now." But the new preemptive and—most critically—unilateral foreign policy America now pursues has squandered much of that international support.

The Bush policy has become one of potentially endless wars abroad and a domestic agenda that mostly consists of tax cuts, primarily for the rich. "Bush promised us a foreign policy of humility and a domestic policy of compassion," Joe Klein wrote in *Time* magazine. "He has given us a foreign policy of arrogance and a domestic policy that is cynical, myopic, and cruel."[6] What happened?

A Mission and an Appointment

Former Bush speechwriter David Frum says of the president, "War has made him . . . a crusader after all." At the outset of the war in Iraq, George Bush entreated, "God bless our troops." And in his 2003 State of the Union speech, he vowed that America would lead the war against terrorism "because this call of history has come to the right country." Bush's autobiography is titled *A Charge to Keep,* which is a quote from his favorite hymn. In Frum's book, *The Right Man,* he recounts a conversation between the president and his top speechwriter, Mike Gerson, a graduate of evangelical Wheaton College. After Bush's successful speech to Congress following the September 11 attacks, Frum writes that Gerson called his boss and said, "Mr. President, when I saw you on television, I thought—God wanted you there." According to Frum, the president replied, "He wants us all here, Gerson."[7]

President Bush has made numerous references to his belief that he could not be president if he did not believe in "a divine plan that supersedes all human plans."[8] As he gained political power, George Bush has increasingly seen his presidency as part of that divine plan. Richard Land, of the Southern Baptist Christian Life Commission, recalls Bush once saying, "I believe

God wants me to be president."[9] And after September 11, Michael Duffy wrote in *Time* magazine that the president spoke of "being chosen by the grace of God to lead at that moment."[10]

Every Christian hopes to find a vocation and calling in the world that is faithful to Christ. But a president who believes that the nation is fulfilling a God-given righteous mission and that he serves with a divine appointment can become quite theologically unsettling. Theologian Martin Marty voices the concerns of many when he says, "The problem isn't with Bush's sincerity, but with his evident conviction that he's doing God's will."[11] The *Christian Century* writes, "What is alarming is that Bush seems to have no reservations about the notion that God and the good are squarely on the American side."[12] And *Christianity Today* writes, "Some worry that Bush is confusing genuine faith with national ideology."[13] Klein writes in *Time* that the president's faith "offers no speed bumps on the road to Baghdad; it does not give him pause or force him to reflect. It is a source of comfort and strength but not of wisdom."[14] A better use of faith would lead to deeper—even self-critical—reflection, but Bush finds in his religion something we humans are too easily tempted by: easy certainty.

I am told by friends of those now running the Defense Department and Bush foreign policy that while many of them are communicants at Washington area congregations, few ever speak of faith as a motivating or shaping factor in their policy choices and decisions. Theology seems not to have an impact on foreign policy, but simply serves to bolster an ideology of U.S. moral supremacy. It seems to be the president who attaches theology to American power.

The Bush theology that is emerging deserves to be examined on biblical grounds. Is it really Christian, or merely American? Does it take a global view of God's world, or does it just assert the newest incarnation of American nationalism in an update of "manifest destiny"? Is it compassionate and humble, as the campaigning George W. Bush said that American domestic and foreign policy should be, respectively? How does the rest of the world view America's imperial ambition? Most important, how does the rest of the church worldwide see it? For when the White House waxes theological, its theology will be viewed as representative of the church.

Getting the Words Wrong

President George Bush uses religious language more than any president in U.S. history, and some of his key speechwriters come right out of the evangelical community. He draws on biblical language and old gospel hymns that, while unknown to many Americans, immediately cause deep resonance among the faithful in his own electoral base. The problem is that the quotes from the Bible and the hymns are too often either taken out of context or, worse yet, employed in ways quite different from their original meaning. For example, in Bush's 2003 State of the Union Speech, the president evoked an easily recognized and quite famous line from an old gospel hymn. Speaking of America's deepest problems, George Bush said, "The need is great. Yet there's *power, wonder-working power,* in the goodness and idealism and faith of the American people" (emphasis added). But that's not what the song is about. The hymn says there is "power, power, wonder-working power *in the blood of the Lamb*" (emphasis added). The evangelical hymn is about the power of Christ in salvation, not the power of the American people, or any people, or any country. It's a complete misuse. And the first verse, from the same hymn, might have been easily recalled in the run-up to the war with Iraq: "Would you over evil a victory win?" A later verse from the hymn might be more appropriately remembered: "Would you be free from your passion and pride?"

On the first anniversary of the 2001 terrorist attacks, President Bush said at Ellis Island, "This ideal of America is the hope of all mankind. . . . That hope still lights our way. And the light shines in the darkness. And the darkness has not overcome it." Those last two sentences are straight out of John's gospel and immediately recognizable to Christians. But again, the light shining in the darkness is the Word of God and the light of Christ. It's not about America and its values. Even his favorite hymn, "A Charge to Keep," speaks of that charge as "a God to glorify," not to "do everything we can to protect the American homeland," as Bush has named our charge to keep.

Bush seems to make this mistake over and over again of confusing nation, church, and God. The resulting theology is more an American civil religion than Christian faith.

The Problem of Evil

Since September 11, hardly a week goes by without President Bush speaking of "evil" and "evildoers." He has turned the White House bully pulpit into a pulpit indeed, replete with "calls" and "missions" and "charges to keep" regarding America's role in the world. George Bush is convinced that we are engaged in a moral battle between good and evil and that those who are not with us are on the wrong side in that divine confrontation.

But who are "we"? Does no evil reside with "us"? The problem of evil is a classic one in Christian theology and in age-old Western philosophy. Clearly, the existence of evil in the world is real in the biblical worldview. Indeed, anybody who could not see the real face of evil in the terrorist attacks of September 11, 2001, is suffering from a bad case of postmodern relativism. To fail to speak of evil in the world today is to engage in bad theology. But to speak of "them" being evil and "us" being good, that evil is all out *there* and that in the warfare between good and evil others are either with us or against us, is also bad theology. Unfortunately, it has become the Bush theology.

But after September 11, Bush's sense of vision and purpose became much more clear and linked to his Christian faith. In the immediate aftermath of the terrorist violence, the White House carefully scripted the religious service in which the president declared war on terrorism from the pulpit of the National Cathedral. Billy Graham also spoke, but the president became the chief homilist and declared to the nation, "Our responsibility to history is already clear: to answer these attacks and rid the world of evil." With almost every member of the cabinet and the Congress present, along with the nation's religious leaders, it became a televised national liturgy affirming the moral and even the divine character of the nation's new war against terrorism, ending triumphantly with the "Battle Hymn of the Republic." War against evil would confer moral legitimacy on the nation's foreign policy and even on a contested presidency.

What is most missing in the Bush theology is any acknowledgment of the truth of one of the biblical passages that George Bush liked to quote before September 11 but hasn't mentioned since. Matthew's gospel says,

"Why do you see the speck in your neighbor's eye, but do not notice the log in your own eye? Or how can you say to your neighbor, 'Let me take the speck out of your eye,' while the log is in your eye? You hypocrite, first take the log out of your own eye, and then you will see clearly to take the speck out of your neighbor's eye."

The question about terrorism that has become unpatriotic to ask in America is "Why do they hate us?" President Bush strongly replies that our enemies hate us because of our values, our freedoms, and our way of life. There is some truth to the fact that some radical Islamic fundamentalists hate Western freedoms, but many people around the world don't hate America or Americans—they hate our policies in many places where they conflict with our best values and ideals. And when those policies are presented with a theological backdrop, it reflects poorly on the American church. Even Richard Land, of the Southern Baptist Convention, was quoted in the *Washington Post* as saying, "We should always remember that [Bush] is commander in chief, not theologian in chief."[15]

A simplistic "we are right and they are wrong" theology covers over the opportunity for self-reflection and correction. It also covers over the crimes America has committed, which lead to widespread global resentment against us. It allows Elliot Abrams, a key member of the Bush National Security Council, to not have to answer questions about his and the Reagan administration's support for death squads in El Salvador, U.S. mercenaries committing atrocities in Nicaragua, or American-backed military dictatorships in Guatemala (and so many other places around the world), which have killed untold thousands. It allows Dick Cheney and Donald Rumsfeld to never have to answer why they and the U.S. government supported Saddam Hussein in his brutal war with Iran, using weapons of mass destruction. It covers over a less than evenhanded U.S. Middle East policy that inflames Arab people everywhere. And it doesn't explain the seeming contradictions of our intervening in Iraq to "liberate" an oppressed people, but not in so many other places, like numerous sub-Saharan African countries, where the death toll and number of human rights violations are as high or higher, but where there is no oil or geopolitical significance.

It is because of the very mixed motives of nations and superpowers that theologian Reinhold Niebuhr wrote that every nation, political system, and politician falls short of God's justice because we are all sinners. He specifically argued that even Adolf Hitler—to whom Saddam Hussein was often compared by George W. Bush—did not embody absolute evil any more than the Allies represented absolute good. When Richard Land said that "Saddam Hussein is evil, and compared to him we are pure and good," he was only half right.[16] Niebuhr's sense of ambiguity and "irony" in history does not preclude action but counsels the recognition of limitations and proscribes both humility and self-reflection.

In Christian theology, it is not nations that rid the world of evil—they are too often caught up in it in complicated webs of political power, economic interests, cultural clashes, and nationalist dreams. The confrontation with evil is a role reserved for God, using imperfect people, churches, and nations as God wills. But God has not given the responsibility for overcoming evil to any nation-state, much less a superpower with enormous wealth and particular national interests. To continue to confuse the roles of God and the church with those of the American nation, as George Bush seems to do repeatedly, is a serious theological error that some might say borders on idolatry or blasphemy.

American unilateralism becomes not just a bad political policy, but bad theology as well. C. S. Lewis wrote that he supported democracy, not because people were good, but rather because they often were not! Democracy provides a system of checks and balances against any human beings getting too much power. If that is true of nations, it must also be true of international relations. The vital questions of diplomacy, intervention, war, and peace are, in this theological view, best left to the collective judgment of many nations, not just one—especially not the richest and most powerful one.

The Theology of Torture

In early 2004, the world was shocked at the release of photos showing the abuse of Iraqi prisoners by American and British soldiers. The painful irony escaped nobody: After going to war to liberate the Iraqi people from the brutality of Saddam Hussein and his "torture chambers," some of the liberators were now accused of brutalizing and torturing Iraqi detainees—in the same Abu Ghraib prison used by Saddam.

The photos showed the horrible details—prisoners severely beaten, stripped naked and humiliated, sexually threatened, deprived of sleep, psychologically intimidated, and perhaps even killed. Pictures of a hooded inmate with wires attached to his body have traveled around the world, especially the Arab world. Forcing nude prisoners to pile up in human pyramids for picture taking and verbal abuse is perceived as especially degrading in the Muslim "honor/shame culture."

In discussing the obscene photos, conservative columnist George Will wrote that we Americans "should not flinch from this fact: That pornography is, almost inevitably, part of what empire looks like. It does not always look like that, and does not only look like that. But empire is about domination. Domination for self-defense, perhaps. Domination for the good of the dominated, arguably. But domination."[17] That is an honest admission.

Christian theology is uneasy with empire, and the pictures from Abu Ghraib prison reveal why. More than just politics is at stake in this scandal; moral theology is also involved, and that is worthy of serious public discussion. Especially when this war's commander in chief speaks often of his Christian faith.

The Christian view of human nature and of sin suggests that we are fallible creatures and thus not good at empire. We cannot be trusted with domination, becoming too easily corrupted by its power and too often succumbing to repression in defending it. Therefore, we should not simply be shocked at the evil we have seen in the horrible prison photos, but also be sobered and saddened by that same potential in ourselves. History teaches that domination can make good people do bad things. The British did horrible things in northern Ireland, the French in Algeria, and the Americans in Vietnam. Brutality is inevitably the consequence of occupation and domination and is an enduring part of the cycle of violence.

In Iraq, young Americans are being shot at and blown up every day. The frustration and anger at being daily targets is enormous. However, to "set the conditions" for the interrogation of prisoners that might yield critical intelligence and perhaps relieve some of that frustration, both soldiers and commanders clearly crossed the line. Now, the detainee scandal is distracting attention from another, equally alarming, consequence of this dangerous occupation: a growing tolerance for civilian casualties in U.S. counterinsurgency military operations. Again the memory of Vietnam haunts.

The fundamental theological issues in the Iraqi prisoner abuse story and the increase of civilian casualties in this war involve the nature of occupation itself and domination as the consequence of empire—the strategy that appears to be the Bush administration's unapologetic choice for fighting terrorism. Brutality is the predictable consequence of domination, the inevitable result of empire, and an enduring part of the cycle of violence. Christian theology suggests that domination is oppressive and corrupting for both the dominated and the dominator. In preferring the virtues of human dignity, justice, and humility, Christianity implicitly teaches that empire is not the best strategy to fight terrorism. In fact, the domination policies of empire often make terrorism worse by producing tragic behaviors that terrorists use to fuel their murderous agendas. The pictures from Abu Ghraib have already become recruiting posters for the next generation of terrorists in the Muslim world.

Truth telling is also central to Christian theology, which teaches that falsehood has consequences. When a war is primarily justified by citing imminent threats from weapons of mass destruction, which are then revealed not to exist, essential trust in political authority erodes. The archbishop of Canterbury, Rowan Williams, recently spoke controversially about the inevitable erosion of trust in political authority when war is justified on erroneous and even manipulated intelligence: "Credible claims on our political loyalty have something to do with a demonstrable attention to truth, even unwelcome truth. A government that habitually ignored expert advice, habitually pressed its interests abroad in ways that ignored manifest needs and priorities in the wider human and non-human environment, habitually repressed criticism or manipulated public media—such a regime would, to say the least, jeopardise its claim to obedience because it was refusing attention."[18]

When "liberators" become "occupiers," greeted not with flowers but with an unexpected and bloody insurgency, the moral ground is further diminished. And when the only arguments left for war and occupation constantly invoke the horrors of "Saddam's torture chambers," American torture in those same chambers deeply undermine the authority of America's arguments and solutions.

The question of moral agency has been much discussed in regard to the Abu Ghraib scandal. Just who is responsible for the horrific pictures and widespread reports of prisoner humiliation, illegal treatment, torture, perhaps even murder, at American hands? From a religious perspective, those soldiers who abused detainees should be held morally and legally accountable, even if they claim to have only followed orders. If there were such orders, commanders should be held even more culpable. Both common sense and the dynamics of how "sin" operates in human beings and their institutions suggest that the "patterns of abuse" reported by the International Red Cross and human rights organizations are most likely true. We are learning that a climate of official toleration and even encouragement may have created pressure for young military police to "soften up" prisoners for interrogation.

In religious terms, the central point is that we always have choices and the responsibility to make ethical judgments based on moral values and established law. Positive moral agency was indeed active in this appalling scandal when Specialist Joseph Darby reported the prison abuses and turned over incriminating pictures to his commanding officer because he "thought it [the abuse] was very wrong."[19] Some of the most disturbing comments in this scandal have come from anonymous military veterans and personnel who have called Darby a "snitch" who should "never get home." Rather, Darby is a moral hero who should be held up to our children as a role model for what to do when your peers are bowing to pressure to do the wrong thing.

But our reflection will be of little worth unless it takes us deeper than revulsion against "bad apples" that taint the reputation of the military. We need more than investigations initiated by the chain of command into prison policies and atmosphere, and more than decisions about how high accountability should go—whether it be to military intelligence, the Secretary of Defense, or even the Oval Office. We must also address the bad theology that contributes to the problems of domination and empire. When the White House promulgates an official theology of righteous empire in which "they" are evil and "we" are good and if you are not with us you are on the side of the evildoers, it contributes to an atmosphere that makes abuse more likely. And when leaders from the American religious Right describe Islam as an "evil religion," they are, however indirectly, helping to set conditions for the abuse of Muslim detainees. Abuse and torture are always more likely when the victims are objectified, made into an "other" that is somehow different and less human than we are. The religious conviction that challenges us to see the image of God in every person is an absolute barrier to the practice of torture. It is also a moral foundation for international accords like the Geneva Convention.

President George Bush is a Christian, but he did not listen to U.S. and world church leaders who overwhelmingly opposed the war in Iraq and who warned about many of the "plagues of war" (to use the language of the Vatican) that have transpired since. Perhaps President Bush should listen to religious leaders now. American domination and empire are both bad policy and bad theology, and they will not succeed in Iraq. Only international initiative and authority have a chance of repairing the damage now. The United States must make the major contribution it clearly owes to reconstruction in Iraq, but only under somebody else's leadership. The domination of empire must be abandoned.

A Better Way

The real theological problem in America today is no longer the religious Right, but the nationalist religion of the Bush administration, one that confuses the identity of the nation with the church, and God's purposes with the mission of American empire. America's foreign policy is more than preemptive, it is theologically presumptuous; not only unilateral, but dangerously messianic; not just arrogant; but rather bordering on the idolatrous and blasphemous. George Bush's personal faith has prompted a profound self-confidence in his "mission" to fight the "axis of evil," his "call" to be commander and chief in the war against terrorism, and his definition of America's "responsibility" to "defend the hopes of all mankind." This is a dangerous mix of bad foreign policy and bad theology.

But the answer to bad theology is not secularism; it is good theology. It is not always wrong to invoke the name of God and the claims of religion in the public life of a nation, as some secularists say. Where would we be without the moral prophecy of Martin Luther King Jr., who held his Bible in one hand and his Constitution in the other as he preached, holding us to our best values? Can anyone deny the prophetic leadership of Archbishops Desmond Tutu in South Africa and Oscar Romero in El Salvador?

In our own American history, religion has been lifted up for public life in two different ways, as we discussed earlier. One invokes the name of God and faith in order to hold us accountable to God's intentions—to call us to justice, compassion, humility, repentance, and reconciliation. Abraham Lincoln, Thomas Jefferson, and Martin Luther King perhaps best exemplify that way. The other way wrongly invokes God's blessing on our activities, agendas, and purposes. Many presidents and political leaders have used the language of religion in such ways, and George W. Bush is falling into that same temptation.

Lincoln regularly used the language of Scripture but in a way that called both sides in the Civil War to contrition and repentance. In perhaps the most famous passage from his second inaugural speech, he said, "Fondly do we hope, fervently do we pray, that this mighty scourge of war may speedily pass away. Yet, if God wills that it continue until all the wealth piled by the bondsman's two hundred and fifty years of unrequited toil shall be sunk, and until every drop of blood drawn with the lash shall be paid by another drawn with the sword, as was said three thousand years ago, so still it must be said 'the judgments of the Lord are true and righteous altogether.'"

Jefferson said, "I tremble for my country when I reflect that God is just." It's easy to vilify the enemy and claim that we are on the side of God and good. But repentance is better. "Demonization can produce hatred, and all of a sudden, we're heading toward a battle of civilizations," says Robert Seiple, former head of World Vision (a Christian relief and development organization) in the *Christian Science Monitor.* As Jane Lampman, who wrote the *Monitor* article, put it, "The Gospel, some evangelicals are quick to point out, teaches that the line separating good and evil runs not between nations, but inside every human heart."[20]

Christians should always live uneasily with empire, which constantly threatens to become idolatrous and substitute secular purposes for God's. As we reflect on our response to the American empire and what it stands for, a reflection on the early church and empire is instructive. New Testament writings contain a description of the Roman Empire and the response of the church.

The book of Revelation, while written in apocalyptic language and imagery, is seen by most biblical commentators as a critique of the Roman Empire, its domination of the world, and its persecution of the church. In Revelation 13, a "Beast" and its power are described. Eugene Peterson's *The Message* puts it in vivid language: "The whole earth was agog, gaping at the Beast. They worshiped the Dragon who gave the Beast authority, and they worshiped the Beast, exclaiming: 'There's never been anything like the Beast! No one would dare to go to war with the Beast!' It held absolute sway over all tribes and peoples, tongues and races." But the vision of John of Patmos also foresaw the defeat of the Beast. In Revelation 19, a white horse, with a rider whose "name is called The Word of God" and "King of kings and Lord of lords," captures the beast and its false prophet.

As with the early church, modern Christians' response to an empire holding "absolute sway," against which "no one would dare to go to war," is the ancient confession that Jesus is Lord. We live in the promise that empires do not last, that the Word of God will ultimately survive the Pax Americana as it did the Pax Romana.

In the meantime, American Christians will have to make some difficult choices. Will we stand in solidarity with the worldwide church, the international body of Christ—or with our own American government? It's not a surprise to note that the global church does not generally support the foreign policy goals of the Bush administration—whether in Iraq, the Middle East, or the wider war on terrorism. Only from inside some of our U.S. churches does one find religious voices consonant with the visions of American empire.

Once there was first-century Rome; now there is a new Rome. Once there were first-century barbarians; now there are many barbarians who are the Saddams of this world. And then there were the Christians who were loyal, not to Rome, but only to the kingdom of God. To whom will the Christians be loyal today?

As the debate over Iraq and the war on terrorism continued in 2004, more than two hundred Christian theologians and ethicists have issued a new "Confession." They express a deep and growing concern about an emerging national theology of war, the increasingly frequent language of righteous empire, and official claims of divine appointment for a nation and president in a war on terrorism. The signers believe such political developments and claims are also seeping into the churches and threatening our very confession of Christ. Therefore, seminary and college professors from many institutions, a lot of them evangelical, have together endorsed a statement titled "Confessing Christ in a World of Violence." We believe that in a climate in which violence is too easily accepted and the roles of God, church, and nation too easily confused, the most important response is a new confession of Christ. No nation-state may usurp the place of God. The statement identifies five points that are central for followers of Jesus, and spells out rejections of current teachings that could undermine our Christian confession.

Here is the confession.

Confessing Christ in a World of Violence

Our world is wracked with violence and war. But Jesus said: "Blessed are the peacemakers, for they shall be called the children of God" (Matt. 5:9). Innocent people, at home and abroad, are increasingly threatened by terrorist attacks. But Jesus said: "Love your enemies, pray for those who persecute you" (Matt. 5:44). These words, which have never been easy, seem all the more difficult today.

Nevertheless, a time comes when silence is betrayal. How many churches have heard sermons on these texts since the terrorist atrocities of September 11? Where is the serious debate about what it means to confess Christ in a world of violence? Does Christian "realism" mean resigning ourselves to an endless future of "preemptive wars"? Does it mean turning a blind eye to torture and massive civilian casualties? Does it mean acting out of fear and resentment rather than intelligence and restraint?

Faithfully confessing Christ is the church's task, and never more so than when its confession is co-opted by militarism and nationalism.

- A "theology of war," emanating from the highest circles of American government, is seeping into our churches as well.

- The language of "righteous empire" is employed with growing frequency.
- The roles of God, church, and nation are confused by talk of an American "mission" and "divine appointment" to "rid the world of evil."

The security issues before our nation allow no easy solutions. No one has a monopoly on the truth. But a policy that rejects the wisdom of international consultation should not be baptized by religiosity. The danger today is political idolatry exacerbated by the politics of fear.

In this time of crisis, we need a new confession of Christ.

1. Jesus Christ, as attested in Holy Scripture, knows no national boundaries. Those who confess his name are found throughout the earth. Our allegiance to Christ takes priority over national identity. Whenever Christianity compromises with empire, the gospel of Christ is discredited.

   We reject the false teaching that any nation-state can ever be described with the words "the light shines in the darkness and the darkness has not overcome it." These words, used in scripture, apply only to Christ. No political or religious leader has the right to twist them in the service of war.
2. Christ commits Christians to a strong presumption against war. The wanton destructiveness of modern warfare strengthens this obligation. Standing in the shadow of the Cross, Christians have a responsibility to count the cost, speak out for the victims, and explore every alternative before a nation goes to war. We are committed to international cooperation rather than unilateral policies.

   We reject the false teaching that a war on terrorism takes precedence over ethical and legal norms. Some things ought never be done—torture, the deliberate bombing of civilians, the use of indiscriminate weapons of mass destruction—regardless of the consequences.
3. Christ commands us to see not only the splinter in our adversary's eye, but also the beam in our own. The distinction between good and

evil does not run between one nation and another, or one group and another. It runs straight through every human heart.

We reject the false teaching that America is a "Christian nation," representing only virtue, while its adversaries are nothing but vicious. We reject the belief that America has nothing to repent of, even as we reject that it represents most of the world's evil. All have sinned and fallen short of the glory of God (Rom. 3:23).

4. Christ shows us that enemy-love is the heart of the gospel. While we were yet enemies, Christ died for us (Rom. 5:8, 10). We are to show love to our enemies even as we believe God in Christ has shown love to us and the whole world. Enemy-love does not mean capitulating to hostile agendas or domination. It does mean refusing to demonize any human being created in God's image.

   We reject the false teaching that any human being can be defined as outside the law's protection. We reject the demonization of perceived enemies, which only paves the way to abuse; and we reject the mistreatment of prisoners, regardless of supposed benefits to their captors.

5. Christ teaches us that humility is the virtue befitting forgiven sinners. It tempers all political disagreements, and it allows that our own political perceptions, in a complex world, may be wrong.

   We reject the false teaching that those who are not for the United States politically are against it or that those who fundamentally question American policies must be with the "evildoers." Such crude distinctions, especially when used by Christians, are expressions of the Manichaean heresy, in which the world is divided into forces of absolute good and absolute evil.

The Lord Jesus Christ is either authoritative for Christians, or he is not. His Lordship cannot be set aside by any earthly power. His words may not be distorted for propagandistic purposes. No nation-state may usurp the place of God.

We believe that acknowledging these truths is indispensable for followers of Christ. We urge them to remember these principles in making their decisions as citizens. Peacemaking is central to our vocation in a troubled world where Christ is Lord.

The statement was initiated by Richard B. Hays, from Duke Divinity School; George Hunsinger, from Princeton Theological Seminary; Richard V. Pierard, from Gordon College; Caryn D. Riswold, from Illinois College; Glen Stassen, from Fuller Theological Seminary; and me, from Sojourners. The signers listed here hail from a broad spectrum of schools. Other supporters come from many other institutions, including Fuller Theological Seminary, Duke Divinity School, Princeton Theological Seminary, and Gordon College; also Wheaton College, Calvin College, Hope College, Harvard Divinity School, North Park Seminary, Asbury Theological Seminary, Associated Mennonite Biblical Seminaries, Baylor University, Westmont College, Pepperdine University, Candler School of Theology, Interdenominational Theological Consortium, Eastern Baptist Theological Seminary, Southern Methodist University, Union Theological Seminary, and numerous others.

In the fall of 2003, the appointment of a controversial general to a high-ranking and sensitive intelligence post caused a brief national debate about the issues of American empire, God, Christianity, and Islam. The following is an open letter I wrote to General William Boykin, whose views on whose side God was on in the clash between American military power and militant Islam caused a furor around the world.

Backward Christian Soldier: An Open Letter to the Christian General

Dear General Boykin,

You've gotten a lot of press this week, General. Perhaps you didn't expect the things you've been saying in churches to go public—about America's "Christian army," the holy war we're waging against the "idol" of Islam's false God, and the "spiritual battle" we're fighting against "a guy named Satan" who "wants to destroy us as a nation, and he wants to destroy us as a Christian army." You call yourself a "warrior for the kingdom of God," but most of your service has been with the Special Forces and the CIA. You say, "We in the army of God, in the house of God, in the kingdom of God, have been raised for such a time as this." You apparently have no doubt that "America is still a Christian nation," while other nations "have lost their morals, lost their values." You think "George Bush was not elected by a majority of the voters in the United States," but that "He was appointed by God." You say, "He's in the White House because God put him there." And maybe you believe God has put you in the new position to which you were just appointed as deputy under-secretary of defense for intelligence.

Because your views sound like a Christian jihad at a time when the U.S. government is sensitive to offending the Muslim world, you have become a controversy. I'm sure you've been under a lot of pressure since the story of your religious views broke in the *Los Angeles Times*.[21] Your critics say your private religious views are your own business, but when you speak with your uniform on, you're a spokesperson for the U.S. military and government. We don't need to make the Arab world angrier at us than they already are and it doesn't help when you say things like, "Why do they hate us? The answer to that is because we are a Christian nation. We are hated because we are a nation of believers." Or when you describe the Muslim warlords you fought in Mogadishu, Somalia as "the principalities of darkness" and a "demonic presence in that city that God revealed to me as the enemy," that "will only be defeated if we come against them in the name of Jesus."

General, I think the hymn "Onward Christian Soldiers" must have been written just for you. I'm sure your superiors have already given you a lesson in politics and public relations. And I've heard you have toned down your opinions and said you didn't mean to offend anyone. Whether you keep your job is a political question, the outcome of which we will know soon enough.

But I want to raise some different issues: biblical theology, bad teaching, and church discipline. General, your theology bears no resemblance to biblical teaching. You utterly confuse the body of Christ with the American nation. The kingdom of God doesn't endorse the principalities and powers of nation-states, armies, and the ideologies of empire; but rather calls them all into question. You even miss the third verse of "Onward Christian Soldiers," which reminds us, "Crowns and thrones may perish, Kingdoms rise and wane, But the Church of Jesus, constant will remain." And let's not misinterpret the famous first verse, "Onward Christian soldiers marching *as* to war, with the *cross* of Jesus going on before (emphasis added)." The cross General, not the Special Forces.

Brother Boykin, I believe you are a product of bad theology and church teaching. Why were you never given sound biblical tools to help you discern the shape of your vocation? Why were you never taught in Sunday school about the real meaning of the kingdom of God, and the universality of the body of Christ? And why have you never heard that only peacemaking, not war-making, can be done "in the name of Jesus?"

General, I really don't want to blame you for the lack of Christian teaching that you have obviously suffered. But there is a legitimate issue of church discipline here. When a high-ranking military officer espouses a zealous religious nationalism which claims the name "Christian" for both his nation and his army, and when he invokes the name of Jesus, not to love our enemies as he instructed, but rather to target them for destruction—the church must discipline that errant brother and name his public statements for what they are—not mere political incorrectness, but idolatry. General you have substituted your nation and your army for God, your faith is more American than Christian, the Jesus you claim is not the Jesus of the New Testament and his kingdom will not be ushered in by the U.S. military.

Whatever happens with your job, I pray that you find a church that offers you the ministry of repentance, forgiveness, and restoration to a more authentic biblical faith.

(A ten-month Pentagon investigation reported by the *Washington Post* found that Boykin failed to obtain clearance for his remarks, failed to clarify that his remarks were personal and not official, and failed to report reimbursement of travel costs from one of the sponsoring religious groups. No action has yet been taken on the report.[22])

CHAPTER 10

Blessed Are the Peacemakers

Winning Without War

IN A WORLD ON FIRE with ethnic conflict, rogue states, and terrorist violence, the traditional Christian doctrines of pacifism and just war are undergoing severe challenges. After years of debates about war, we now confront new adversaries much more difficult to deal with. The new enemy of terrorism is revealing the limits of both our traditional pacifist and just-war positions. Standing above or apart from terrible violence that costs so much in human lives is not morally tenable. And a pacifism that objects to war without having answers to the very real threats of horrific violence, sponsored either by international networks of terror, or states, isn't very helpful in today's world. Nor is classical just-war doctrine particularly relevant in this battle; it simply isn't enough to offer endless academic scrutiny of criteria for fighting wars but not actually preventing any. Just-war thinking has few answers to new forms of violence that are not as vulnerable to traditional war-fighting methods. And just-war arguments are becoming increasingly suspect when critical judgments become more dependent on technology than morality.

The real question today is how to resolve the conflicts, how to reduce the violence, how to heal the causes of war, and, most critically, how to defeat terrorism. Maybe a new approach is called for. Will the world just see the churches arguing about war or actually doing the things that make for peace?

The challenge to nonviolence today is clear: If nonviolence is to have any credibility, it must answer the questions that violence purports to answer, but in a better way. Saying no to violence is good, but having alternatives is better. What does real peacemaking look like in our world, and what are the initiatives and institutions that could become alternatives to both war and terrorist violence by resolving the inevitable conflicts that lead to them? Where are the most hopeful signs today of peacemaking that is actually making a difference?

I remember a debate on the problem of war in which I participated at Fuller Theological Seminary. I was put in the "pacifist pulpit" and looked across to my dialogue partner in the "just war pulpit." Then I looked down at the audience far below in the pews, and the problem struck me right away. The debate between the pacifist and just war traditions in the church was taking place high above "the ground" and way over people's heads. I suggested we both get back down to earth and together work with everybody else in the room to reduce the growing violence in our world by actually trying to *solve* the problems that create the conflicts.

Sitting in the seminary audience was Fuller Professor Glen Stassen, who speaks of "just peacemaking," instead of either pacifism or just war. As a teacher of ethics, Stassen encourages his students to find "transforming initiatives" that work to create practical, peaceful answers to injustice and violence. "Peace building" is a critical concept and process in many nations of the world today, and aims to solve problems before we get into situations where there are fewer and fewer alternatives to violence or war.

We always should be reminded that Jesus did not say, "Blessed are the peace lovers" (which, of course, everybody claims to be). He rather said, "Blessed are the peace*makers*," which is always a much more difficult task. History, especially recent history, is demonstrating an abundance of practical, nonviolent strategies, techniques, and practices that are yielding positive results. Nonviolent victories against the tyrannies of communism in Eastern Europe, overthrowing military regimes in Latin America, and bringing down powerful dictators like Ferdinand Marcos in the Philippines and Slobodan Milosevic in Serbia, have shown great promise and possibilities for the future. Various strategies of nonviolent resistance have actually proved to be more effective than counterviolence and war in bringing down oppressive regimes. The idea that such powerful states are simply too brutal for nonviolence to work has been proved false.

Several months before the war against Iraq, Peter Ackerman and Jack DuVall, the co-authors of *A Force More Powerful: A Century of Nonviolent Conflict,* a PBS documentary and book, wrote:

> The reality is that history-making nonviolent resistance is not usually undertaken as an act of moral display; it does not typically begin by putting flowers in gun barrels and it does not end when protesters disperse to go home. It involves the use of a panoply of forceful sanctions—strikes, boycotts, civil disobedience, disrupting the functions of government, even nonviolent sabotage—in accordance with a strategy for undermining an oppressor's pillars of support. It is not about making a point, it's about taking power.
>
> Regimes have been overthrown that had no compunction about brutalizing their opponents and denying them the right to speak their minds. How? By first demonstrating that opposition is possible, peeling away the regime's residual public and outside support, quashing its legitimacy, driving up the costs of maintaining control, and overextending its repressive apparatus. Strategic nonviolent action is not about being nice to your oppressor, much less having to rely on his niceness. It's about dissolving the foundations of his power and forcing him out.[1]

The path of active peacemaking is where we are likely to find the alternatives to war and violence that are so desperately needed in a world full of terrorists, terrorist states, unilateralist superpowers, and weapons of mass destruction.

This has been a very difficult time for Christian peacemakers who believe that following Jesus leads to the path of nonviolence. Despite the great challenges to that commitment in light of new terrorist violence, I am still committed to active nonviolence. But since September 11, we've all had to go deeper to understand what real peacemaking might require of us.

I have been part of the peace movement for more than three decades. But the U.S. government's "war on terrorism" presents far more difficult challenges and harder questions than any of the other wars and interventions many of us have fought against. In those other wars, declared and otherwise—from Vietnam to Central America, from Chile to the Congo—there was no worthy goal to be pursued, and any notion of "defending" America was only Cold War propaganda. In the name of anti-communism, the United States violated its own professed values again and again, by backing a succession of ugly regimes that killed tens of thousands of their own people, trampling on every human right Americans hold dear. Our government backed the wrong people over and over in Latin America, Asia, and Africa—even in South Africa (until the very end), where the issues of justice and injustice were so clear. We have never really stood up for Palestinian human rights and self-determination or served as an honest broker for Middle East peace with our ally Israel. After the first war against Saddam Hussein, we ultimately made the Persian Gulf safe, not for democracy, but for our own oil interests. And for fifty years, U.S. nuclear weapons policy has been based on a willingness to exterminate hundreds of millions of people. The United States has led the world in weapons sales, which have fueled conflicts around the globe. Under both Republican and Democratic presidents, U.S. foreign policy has been morally flawed at its core. I have protested it with twenty arrests in thirty years, all for nonviolent civil disobedience.

But the current challenge is much more complicated. The September 11 terrorists murdered three thousand people in one day, and they did so with a cruel intentionality. That those people were civilians mattered nothing to the mass murderers. While President Bush's morally simplistic "good versus evil" rhetoric is spiritually dangerous (America has hardly been "good," given the above litany of wrongdoing), an inability to see the stark face of evil in the events of September 11 is indeed a moral failure. Our postmodern and politically correct world has a hard time naming evil, but Christians shouldn't. This was a horrific crime against humanity.

Although I opposed the language and tactics of war in this campaign against terrorism, the task of preventing further terrorist violence against innocent people is a very worthy goal, and the self-defense of Americans and other people is clearly at stake here. If there is a good and necessary purpose in defeating terrorism, and if the lives of my neighbors and my family are indeed at risk, how do I respond?

While terrorists use and manipulate American global injustices to justify their crimes and to recruit the angry and desperate for their violent purposes, they have no interest in the international justice and peace that many of us have lived and fought for—indeed, they are its mortal enemies. The terrorists' vision for the world is oppressive; they would destroy democracy, deny human rights, repress women, and persecute people of other faiths and even those of their own religion who disagree with them. Even worse, they blaspheme the name of God by doing their violent work in the name of religion.

To dismiss them as a small group of Islamic fundamentalists or marginal extremists is not enough; these terrorists are educated, well financed, and coldly calculating ideologues who will kill quickly and massively whenever it suits their clear purpose—to take power over Islam and the entire Muslim world. We must be realistic at this moment and confront the fact that terrorists are even now planning further violence against innocent people, on as large a scale as their weapons and capacities will allow. They have shown themselves to be people not bound by conscience or by any limits on the destruction they seek.

Therefore, stopping the terrorist violence is a moral cause. But how? How do we prevent them from killing more innocents? Most poignantly, how do advocates of nonviolence try to stop terrorism? We must remember the challenge: For nonviolence to be credible, it must answer the questions that violence claims to answer, but in a better way. I've opposed a widening "war against terrorism" that bombs more people and countries, kills other innocents, recruits even more terrorists, and fuels an unending cycle of violence. But those who oppose the logic of war must have a better alternative.

We are in a new situation. The fight against terrorism is indeed a moral cause, unlike many U.S. foreign policy goals in the Cold War period. Yet this battle will not likely be won with the traditional methods of war; indeed, fighting a succession of endless wars against Muslim countries will only exacerbate the spread of terrorism and claim many innocent lives at the same time. What, then, shall we do?

We must advocate the mobilization of the most extensive international and diplomatic pressure the world has ever seen against the bin Ladens of the world and their networks of terror—focusing the world's political will, intelligence capabilities, security systems, legal intervention, multinational policing, and, finally, swift and sure international law enforcement against terrorism. The international community must also dry up the terrorists' financial networks, isolate them politically, discredit them before international tribunals, and expose the ugly brutality behind their terror.

International law must be empowered to act as never before and authorized to be the primary global instrument against terrorism. The adjudicatory capacities and enforcement mechanisms of international law are the best alternative we have to the dangerous and counterproductive strategy of endless war. That may now be our choice: either multinational cooperation under the rule of international law or the unilateral preemptive wars of the world's superpower. Choosing the first course over the second is not only the best political wisdom; it is also a theological imperative. Unilateral American wars against terrorism cannot conform to the ethics of just war; aggressive multinational legal enforcement could. And such international legal action could also be the alternative to war that pacifists have always sought. Such international and collective action should not be regarded as weak, but rather may be the only thing strong and effective enough to defeat terrorism.

But when the international community has spoken, tried and found terrorists guilty, and authorized their apprehension and incarceration, we will still have to confront the ethical dilemmas involved in enforcing those measures. Terrorists must be found, captured, and stopped. And their violent plans must be disrupted *before* they are carried out. All this involves using some kind of force.

To accept any use of force is a very difficult thing for those committed to nonviolent solutions. Is any kind of force consistent with nonviolence? If so, what kind? What limitations are required? What ethical considerations must be brought to bear?

Most nonviolence advocates, even pacifists, support the role of police in protecting people in their neighborhoods. We would call the police if we or our families were under attack. Can we effectively apply this to the international situation? Perhaps it is time to explore a theology for global police forces, including ethics for the use of internationally sanctioned enforcement—precisely as an alternative to war.

Several theologians have explored this possibility. In his seminal book, *The Politics of Jesus,* Mennonite theologian John Howard Yoder wrote, "The distinction made here between police and war is not simply a matter of the degree to which the appeal to force goes, the number of persons killed or killing. It is a structural and a profound difference in the sociological meaning of the appeal to force. In the police function, the violence or threat thereof is applied only to the offending party. The use of violence by the agent of the police is subject to review by higher authorities. The police officer applies power within the limits of a state whose legislation even the criminal knows to be applicable to him. In any orderly police system there are serious safeguards to keep the violence of the police from being applied in a wholesale way against the innocent. The police power generally is great enough to overwhelm that of the individual offender so that any resistance on the offender's part is pointless. In all of these respects, war is structurally different."[2]

More recently, biblical scholar Walter Wink wrote in *The Powers That Be*, "Perhaps it would be helpful to distinguish between *force* and *violence. Force* signifies a legitimate, socially authorized, and morally defensible use of restraint to prevent harm being done to innocent people. *Violence* would be a morally illegitimate or excessive use of force. A police officer who must arrest a killer may have to use force to restrain him. Such a use of force falls within the definition of his or her office as spelled out by society and Scripture."[3]

Following the beginning of the Iraq War, Gerald Schlabach, professor of moral theology at St. Thomas University, explored the question in *America* magazine.[4] He noted several ways in which war is significantly different from policing.

- "Political leaders draw on the rhetoric of national pride, honor and crusading to marshal the political will and sustain the sacrifices necessary to fight wars. . . . Police officials by contrast appeal to the common good of the community to justify their actions."
- "Police officers are expected to use the minimum force needed to achieve their objective, and are judged harshly if there is 'collateral damage' of the kind that routinely occurs in warfare."
- "War can never be subject to the rule of law in the way that policing is."

All of these are asking whether those committed to nonviolence might support the kind of necessary force utilized by police, because it is (or is designed to be) much more constrained, controlled, and circumscribed by the rule of law than is the violence of war, which knows few real boundaries. If that is true for the function of domestic police, how might it be expanded to an international police force acting with the multinational authorization of international law? As Schlabach concludes, "Viewing the enforcement of international law as 'just policing' would allow pacifists to integrate the contributions they have made to international peacemaking into a process of 'just policing,' without requiring them to condone warfare even in exceptional cases." New Testament theologian Tom Wright has provocatively suggested that the ethics for global policing might be extrapolated from Paul's words in Romans 13 on the responsibility of political authority.

Since September 11, I've talked to a wide range of Christian peacemakers. Some are delving into Dietrich Bonhoeffer's painful decision, as a pacifist, to join the plot to assassinate Hitler. Others are rereading French theologian Jacques Ellul, who explained his decision to support the resistance movement against Nazism by appealing to the "necessity of violence" but wasn't willing to call such recourse "Christian." Many are going back to Gandhi and asking what he meant when he said that nonviolent resistance is the best thing but that violent resistance to evil is better than no resistance at all.

Some believe that there can be no forceful resistance to terrorism, either because of American foreign policy sins or because of their principled pacifism. Others are only willing to deal with "root causes" and continue to oppose the American foreign policy that, in their view, is behind this terrorism. They point out the true fact that the United States has itself been guilty of sponsoring or supporting "state terrorism"—a painful reality I've observed in the Palestinian West Bank and Gaza, occupied by Israeli Defense Forces.

But many practitioners of Christian peacemaking, including myself, can't accept such a nonresponse to horrific terrorism, despite the history of U.S. foreign policy. I believe we must find a way to deal with the threat of terrorism—a threat that must not be avoided or minimized by those committed to nonviolence. We cannot turn away from this.

The just-war theory has been used and abused to justify far too many of our wars. This crisis over terrorism should turn us to a deeper consideration of what peacemaking means. In the modern world of warfare, where far more civilians die than soldiers, war has become increasingly obsolete as a way of resolving humankind's inevitable conflicts. And war is an especially ineffective response to the new challenges of terrorism.

I am increasingly convinced that the way forward may be found in the wisdom gained in the practice of conflict resolution and the energy of a faith-based commitment to peacemaking. Indeed, the number of people, projects, and institutions experimenting in nonviolent methods of conflict resolution has been growing steadily over the past decade and with the promising results suggested above.

Theologian Stanley Hauerwas, author of the seminal *The Peaceable Kingdom,* told me shortly after September 11, "I just don't feel like I've found a voice about all this yet." Hauerwas doesn't like it when people tell pacifists to "just shut up and sit down" during a time like this. He believes that pacifists cannot be expected to have easy policy answers for every difficult political situation, especially when they are often created, at least in part, by not listening to the voices of nonviolence in the first place.

Nevertheless, he believes the advocates of nonviolence can and should offer alternatives that reduce the violence in any conflict. As a professor of ethics, he is quite willing to call governments to observe the principles of a just war, such as the recognition that soldiers killing each other is morally preferable to soldiers murdering civilians. And Hauerwas favors the use of international courts and global police to resolve conflicts. But he doesn't agree with the conventional wisdom that says, "The world changed on September 11." Hauerwas says, "No, the world changed in 33 A.D. The question is how to narrate what happened on September 11 in light of what happened in 33 A.D."

Walter Wink offers a crucial critique of how—in the war against terrorism—the "myth of redemptive violence" is again being used to try to prove to us how violence can save us. He remains convinced that it cannot. Nonetheless, he admits to being glad when the "bad guys" lost in Afghanistan and women, among others, were liberated from Taliban tyranny. He too would greatly prefer the course of international law and police. We simply haven't trained the churches, or anybody else for that matter, in the crucial theology and practice of active nonviolence, says Wink. That must now become our priority. Wink would no doubt agree with Glen Stassen's "transforming initiatives" that can be taken to reduce violence in any situation of conflict. Exploring what practical nonviolent initiatives can be undertaken to open up new possibilities is more important than merely reiterating that one doesn't believe in violence.

John Paul Lederach, who teaches at Notre Dame, is helping us to open up those possibilities as a practitioner of nonviolence in many situations of global conflict. In this terrorism crisis, he has insights into how a network like bin Laden's might be defanged and defeated without bombing an entire country. In particular, Lederach speaks of the need to form "new alliances" with those closest to the "inside" of a violent situation. In this case, he feels that Islamic fundamentalists who don't share the terrorist's commitment to violence might be the most instrumental group in defeating them. Undermining violence from within, Lederach feels, can often be more effective than attacking it from without.[5]

If the aggressive application of international law is to be an effective strategy in countering terrorism, an important question must be answered, "What is terrorism?" Is the definition of terrorism simply in the mind of the beholder, as some say? Is it finally just a political definition, controlled ultimately by who wins bitter conflicts? The events of September 11 and the cycle of violence in the Middle East make preventing terrorism more urgent than ever. Yet the nations of the world still cannot even agree on the definition of terrorism. Two events in 2002 highlighted the problem.

At its first gathering after September 11, in April of 2002, the Organization of the Islamic Conference met in Malaysia. In his opening speech, Malaysian Prime Minister Mohammad stated, "I would like to suggest here that armed attacks or other forms of attacks against civilians must be regarded as acts of terror and the perpetrators regarded as terrorists."[6] He concluded that Palestinian suicide bomb attacks against Israeli civilians and Israeli Defense Force counterattacks against Palestinian society should both be condemned as terrorism.

But his attempt at a clear and moral definition gained little support from the other delegates present, many of whom regard the Palestinian bombings as legitimate acts of resistance against the Israeli occupation. Three days later, the final declaration of the conference stated, "We unequivocally condemn acts of international terrorism in all its forms and manifestations, including state terrorism, irrespective of motives, perpetrators and victims, as terrorism poses a serious threat to international peace and security." So far, so good. But then the statement declared, "We reject any attempt to link terrorism to the struggle of the Palestinian people." So much for moral consistency.[7]

A *Los Angeles Times* story reported that "the U.N. General Assembly, meanwhile, stymied by a deep rift between the Western powers and leading developing nations, has quietly suspended its quest for a consensus definition of terrorism. Such a definition is a prerequisite for a long-delayed international convention that would give the U.N. efforts legal teeth by criminalizing terrorist activity anywhere in the world."[8]

The United States does not want the definition of terrorism to include acts of violence against civilians that are committed by the military forces of recognized states, but only those by individuals or nonstate organizations. Acts by states might be violations of international law but are not terrorism, according to the United States.

Two of the clearest examples of the clash of definitions came from the Israeli and Iraqi ambassadors. Israel said, "Since its inception, Israeli citizens have been the targets of countless terrorist attacks. This past year, Israel has been compelled to engage in legitimate self-defense." Iraq said, "The acts carried out by the Israeli authorities in Palestine and the occupied Arab territories . . . are considered, according to all standards, organized terrorism against a whole population." Both points of view come under judgment by the famous words of Jesus, "Why do you see the speck in your neighbor's eye, but not the log in your own?"

Is the violence we now witness each night on our television screens legitimate self-defense or organized terrorism? The old cliché says that one person's terrorist is another person's freedom fighter. But that is a very bad and dangerous cliché.

In the General Assembly debate, Secretary General Kofi Annan, noting these difficult issues of definition, said, "I understand and accept the need for legal precision. But let me say frankly that there is also a need for moral clarity. There can be no acceptance of those who would seek to justify the deliberate taking of innocent civilian life, regardless of cause or grievance. If there is one universal principle that all peoples can agree on, surely it is this. Even in situations of armed conflict, the targeting of innocent civilians is illegal, as well as morally unacceptable."[9]

Where are the moral lines? Where will the violence and counterviolence end? If Israeli helicopters firing rockets into a crowded refugee camp is justified in the name of self-defense, can we also justify a suicide bomber at a seder in the name of the struggle against occupation? It may be the particular obligation of religious communities to make clear that neither is morally justifiable.

Fuzzy and ideological definitions of terrorism just make it easier to kill people. When you know your actions will kill innocent noncombatants, that's terrorism. And it must be clearly named as unacceptable—no matter who does it (individuals, groups, or states), whatever the weapons, the expressed intentions, or political justifications. Deliberately taking the lives of innocent civilians simply must be morally condemned. That's a clear definition of terrorism and a beginning of resistance to it.

In the crisis over terrorism, Christians must continue to defend the innocent from military reprisals, prevent dangerous and wider wars, and oppose superpower unilateralism. But we must also help stop the bin Ladens, their networks of violence, and the threat they pose to everything we love and value. All that presents difficult questions for peacemakers, but it is a challenge we dare not turn away from.

The Bush administration chose the course of unilateral war instead of international law in its war on terrorism. In Afghanistan, more civilians were killed than the number of Americans who died on September 11, and outside of Kabul, the country is reverting back to control of warlords and even Taliban resurgence. Iraq is now a place of violent insurgency, political chaos, and concentrated terrorism, while even administration officials admit that the terrorist threats to the United States may now be even greater than before September 11. Bin Laden is still at large, his networks are more dispersed, and close observers report that a new generation of terrorists has come of age and is now operational. Indeed, terrorists from other countries are now in Iraq, where their bombings of civilians are exacerbating the already violent situation. War as a tactic in the war against terrorism has not succeeded.

It can now be argued that the Iraq war may ultimately make the defeat of terrorism more difficult because of the division it caused among key allies, the deeper resentment it has triggered in Muslim countries, and the failure of the war to produce the promises of democracy in Iraq or the beginning of a Middle East peace agreement, which seems further away than ever. The war in Iraq has proved to be a great distraction and diversion from the fight against terrorism, rather than a necessary component.

Pope John Paul II vigorously opposed the U.S. war in Iraq. He said it was the wrong way to defeat terrorism. Yet on Easter Sunday 2004, the pope called upon the world "to unite" in overcoming terrorism and to resist its "logic of death." Reject war, but unite to defeat terrorism. That is a message for our time.

No one has all the answers. But humility is a good trait for Christian peacemakers, while self-righteousness is both spiritually inappropriate and politically self-defeating. This much is clear: Jesus calls us to be peace*makers,* not just peace*lovers*. That will inevitably call us to face hard questions with no easy answers. In the end, Christian peacemaking is more a path than a position.

CHAPTER 11

Against Impossible Odds

Peace in the Middle East

"FROM ASHES YOU HAVE COME, to ashes you shall return. Repent and follow Christ." With these traditional words from the Ash Wednesday liturgy, the sign on the cross was placed on the foreheads of Sojourners staff members at our midday prayer service. But these special ashes came from a burned-out house on the West Bank, recently shelled by the Israeli army.

On the altar were many reminders of the violence and pain of Lent in the Middle East, where I had just spent eight days. There were pictures of a Palestinian family whose home had been hit by the bombs that now fall most nights, and photos of the inside of what used to be their house. There were rubber bullets and live ammunition I found on the ground in the conflicted city of Hebron, where members of the Christian Peacemaker Team practice daily nonviolence at their own risk. There was a newspaper column by Amira Hass, a courageous Israeli journalist who writes passionately and eloquently about how her own government's policy of "closure" in the West Bank in Gaza has become a "great robbery" of the Palestinian people's ordinary life. There were clippings about horrendous suicide bombings resulting in the deaths of countless Israeli civilians. Along with other pieces of broken and charred buildings there was a plaque of the Sabeel Center, the Palestinian Christian organization that sponsored the conference that brought me to Jerusalem and the surrounding areas. Symbolically, the beautiful plaque had cracked right down the middle during my return trip home.

After clearing Customs at the Tel Aviv airport, I rode the thirty-five miles to Jerusalem, where I attended the international peace conference convened by the Sabeel Ecumenical Liberation Theology Center. Everywhere I looked were enormous Israeli settlements, always on the highest ground, with the most modern, first-world living conditions anywhere, towering over the much-poorer Palestinian villages down in the valleys.

I saw on this trip how Israeli settlements now loom over the West Bank and Gaza—and likewise loom over the chances for peace in the Middle East. They are the "facts on the ground" that shape virtually everything about Middle East politics today. Long before he became prime minister, Ariel Sharon controlled the future as the chief architect of the settlement policy. Settlements are aggressive forays into Palestinian territory by people who believe that God has given them all the land. Each one makes lasting peace that much more difficult. It's obvious now that this was the intent of the policy from the beginning.

Many of the settlers are American Jewish immigrants to Israel. I've heard stories from Palestinians about an SUV pulling up alongside a Palestinian family whose roots go back ten generations. An American Jew from New York City, here only two weeks, screams at the Palestinian family: "Get off this land! God gave it to us!" With the settlements policy, that's now happening. Israeli soldiers are in the West Bank and Gaza, not to keep law and order, not to protect Palestinians from violence or crime, but only to protect the settlements and the settlers. Control the roads, control movement, control the daily life of the entire Palestinian population—that's the consequence of the settlements policy.

Most people in the Middle East and elsewhere have accepted the logic of a two-state solution and the formula of land for peace. But after days of watching, listening, and talking to both Palestinians and Israelis, I began to feel that the "peace process" had now become, as Palestinian leader Jean Zaru described, "the structure for our domination." The more I saw, the more it reminded me of apartheid in South Africa. There is no contiguous Palestinian territory in the West Bank or Gaza, no such thing as a Palestinian state, nor one in the making. There are only pieces of Palestinian territory, with Israel controlling everything in between.

American Rabbi Michael Lerner says the situation is as if somebody took your house away (the way Israel took Palestinian land in the 1967 war), and then says they'll give you parts of your house back. But they still control the hallways and the bathrooms! Palestinians can't get from one bedroom to another without going through a hallway controlled by the Israelis. As long as the settlements remain, the only possibility is disconnected territories housing the Palestinian workers who service the Israeli state.

The Israeli policy is called "closure." Everything gets closed down in the West Bank and Gaza, and Palestinians are not allowed to move freely—to go to school, to work, or even to visit family. All Palestinians are required to have permits and pass through interminable checkpoints. Our group was stopped at every checkpoint, even though we were an international delegation in large buses. We had some clout and were no threat to the Israelis, and they still held us up for hours. If you are a Palestinian, you wait. And you wait. I heard many stories—for example, of a woman in labor, stopped at a checkpoint on the way to the hospital. She was forced to deliver her baby in the backseat of her car, waiting at the checkpoint. The soldiers then ordered her outside the car, where she collapsed on the ground in utter exhaustion, with the umbilical cord still attaching her to her baby, while Israeli soldiers laughed. In July, another baby born at an Israeli checkpoint died before reaching the hospital. These women experienced the extreme of the type of indignities visited on virtually all Palestinians every day.

One of the most passionate and visible critics of the closure policy is Israeli journalist Amira Hass, whom I mentioned earlier. When I met her, she admitted to being "obsessed" with her government's practice. She, like many other Jews both in Israel and America, believe the policies of settlements and closure are as morally damaging to the Israelis as they are oppressive to the Palestinians. Hass describes the closure policy as "the theft of spontaneity."

If you were an activist in apartheid-era South Africa, you could be pulled out of bed in the middle of the night and killed. But ordinary South Africans, though poor and oppressed, could still visit their mothers or join their buddies to play soccer, and generally they were able to move freely around the country. Palestinians, however, can't just wake up in the morning and decide to go visit a friend, or at the end of the day go see the sunset at the water's edge—the theft of spontaneity. Jean Zaru told me she hadn't worked with her assistant face to face for three months because they couldn't get in the same room at the same time. It was easier for international visitors to come to the Sabeel conference in Jerusalem than for local Palestinians to get there from their own villages and cities.

There is indeed Palestinian violence against Israeli settlements. Shots and mortar shells have been aimed into them. Many Jews have been killed, and the fear is very high. There have even been casualties among Israeli children. Two fourteen-year-old boys were found dead in a cave near their settlement, their bodies battered and mutilated with rocks, killed by Palestinians. And the world has seen the results of suicide bombers in Tel Aviv restaurants and discos, suburban shopping malls, and even Israeli school buses. Israeli parents, no matter what their views on peace, now live in fear each time their children go to school or their teenagers go out at night. We have been horrified by graphic pictures of the bodies of Jewish children blown to bits. From a moral perspective, attacks against civilian populations are terrorism, no matter their source and regardless of their political motivation or grievances. Such terror can never be morally justified. Never.

But the Israelis regularly use such incidents to justify shelling Palestinians in massive, disproportionate retaliation. They've even resorted to bombing Palestinian targets with F-16 fighter planes. The casualties are enormous, including Palestinian children and infants caught in the middle of attacks in retaliation for terrorism. Such deliberate and lethal violence against civilians must also be called terrorism.

The Israeli army is shelling the most exposed houses in Palestinian villages directly from the settlements, knowing they're attacking unarmed civilians with families and children. I went into Palestinian homes that had been shelled and met the families most directly affected. In one home, I saw the huge shell hole in the wall of the children's bedroom. The kids were scared that night, cowering in their parents' room down the hall, and otherwise would surely have been killed.

At this writing, according to B'Tselem, the Israeli Information Center for Human Rights in the Occupied Territories, the death toll among both Palestinians and Israelis since the start of the second intifada in September 2000 is

3,168 Palestinians killed, including 596 children under eighteen, and 639 Israeli civilians, including 112 children under eighteen, as well as 291 members of the Israeli security forces.[1] In a very moving moment at the start of the Sabeel conference, we named each victim of the violence, from all sides. Every individual life counts in God's eyes.

Movements are responsible for the images they project. When the Israeli military shot and killed twelve-year-old Mohammed Dura in his father's arms as they cowered in fear against a wall in Gaza, the powerful images went around the world. During my visit, I drove through that fatal intersection, being told that if we stopped the soldiers would shoot. Forty-five minutes later, a bomb went off at the very spot we had just passed. But three days after the death of little Mohammed, two Israeli soldiers were captured and lynched by angry Palestinians in the city of Ramallah in the West Bank. The image flashed around the world was that of bloody hands raised by an angry Palestinian mob over the lynched soldiers' mutilated bodies. If the images from Birmingham and Selma had been dead cops instead of peaceful marchers being beaten, we wouldn't have won the civil rights struggle in America.

There is no "symmetry" in the violence of the Middle East today. Israeli violence is enormously disproportionate to Palestinian violence. That includes the violence of the settlements and closure policies themselves and the violence of Israeli military practices, especially in their retaliation against Palestinian attacks. Despite this lack of proportionality, there is no moral or strategic justification for the terrorist Palestinian violence targeted against civilians in response to Israeli domination. No argument, even lack of symmetry, will suffice.

But there still are courageous voices in the Middle East calling for nonviolence. Amid the downward spiral of violence and counterviolence, a growing conversation in the Middle East, the United States, and Europe suggests that only new initiatives of nonviolence may now be able to redress the deadly despair of the current deadlock. The violent extremists on both sides of the Palestinian/Israeli conflict seem to have a symbiotic relationship: The Hamas terrorists and the right-wing Likud Party ideologues seem to need each other. And many have pointed out that a disciplined Palestinian movement of nonviolent civil disobedience could be Ariel Sharon's worst nightmare. But as Gandhi often pointed out, nonviolent movements still have casualties—a painful fact that could be especially likely in any confrontation with the Israeli Defense Forces.

It is far too early to say whether such discussions will lead to significant action that might make a difference for peace. But in the Middle East, one discovers a growing number of people and organizations committing to nonviolence. These are the fragile signs of hope.

I visited the Christian Peacemaker Team (CPT) in Hebron, one of the most conflicted areas in the West Bank. The Jewish settlements are actually inside the city of Hebron, so there are street confrontations nearly every day. The only force between the warring factions is the CPT, which consists mostly of Americans and Canadians from the historic peace churches. Days are spent offering presence, relationship, and accompaniment for those who need it. For example, a group of Palestinians were on their way to the mosque for worship, and Israeli soldiers stood in their way. The Muslims went to their knees to pray, refusing to disperse, and the Israelis trained their guns, apparently about to open fire on an unarmed crowd. A twenty-three-year-old American woman and a young man from Canada, both from the CPT, jumped in front of the soldiers with their arms spread and cried, "Please, these are unarmed people, do not shoot them!" That stopped the soldiers from shooting, but the Israeli military put the two CPTers in jail for the night. But when they came back to the city the next day, townspeople said "it was like welcoming Jesus" because the young nonviolent activists had saved countless lives. The Christian Peacemaker Team is both a heroic and a practical project of nonviolence in the Middle East, but it's very, very small. Expanding the CPT style of presence in several other areas could make a real difference.

Interfaith teams from the United States and Europe—Christian, Jewish, and Muslim—are now being sent to the Middle East in small numbers, both for critical moments and situations as well as for long-term presence. What comes to mind for me were invitations from South African anti-apartheid leaders asking Americans to join them for crucial periods in their struggle, and Witness for Peace in Nicaragua, which sent more than five thousand North Americans to conflicted war zones.

In 2003, two international activists paid the price for their nonviolent solidarity. On March 16, twenty-three-year-old Rachel Corrie, an American peace activist, was crushed to death by an Israeli bulldozer as she defended Palestinian homes in Gaza Strip. The army described the incident as a "regrettable accident" and said the protesters were acting "irresponsibly." Less than a month later, on April 11, Tom Hurndall, twenty-one, a British photography student and member of the International Solidarity Movement, was shot in the head by an Israeli soldier in Gaza as he tried to guide two young girls to safety. On January 13, 2004, he died without regaining consciousness.

I was most impressed with people—Palestinian and Israeli—who hadn't given up on peace. The peacemaking conference in Jerusalem was sponsored and attended by some of the most committed and influential Palestinian activists for peace, many of them Christians. And the Jewish peace activists I met each day also made a deep impression on me.

Nonviolence must address what's happening to Jewish souls as well. Not surprisingly, given all the daily stress and fear, spousal abuse is at an all-time high among the Israeli population. Jewish therapists say that they're treating many young men who scream at night, unable to sleep because of what they've done as soldiers to Palestinians. Any movement for nonviolence aims at the souls of all those in conflict.

In the Middle East, we must turn from an ineffectual peace process to an effective peace strategy. The Jerusalem conference I attended created real solidarity for a Palestinian/Israeli peace settlement. But solidarity is not the same thing as strategy, and it's time to move from solidarity to strategy. Most of the conference participants believed the most effective strategy would focus on the settlements and on ending the occupation of the West Bank and Gaza. A parallel nonviolent movement in the Middle East and in the United States was proposed by many. The international participants in the Sabeel conference agreed that they could not call for nonviolence in the Middle East unless they were prepared to mobilize nonviolent campaigns on their own turf as well.

Christian Palestinians now make up only 2 percent of the population, and a nonviolent movement that seems to be coming mostly from the Christians might feel like a Western movement to the Muslim Palestinians. But there are also Muslims who believe in nonviolence, who believe the Qur'an forbids the violence and terrorism that others pervert their faith to justify. Any successful movement based in nonviolence will have to be Christian, Muslim, and Jewish.

There are some hopeful signs of that, both in the Middle East and internationally. And we are beginning to see a coming together of people and organizations in the United States for a potential new campaign for Middle East peace. Jews United for a Just Peace, or "Junity," has held several meetings. Olive Tree Summer, sent North American Jews and others to the Middle East for a series of high-profile protest activities. Churches for Middle East Peace is mobilizing mainline and Catholic churches, and a group of over forty moderate evangelical leaders have issued a very significant statement against "Christian Zionism." (See the sidebar on page 185.)

I have been most encouraged by the tremendous determination of my Palestinian friends who persevere in the midst of an almost impossible daily situation. Their energy, faith, passion, and determination have deeply impressed me. But they know that they can't win their freedom by themselves. It will take an international movement to press for a just and lasting solution between Palestinians and Israelis.

While Middle East leaders and American politicians debate "road maps" and timetables for cease-fires and cooling-off periods—all very important—a public momentum has to build for an end to the violence and for a just peace. We have friends, both Palestinian and Israeli, who are putting their lives on the line for that kind of peace. And we can't continue to let them suffer or struggle alone.

My wife, Joy, my son Luke, and I had dinner recently with our friend Michael Lerner and his wife, Debora, in their Berkeley, California, home. Two nights later, we shared in a Shabbat service in their living room. Luke especially loved the lively singing and dancing of the Beyt Tikkun congregation, which brings together Jews and other spiritual seekers each week from around the Bay Area. Michael is a rabbi, and it was a delight to see him in his element—leading prayers, teaching Torah, and joyously moving around the room with his hands clapping high over his head.

As I looked out his window over the San Francisco Bay, I remembered that the location of this very house, complete with address and directions for how to get there, has been posted on a right-wing Jewish website that labels Michael as a traitor to Judaism. Why? Because he has defended the human rights of the Palestinian people, protested the building of Jewish settlements on the West Bank and in Gaza, and says that taking other people's land and bulldozing their homes is wrong. He says it violates Jewish law and ethics.

Michael is committed to the state of Israel and also has been very outspoken against Palestinian violence. But he has shown the courage to challenge the policies of the Israeli government and the attitudes of many Jews toward the Palestinians. Michael Lerner calls for a new movement of nonviolence in the Middle East, and in the United States, toward a just peace in the embattled Holy Land. Because of his courage, his *Tikkun* magazine has lost many subscribers and, most critically, many donors. Michael has received death threats and a torrent of criticism from defenders of Israeli government policy. But Lerner's prophetic voice has also struck a chord, both with Jews looking for a voice of conscience and with others seeking a voice for peace.

Other American Jewish voices are rising up to protest the Israeli occupation of Palestinian land. Arthur Waskow, another rabbi and friend of Sojourners, has been an articulate spokesperson for nonviolent alternatives to both occupation and terrorism. In February 2002, he helped lead a delegation of ten rabbis and about sixty other American and European Jews to Israel and Palestine to help Rabbis for Human Rights plant olive trees. With this symbolic gesture they began the replacement of the thousands of Palestinian olive trees that have been destroyed by Israeli settlers and military forces. Arthur wrote, "With our own hands—hands more used to computer keyboards and the social workers' handshake and the rabbi's turning of pages in an ancient book—we dig with a pickax, carry trees in tender arms, pick our way through foot-deep mud, place trees in the holes, pat the earth into shape around them, unbend weary backs to grin and sing and pray with each other."[2]

But it is even more dangerous for those in the Middle East to speak and act for peace. When I was there, I met with Jeremy Milgrom of Rabbis for Human Rights. Soft-spoken but passionate for a peaceful solution to this terrible conflict, Milgrom and this little band of rabbis in Israel has had an impact far beyond their numbers. The group's executive director, Rabbi Arik Ascherman, visited Sojourners and told us of the tremendous opposition they receive for simply standing up for Palestinian human rights. Days after his return to Israel, we saw in the *New York Times* a photograph of Ascherman being arrested by Israeli police during a solidarity march to a "closed military zone" where Palestinian lands had been seized for a settlement outpost. In the spring of 2003, he was again arrested on two charges of standing in front of bulldozers to try and prevent the demolition of Palestinian homes. He faces a possible three years in prison for these "offenses."

On my last night in Jerusalem, I had dinner with Jeff Halper of the Israeli Committee Against House Demolitions. I had seen Jeff in a BBC documentary, sitting in front of an Israeli bulldozer that was about to destroy a Palestinian home. Halper and his colleagues have been arrested and jailed many times for such actions, bringing embarrassing attention to Israeli policies that otherwise would go virtually unnoticed by the outside world. His insights into what's behind Israeli policy are some of the most penetrating I've heard, and he has helped me understand Jewish feelings today more than anybody.

"Israel is strong," said Halper. Indeed, it's the fifth largest military power in the world, economically dominant, deep in leadership cadres, healthy in civil society and culture. "But we don't know we're strong," he said. "We still believe we are victims. As long as you believe you're a victim, you are not accountable." The truth of Halper's words hit me like a bolt of lightning, applying to many situations far beyond the Middle East. Israelis do not feel accountable for what they are doing to Palestinians because they believe they are still victims. And as "victims" they must defend themselves whatever the cost, whatever the consequences.

For many, the situation in the Middle East seems utterly hopeless. Violence, counterviolence, suffering, and more suffering have become the lifecycle for both Palestinians and Israelis. The abysmal failure of the political leadership of both sides adds to the feeling of hopelessness and despair. But what is most often missing from the media coverage of the Middle East and what is most promising for the future is the number of sane and courageous voices for peace that *are* there—on both sides. That's what I have had the blessing to see and to hear, and that is, I hope, the best contribution of this chapter. Those voices could conceivably lay the foundation for the new strategy that is undeniably needed for any resolution of the Israeli/Palestinian conflict.

I was invited to the Middle East to speak at the Sabeel Ecumenical Center's Alternative Assembly, sponsored by Palestinian Christians who are, in my view, among the most unsung heroes of the Middle East conflict. Identifying deeply with the suffering of their own people and courageously outspoken against the injustice of Israeli government policies, Palestinian Christian leaders are consistently calling for a nonviolent alternative to the cycle of revenge and retaliation.

Reverend Naim Ateek, known as a liberation theologian among Palestinians, is an Anglican priest and president of the Sabeel Center. Naim, one of the most articulate voices for justice and peace in the Middle East, was unequivocally clear at the Sabeel conference in pointing to the way of Jesus as both a spiritual and strategic imperative. He functions as both a pastoral and prophetic leader in the midst of the present conflict, both comforting the afflicted and afflicting the comfortable. Jean Zaru, Sabeel's vice president, is the presiding clerk of the Ramallah Friends Meeting and a Palestinian Quaker. Jean exposes the "structural violence" of the Israeli occupation and the necessity of nonviolent resistance to it. She has committed her life to the struggle for liberation.

My old friend Jonathan Kuttab has been practicing law and nonviolence for almost two decades. As an internationally recognized human rights lawyer, he has cofounded several human rights organizations and trained many other lawyers. It is the courage of such friends that keeps drawing me back to the Middle East. Their daily experience is one of long-suffering, and all they ask of us is the solidarity of friendship.

The sun was shining over the Holy City as I enjoyed a panoramic view from the top of the Mount of Olives. One can imagine Jesus weeping over the city again because of what is happening down below in the contested streets of conflict. Day after day, I met and listened to courageous people, both Palestinian and Israeli, whose hearts are breaking but who are still working to end the terrible cycle of violence.

I went into the areas of greatest conflict to listen, learn, and seek to understand people's anger and fear. I saw demolished houses, confiscated land, bullet holes, broken windows, and, of course, military checkpoints staffed by young Israeli men who would later become haunted in sleepless nights. This violence needs healing, not fueling, but the political leaders seem incapable of that. The Sabeel peace conference called for a Palestinian Mandela to shake hands with an Israeli De Klerk (referring to the historic leadership offered in South Africa by the black African National Congress leader and the white prime minister in Pretoria). But it must be asked, what if neither is forthcoming? At the very core of what we face in the deepening Israeli/ Palestinian conflict is a leadership crisis in both communities.

But I did meet leaders in both communities. We prayed, cried, and strategized together, believing against present political realities that violence can end. We talked about moving from a peace "process," dominated by negotiations, that doesn't produce promised results, to a peace "strategy" based on nonviolent resistance both in the Middle East and internationally. Together, we remembered how that happened in South Africa.

Many in the region have seen their hopes dashed time and time again. Hope is a fragile thing in this land, despite the abundance of holy sites. Yet there is something about seeing trees that were here when Jesus was alive that brings his presence to mind again. We held candles by the gate to the Old City to invoke his presence in the midst of a deep crisis in the Holy Land. It was Naim Ateek, the host of our conference, who passionately called all of us again to the path of peace—by following the way of Jesus in the land of his birth and death.

The Church of the Nativity in Bethlehem, traditional site of the birth of Jesus, has often been under siege. "Terrorism!" one side shouts. "Occupation!" the other side shouts back. Each side seems to have only one message, never hearing the other. "Occupation!" "Terrorism!" The competing claims fly through the air while innocent civilians die. Both realities are true. The Israeli occupation of the West Bank and Gaza is illegal and immoral, and it must end. Palestinians are entitled to live in peace and security without blockades, closures, and the daily harassment of their entire population. But suicide bombings of innocent Israeli civilians are not the way to end the occupation. The moral truth that condemns both is that there is nothing—no cause, no ideology, no true religion—that can ever justify the deliberate killing of civilians. That is the definition of terrorism.

Whether it is a Palestinian with an explosive belt blowing up a Seder celebration or an Israeli pilot in an Apache gunship firing rockets into a refugee camp—it is terrorism. Elderly people and children, women and men, deliberately killed for political objectives is terrorism; and it is wrong. The Israelis have the superior firepower; Palestinian deaths greatly outnumber Israeli deaths. But the mothers and fathers of dead children take no interest in talk of relative political power or symmetry. Dead children simply rend the souls of their parents and cause the God who created those children to weep.

The immediate question is how to stop the current violence. It will take immediate action by the United States and the world community to achieve a situation in which a secure state of Israel and a viable state of Palestine live side by side in peace. The United States should immediately work to bring about the creation of an international protection force to shield both Israelis and Palestinians from further violence and should call a regional and international peace conference—including Israel, the Arab states, religious leaders, and civil organizations—to finally establish, and then enforce, the plan for peace.

There Has Been Enough Killing—It's Time for Peace

The voices of peace and nonviolence, though marginalized by both sides, continue to challenge their respective authorities. Let us listen to them.

Hanan Ashwari (Palestinian Legislative Council): "Why and when did we allow a few from our midst to interpret Israeli military attacks on innocent Palestinian lives as license to do the same to their civilians? Where are those voices and forces that should have stood up for the sanctity of innocent lives (ours and theirs), instead of allowing the horror of our own suffering to silence us?"

Jonathan Kuttab and Mubarak Awad (lawyers and human rights activists): "The Palestinian people have a genuine chance to achieve their national goals, in spite of the enormous gap between them and their foes, if they pursue a conscious, organized strategy of nonviolent resistance to the occupation on a massive scale."

Neta Golan (from Israel's Gush Shalom): "Inside the pockmarked building surrounded by Israeli tanks and snipers, there is one question on everyone's mind: how many international laws does Israel need to break before the U.N. demands a full and immediate withdrawal? The list of violations is reaching unprecedented levels."

B'Tselem (the Israeli Information Center for Human Rights in the Occupied Territories): "Endangering the lives of innocent civilians constitutes a flagrant violation of the most basic principles of international humanitarian law. Such acts cannot be justified based on 'military necessity' as the IDF [Israeli Defense Forces] has frequently claimed in regard to many other violations."

Rabbi Arik Ascherman (Rabbis for Human Rights): "These are the kind of actions which we must oppose, even when we are suffering from the terror attacks which we condemn. The true test of our humanity and commitment to human rights is whether we can stand up at moments like this and say, 'This crosses the line.'"

A Letter to President Bush from Over Forty Evangelical Christian Leaders

July 2, 2002

President George Bush
The White House
1600 Pennsylvania Avenue N.W.
Washington, D.C. 20500

Dear Mr. President,

We write as American evangelical Christians concerned for the well-being of all the children of Abraham in the Middle East—Christian, Jewish and Muslim. We urge you to employ an even-handed policy toward Israeli and Palestinian leadership so that this bloody conflict will come to a speedy close and both peoples can live without fear and in a spirit of shalom/salaam.

An even-handed U.S. policy towards Israelis and Palestinians does not give a blank check to either side, nor does it bless violence by either side. An even-handed policy affirms the valid interests of Israelis and Palestinians: both states free, economically viable and secure, with normal relations between Israel and all its Arab neighbors. We commend your stated support for a Palestinian state with 1967 borders, and encourage you to move boldly forward so that the legitimate aspirations of the Palestinian people for their own state may be realized.

We abhor and condemn the suicide bombings of the last 22 months and the failure of the Palestinian Authority in the first year of the intifada to stop the violence against Israeli citizens. We grieve over the loss of life, particularly among children, and the suffering by Israelis and Palestinians. The longer the bloodletting continues, the more difficult it will be for both sides to reconcile with each other.

We urge you to provide the leadership necessary for peacemaking in the Middle East by vigorously opposing injustice, including the continued unlawful and degrading Israeli settlement movement. The theft of Palestinian land and the destruction of Palestinian homes and fields is surely one of the major causes of the strife that has resulted in terrorism and the loss of so many Israeli and Palestinian lives. The continued Israeli military occupation that daily humiliates ordinary Palestinians is also having disastrous effects on the Israeli soul.

Mr. President, the American evangelical community is not a monolithic bloc in full and firm support of present Israeli policy. Significant numbers of American evangelicals reject the way some have distorted biblical passages as their rationale for uncritical support for every policy and action of the Israeli government instead of judging all actions—of both Israelis and Palestinians—on the basis of biblical standards of justice. The great Hebrew prophets, Isaiah and Jeremiah, declared in the Old Testament that God calls all nations and all people to do justice one to another, and to protect the oppressed, the alien, the fatherless and the widow.

Finally, Mr. President, be assured of our prayers for you and your cabinet as you lead our nation in this troubled time. May the strength and peace of the Lord be with you.

Sincerely,

CHAPTER 12

Micah's Vision for National and Global Security

Cure Causes, Not Just Symptoms

NEW OPTIONS FOR PUBLIC LIFE, and even political policy choices, can be inspired by our best moral and religious traditions; especially when present options are failing some fundamental ethical tests. For example, the biblical prophet Micah offers some deep insights that could be most critically applied to issues of national and global security. While there are no easy religious answers to hard political questions, some people are turning to ancient biblical writers for the visions to shape new political answers.

The eighth-century Micah has become my favorite prophet of national security. Listen to his prescriptions:

> In days to come the mountain of the Lord's house shall be established as the highest of the mountains, and shall be raised up above the hills. Peoples shall stream to it, and many nations shall come and say: "Come, let us go up to the mountain of the LORD, to the house of the God of Jacob; that he may teach us his ways and that we may walk in his paths." For out of Zion shall go forth instruction, and the word of the LORD from Jerusalem. He shall judge between many peoples, and shall arbitrate

> between strong nations far away; they shall beat their swords into plowshares, and their spears into pruning hooks; nation shall not lift up sword against nation, neither shall they learn war any more; but they shall all sit under their own vines and under their own fig trees, and no one shall make them afraid.

In chapter 7, we discussed the dangers of a foreign policy based primarily on fear; and in chapter 9, one that believes that God is on our side in the struggle between good and evil. Prudent and strategic action is clearly called for in response to real threats in the world, but focusing exclusively on fear as the central motivation for U.S. foreign policy is both a moral and a practical mistake. And to believe that your own nation is "the greatest force for good in history," as President George Bush proclaimed from his Texas ranch in August 2002,[1] and that those who oppose us are "evil" is indeed a dangerous religion for the world. Yet this is exactly the course that the White House has chosen.

As has already been said, September 11 shattered the American people's sense of invulnerability. My English wife, Joy, has a hard time understanding this, coming from a country more like the rest of the world, where vulnerability is simply part of the human condition. For the people of the Middle East today, that vulnerability is a daily fact of life; for the citizens of Sarajevo, San Salvador, and Cape Town, it is still a vivid memory; and even for the people of London, Paris, and Berlin, the more distant memory of life under attack is kept alive by people of Joy's parents' generation, who lived through World War II. But for most Americans, two great oceans and the history of never suffering a foreign attack on our own territory created the illusion of safety. September 11 suddenly changed all that and shook our national psyche to its very foundations, as we have become a nation ruled by the fear of terror—and the terror of fear.

We have suggested that September 11 could have been a teachable moment, the time when America joined the rest of the world in the inevitable vulnerability that comes from being human—accepting the truth that to be human in this world is to be vulnerable. Strengthening international law and institutions, with a much greater capacity to intervene and respond decisively to real threats and dangers to the world community, would have been a wise choice for the United States. Radically reducing the world's arsenals (having squandered the opportunity to do so immediately following the Cold War) and committing to abolish all weapons of mass destruction, including our own, would have been the smartest decision. The worldwide sympathy for the United States following September 11 would have put us in a clear position to lead the other democracies in a fresh determination to ensure human rights, seriously address the scandal of global poverty, protect the environment, and finally resolve the most stubborn conflicts that continue to undermine global security. The rest of the world would have welcomed American leadership, but what they got was U.S. domination.

Instead of recognizing the vital nature of multinational cooperation and the wisdom of confronting the root causes of global unrest, including terrorist violence, the Bush administration asserted the right to pursue American security by any means it saw fit—including unilateral and preemptive war. The only path to peace and security would be through U.S. military supremacy. In a dangerous world, the United States chose to shed the constraints of international relationships and institutions, even those of former allies. And as we described in chapter 9, the mission of American triumphalism was named as a righteous one, cloaked in the symbols of a nationalist religion.

It soon became clear that the political leaders who run the foreign policy of the United States now intend to control the rest of the world too. Their intention is clear in the language they now employ. But their ambition is as dangerous as it is illusory. During the run-up to the Iraq war, another head of state privately said, "The question now is whether America will listen to any of us, or if the rest of us just have to do all the listening."

When we do listen to American officials, one thing becomes completely clear—they have a vision. They are not confused, ambiguous, or equivocal; rather their vision is coherent, compelling, and very aggressive. Their definitions of *peace* and *security* are based almost entirely on American military might and economic power. The peace of America is now like the peace of Rome. Pax Romana has been replaced with Pax Americana, a phrase being offered quite openly in the aftermath of the breathtaking military conquest of Iraq. The positive use of the word *empire* is becoming more and more acceptable.

The political leaders who have ascended to power—like Vice President Dick Cheney, Defense Secretary Donald Rumsfeld, Undersecretary Paul

Wolfowitz, Pentagon adviser Richard Perle, and, of course, President George W. Bush—are clear in their words, can be taken at their word, and should be judged by what they say. Presidential words from battleships point the way. Their plans are not secret. Those who now lead this nation believe, I think sincerely, that peace comes only through unquestioned military superiority and that security depends primarily, and almost exclusively, on utterly dominant military and economic power for the United States.

It is important to recognize that the assertion of American domination in the world is a "moral" claim. It goes much deeper than merely accusing our leaders of shedding blood for oil. Certainly, economic self-interest is a real factor here. But if crude imperialism were all this was, the policy would be much easier to challenge. Rather, I think the proposition for American domination must be taken seriously as a moral assertion about the things that make for peace—then challenged at that level.

The real question becomes, Is there another vision? In particular, can the religious community help to offer a better way? Do the people of God have a vision?

When political leaders speak and act this way, the issues at stake are more than just political. When you say peace depends solely on unquestioned military preeminence, you're contradicting the biblical prophets. The prophets do not agree.

Micah, in particular, has another view. He has another vision. Micah is making a contradictory moral assertion: "He shall judge between many peoples, and shall arbitrate between strong nations far away; they shall beat their swords into plowshares, and their spears into pruning hooks; nation shall not lift up sword against nation, neither shall they learn war anymore; but they shall all sit under their own vines and under their own fig trees, and no one shall make them afraid." Micah is saying, you simply cannot and will not beat "swords into plowshares" (remove the threats of war) *until* people can "sit under their own vines and fig trees" (have some share in global security). Only then will you remove the fear that leads inextricably to conflict and violence. Military solutions are insufficient to bring peace and security, claims the prophet. In fact, Micah's moral assertion directly counters the claims of political leaders who say that peace and security can be found only in military supremacy.

Several millennia later, Pope Paul VI paraphrased Micah when he said, "If you want peace, work for justice." The prophet's insight is that the possibilities for peace, for avoiding war, depend upon everyone having enough for their own security—having a little vine and fig tree. The wisdom of Micah is both prophetic and practical for a time like this. If the tremendous gaps on our planet could be leveled out just a little, nobody would have to be so afraid. Micah understood it was the great imbalances and ambitions that lead to war. Anglican Archbishop Rowan Williams says it well, "There is no security apart from common security." The Israelis will never be secure until the Palestinians feel secure; the developed world will never be secure until the developing world also achieves some economic security; America will never be secure until the injustice and despair that fuel the murderous agendas of terrorists have finally been addressed. This isn't just a political idea, it's a principle rooted in the prophets: There is no security apart from collective security—no national security apart from more common global security.

I have been moved by the stories of American families reunited after the Iraq war—soldiers coming home, loved ones full of gratitude. I talked with many of them, especially the many young kids who bore the brunt of the war effort. And I have listened carefully to the words of both the families and their young soldiers, both before they went over and after they came back. Based on their own words, I've concluded that those who supported and fought this war did so for different reasons than the political leaders who sent them. I was often impressed with the reasons the servicemen and women gave for going to war and risking their lives: Many believed their families were in danger from a terrorist attack with weapons of mass destruction—the chemical, biological, and even nuclear weapons the Bush administration falsely claimed that Saddam possessed—and they wanted to protect their loved ones. Soldiers also always fight for each other—their brothers and sisters in the line of fire. And many told me they wanted to liberate the people of Iraq.

I also wanted to see the people of Iraq liberated from Saddam Hussein, and I am grateful to see "the butcher of Baghdad" gone. I was in Iraq a decade ago, as a delegation of church leaders tried unsuccessfully to persuade Saddam to withdraw from Kuwait and avoid a war. I have witnessed how utterly brutal a dictator and human rights abuser he was—every bit as cruel as

U.S. government leaders painted him. Saddam Hussein had oppressed his people, and some of these American and British kids wanted to help liberate them. Most didn't know how the U.S. government had previously supported and even armed Saddam's Iraq, had far too often backed other regimes as brutal as his, and had more recently failed to intervene in clear situations of genocide in Rwanda, Sudan, and the Congo, just to name a few. But there was no oil in Rwanda, no other important geopolitical interests.

But those who were the architects of the war with Iraq sent those young people off to war for different reasons. Neither (what proved to be) the false claims of weapons of mass destruction nor the liberation of the Iraqi people was their primary motivation. Instead, Iraq was to be the *case study* of American power and dominance, and (we now know) the war was planned even before September 11 by the American "neoconservatives" who control U.S. foreign policy. The Project for the New American Century, led by many who later assumed top positions in the Bush administration, wrote in September 2000, "The United States has for decades sought to play a more permanent role in Gulf regional security. While the unresolved conflict with Iraq provides the immediate justification, the need for a substantial American force presence in the Gulf transcends the issue of the regime of Saddam Hussein."[2]

They wanted a vivid demonstration of the raw military power the United States possesses and is able to exercise at will, so convincing that others, in the region and around the world, would be more submissive to American interests and goals. The war in Iraq was aimed also at the other nations President George Bush had identified as part of the "axis of evil"—such as Iran and North Korea–and at the many countries, especially in the Middle East, that the United States wished to influence. Look at Iraq, and see your own future—that was the message. This was a war fought for American empire.

But the young soldiers did not fight this war for American empire. And I don't think most of their parents, or most of the American people, believe in American empire. Our leaders are taking us down the wrong road.

Micah can help those kids who supported the war and their families. Micah can help us all. Micah knew that we will not beat our swords into plowshares; we will not overcome war, will not be safe, will not protect our families, and will not prevent further wars—or further terrorism—until more people have their own vines and fig trees. You want to be safe and keep your family safe? This is the way. Everyone needs their own little piece of the global economy, their own small stake in the world, their own share of security for themselves and their families. Because when you have a little patch on which to build a life for yourself and your family, nobody can make you afraid—or at least it's much harder. It's both a prophetic and a practical prescription for peace and security. What eighth-century Micah understood is that there is no security for us until others also feel secure. Micah knew our weapons cannot finally protect us; only a world where most people feel safe can do that.

I live twenty blocks from the White House—Joy, Luke, Jack, and I. Whenever I fly away from home for a speaking engagement, I'm aware they're left behind in what is probably one of the world's top ten terrorist targets. Anything that hits the White House, or anywhere close, hits us. I'm aware of that all the time. Do I want to stop further terrorist violence against innocent people, maybe even my own family? With all my might, I do. But, as I always say to audiences of people around the country, "not in a way that takes out other people's five-year-olds." How do we do that?

Micah is pleading with us to go deeper, to the resentments and the angers, the insecurities and injustices, embedded in the very structures of the world today. We must go deeper than war. And Micah knew the cruel connection between poverty and war.

Part of my job at Call to Renewal is to read federal budget resolutions. It hasn't been pleasant reading these last few years because the cost of the war in Iraq will now be measured in the loss of health care, in our inability to provide education, and in the alarming percentage of people who will be hungry. Yes, there are still hungry people in America, though the rest of the world can hardly believe it. The American occupation of Iraq now costs about $1 billion per week. Yet American religious leaders cannot succeed in getting our government to approve $5 billion for child care over five years. We warned about the cost of the war for the poor, and our worst expectations have come true.

War and security, and people being afraid, have pushed the poor off the agenda. There is simply no space in the budget or in the national public discourse to talk about poor people. What about the faith-based organizations that do so much work on the ground in soup kitchens and shelters? They are all struggling to survive because there is no money. And the government is telling us, in effect, there are no resources, so you do it. You're so good at this, you change lives. They are asking faith-based organizations to make bricks without straw. And we're all struggling.

A modern American prophet, like Micah, once said, "A nation that continues year after year to spend more money on military defense than on programs of social uplift is a nation approaching spiritual death."[3] He was Reverend Martin Luther King Jr., and he also made the connection between war and poverty.

We had only a few weeks to organize "Pray and Act: A Service for Peace and Justice" on January 20, 2003, the holiday of Martin Luther King Jr.'s birthday. We knew the service in the Washington National Cathedral and the procession to the White House for a candlelight vigil would succeed only if "the moment" carried the day. John Chane, the Episcopal bishop of Washington, felt strongly that this could be a historic event, that the cathedral was hosting a service that could challenge the nation's rush to war.

Almost twenty-five leaders from denominations, religious orders, and national faith-based organizations appeared, and the middle of the sanctuary began to fill as the service participants gathered, many in their clerical robes and stoles. This was a genuinely prayerful event, not a political rally, so we joined hands to pray that God's will be done in this service and in our nation. As we processed to the front of the cathedral, the aisles nearest the pulpit seemed full, but from where we were sitting, I couldn't see beyond the first twenty rows. From the start, the music, the readings from Dr. King's last Sunday sermon (preached in the National Cathedral), the biblical texts, and the prayers and meditations created a profoundly spiritual environment.

It wasn't until I climbed up into the pulpit to preach that I could see all the people. The National Cathedral was packed to capacity; an estimated thirty-two hundred people had come to pray for peace and justice. I was amazed. This was clearly a response to the critical moment. Some who were there said they had been to the big demonstration the previous Saturday and felt mixed about the messages there. They didn't want to attack or demonize the president, and their most enthusiastic response came when I called on

George W. Bush to undertake a new "faith-based initiative" for peace. Many expressed a deep gratitude for the opportunity to express their faith in relation to war with Iraq.

The media were there too and could easily see the strength of the churches' response. It was one of those unique times when our message as Christians was clear, not mingled with many others. Broad coalitions are often a good thing, but there is also a need to witness as people of faith. The cathedral service became that time. I preached from Micah 4, the call to go beyond "anti-war" to the deeper peace that can only be based on justice. The sermon summarized the insights from Micah in this chapter and applied them to the impending war on Iraq.

Here is the text from the January 2003 Martin Luther King Jr. remembrance service in the National Cathedral, and my sermon on it.

[The Lord] shall judge between many peoples, and shall arbitrate between strong nations far away; they shall beat their swords into plowshares, and their spears into pruning hooks; nation shall not lift up sword against nation, neither shall they learn war any more; but they shall all sit under their own vines and under their own fig trees, and no one shall make them afraid. (Micah 4:3–4)

We gather as our nation moves closer to a decision about whether to go to war with Iraq. It will be a momentous choice, with great consequences for the life of the world. I just returned from England, where the debate about this war is also raging. Most there hope Prime Minister Blair will not commit Britain to a war without genuine international support.

Virtually every church body which has spoken, internationally and in the United States, has concluded that a war on Iraq would not be a just war. Never before have the churches in America been so united on the issue of peace. Never before has the House of Bishops of the Church of England spoken out so clearly and strongly against the direction the British government is taking.

The churches have warned about the risk and cost of a potential war which could easily result in unpredictable and unintended consequences—high numbers of civilian casualties, the death of many American and British servicemen and women, more instability and violence in the already volatile Middle East, more anti-American sentiment around the world, and perhaps even more terrorism against our people. But at a deeper level, the churches are witnessing to the need for a new "world perspective" of which Martin Luther King Jr. spoke.

Today we remember the birthday of our brother Martin Luther King Jr. We have heard words from his last Sunday sermon on earth, given from this pulpit, where he called for an alternative to war and bloodshed, for a refocusing of our attention on the most dangerous enemies of our age—poverty, racism, and hopelessness—and for the development of a new perspective.

Martin Luther King Jr. was a modern day Micah, who knew that we will not beat our swords into plowshares until everyone has their own vine and fig tree—their own little piece of the global economy, their own small stake in the world, their own share of security for themselves and their families. Because when you have a little patch upon which to build a life, nobody can make you afraid. And it is fear that leads to violence. We must learn, as both the twentieth-century Martin Luther King Jr. and the eighth-century Micah understood, that there is no security apart from a common security—a global security. That spiritual reality is truer today than ever before. Our weapons cannot finally protect us; only a world where most people feel secure will truly be safe for us and our children.

Both of our prophets, Micah and Martin, urge that we go deeper—to the resentments, angers, insecurities, and injustices embedded in the very structures of today's world. We must go deeper than war.

Micah and Martin knew well the cruel connection between

poverty and war. The cost we will pay for war in Iraq will also be measured in the loss of health care for millions of poor American children, our inability to provide the education that frees inner-city youth from the prison of poverty, the shame of women and children forced to live in homeless shelters, and an alarming percentage of people going hungry in the richest nation in the history of the world. We fear that the urgent need to overcome poverty, at home and around the world, will literally be pushed off the agenda in favor of the resources, attention, and priority that war demands.

So today we don't just say no to a war with Iraq, we say yes to the biblical prophet Micah, yes to the American prophet Martin Luther King Jr., yes to our own church leaders, yes to the international body of Christ, yes to the millions of our fellow citizens across the country and across the political spectrum who don't want this war, yes to our own poorest and most vulnerable citizens, yes in urging our national leaders to find another way.

Today we pray to God and plead with our national leaders to avoid the destructiveness of war and find a better way to resolve the very real threats involved in this conflict with Iraq. We believe that is possible, and we believe we can still stop this war before it starts.

From this National Cathedral and then in our candlelight vigil at the White House, we appeal to President George W. Bush today, not in anger, but in hope, to a fellow brother in Christ, to heed the words of the prophets, the words of our brother Martin Luther King Jr., the words of Jesus the Prince of Peace—to win this battle without war, to transform our swords into plowshares and, yes, to persevere in disarming the world of weapons of mass destruction—all of them, including our own—but without the killing of more innocents. Provide us, Mr. President, with a leadership for peace that would sow the seeds of justice. Mr. President, the hour is late, we stand at a midnight in history, and what we need from you is a faith-based initiative for peace.

May God bless our prayers for peace, and, at this critical hour, God bless America, and God bless the world.

Almost half the world—three billion people—now lives on less than $2 per day, and one billion people live on less than $1 a day, including half the population of sub-Saharan Africa. Bread for the World, the Christian hunger organization, does some of the best work on Capitol Hill, focusing the energy of faith communities on budget priorities for hungry people around the world. My friend David Beckmann leads that organization and courageously reminds his affluent American audiences that thirty thousand children die every single day from utterly preventable causes—hunger, diseases due to hunger, unsafe drinking water, etc.—things that we *could* change if we just wanted to. When I preach, I sometimes remind congregations that during the service we're sharing together, and the time it takes for lunch afterward, more children will have died needlessly than all the people who perished in the attacks of September 11. Yet there won't be any headlines, emergency newscasts, or memorial services for them. Micah is saying that until all the children who died from hunger on September 11 are as important to us as those who died in the terrorist attacks, we will not be safe or secure. As we have said, there is no simplistic causality between global poverty and global terrorism, but most people sense that there is a spiritual and political connection. It is both a moral issue and a practical matter of national security.

There are voices rising up in our world that speak directly to the challenge of Micah. I believe they are modern-day prophets, often coming from unexpected places. One is the most famous rock singer in the world, the leader of the Irish band U2. Of course, I'm speaking of Bono, who has become a serious and well-informed activist, talking always about Africa and HIV/AIDS. Bono is a spiritual man, though not a churchy person, and often comes to Washington, D.C. Over dinner together one night, Bono told me he had started the day "on my knees, begging God for the right words to say." On that day alone, he spoke with Secretary of the Treasury Paul O'Neill, National Security Adviser Condoleezza Rice, and President Bush. Many people, including political figures, are listening to Bono as he passionately pleads the cause of millions of people afflicted by HIV/AIDS and implores the world to respond. But the image of this rock superstar on his knees, quietly, privately, "begging" God for the words to say to some of the world's most important leaders was both humbling and heartening.

I talked to Bono that night about a subject surprisingly pertinent to his mission—biblical archeology. He smiled and confessed that he'd never studied it, but he became very interested when I shared what some of the biblical archeologists had found. When they dig down into the ruins of ancient Israel, they find periods of time when the houses were more or less the same size, and the artifacts show a relative equality between the people, with no great disparities. Ironically, during those periods, the prophets were silent. There was no Micah, Amos, Isaiah, or Jeremiah because there was nothing to say. But then they dig down into other periods, like the eighth century, and find remains of huge houses and small shacks, along with other evidence of great gaps between the rich and poor. And it was during those periods that the voice of the prophets rose up, to thunder the judgment and justice of God. I told Bono that he might be one of those prophetic voices today.

The next night Bono spoke at the Africare dinner in Washington to fifteen hundred of the capital's leaders and media. "Excuse me if I'm a little nervous," Bono apologized, "but I'm not used to speaking to less than twenty thousand people!" Then he looked at the delighted crowd of Washington's elites and spoke like a preacher:

> So you've been doing God's work, but what's God working on now? What's God working on this year? Two and a half million Africans are going to die of AIDS. What's God working on now? I meet the people who tell me it's going to take an act of God to stop this plague. Well, I don't believe that. I think God is waiting for us to act. In fact, I think that God is on His knees to us, to the Church. God is on His knees to us, waiting for us to turn around this supertanker of indifference, our own indifference a lot of the time. That God Almighty is on His knees to us—I don't know what that means. Waiting for us to recognize that distance can no longer decide who is our neighbor. We can't choose our neighbors anymore. We can't choose the benefits of globalization without some of the responsibilities, and we should remind ourselves that "love thy neighbor" is not advice: it is a command.[4]

It was quite an image in Washington, D.C., a very compelling and powerful picture of God, on his knees, begging *us* to respond. Bono also reminded them how the United States invested in Europe after the Second World War, as a bulwark against communism. Now, he said, Africa is in the same kind of position, vulnerable to extreme ideologies, that Europe once was. The pragmatic prophet said it would be both right and smart for the West to invest in preventing fires rather than putting them out, which is always a lot more expensive. I could hear the tones of Micah in the voice of Bono. The cost of five days of war in Iraq would have eliminated illiteracy worldwide. Which will finally make us more secure?

Another Micah-like voice is from Gordon Brown, the chancellor of the exchequer in Great Britain—the most loyal U.S. ally. Brown has called for a boost in global aid from the developing countries of $50 billion a year over the next decade to provide a surge of funds to build schools, hospitals, and sanitation facilities in the developing world to meet the 2015 U.N. deadline world leaders have set for cutting "extreme" global poverty in half. The chancellor says, "What has happened in Afghanistan and elsewhere raises global issues on terror on which we must respond with resolution, but also about the integration of the poorest countries into our economy." Gordon Brown grew up in a manse, his father being a Church of Scotland minister. He's paying attention to the world, he's troubled, and he's listening to the message of the biblical prophets.

In a November 2001 speech in New York City, Brown said,

> I want this generation to be remembered as the first generation in history that truly made prosperity possible for the world and all its people. I want us to be remembered not only as the generation which—in the face of terrorism—freed the world from fear, but as the generation which—in the face of deprivation and despair—finally freed the world from want. This is a great ambition—a grave responsibility—but a genuine possibility given to no other generation at any other time in human history.
>
> The challenge is as new as today's debt crisis, but it is as old as the call of Isaiah to "undo the heavy burdens and let the oppressed go free." The difference is that thousands of years after those words were first written, we now hold in our hands the power to obey that ancient command.[5]

Brown knows that poverty is not the only cause of terrorism; it's more complicated than that, with roots that are also religious, cultural, and ideological.

An American-controlled new world order, based on unilateral instead of multilateral action and military power rather than international law, will not create a framework the world will trust for peace or security. Unresolved grievances—like the Israeli-Palestinian conflict, corrupt relationships with Arab regimes protected by oil, and globalization policies that give advantage to wealthy countries over poor nations—not only undermine justice, but also our security. And these enormous problems will not simply be overcome by the imposition of U.S. military superiority. At a recent World Trade Organization meeting, analysts bluntly admitted that the objective of the United States was to advance the interests of multinational corporate interests over the interests of developing countries. That is not the path to peace.

We need a leadership for peace that will sow the seeds of justice. Listen to Micah. We do stand at midnight in history, as Dr. King said, and we need nothing less than a faith-based initiative for peace and global security. A rock star, a chancellor, and young people across the world are all talking about globalization, HIV/AIDS, and reducing global poverty—and all in the prophetic voice of Micah. I am convinced that global poverty reduction will not be accomplished without a spiritual engine and that history is changed by social movements with a spiritual foundation. That's what's always made the difference—abolition of slavery, women's suffrage, civil rights—they were social movements, but they all had spiritual foundations.

This will be no different. That's why religious leaders are engaging political leaders on critical issues of national and global security, pushing the agenda deeper with the insights of Micah. In doing so they are bringing spiritual energy to key political questions and engaging global politics with the moral authority of religion. That could help move the sometimes deadlocked international discussions to a new and deeper level. What if the religious community could become a factor in the G–8 discussions with heads of state and finance ministers, offering a voice and a distinctive influence that crosses national boundaries? The smart politicians who care about these issues understand that without a new moral energy, we will not succeed with reducing global poverty or making the world more secure.

Speaking in Britain during the summer of 2003, and right after the war with Iraq, I contrasted the vision of Micah with that of Donald Rumsfeld, the U.S. secretary of defense and the chief architect of the war. I was careful neither to caricature nor demonize Rumsfeld but to lay out his stated plan for peace through unilateral military might—using the secretary's own words. Then I compared Rumsfeld's agenda for American military dominance with the vision of common security offered by Micah and emphatically stated that the two visions were not compatible. We have to choose one or the other, I told the audience of eleven thousand British Christians. Their tumultuous response indicated a spiritual hunger for Micah's better vision.

After I spoke, a man from the audience wanted to have coffee. He introduced himself as a major general in the British army and said he had just returned from Iraq. In our discussion, he stunned me by saying he had been helping to oversee the occupation of Iraq, working with Americans Jay Garner and then Paul Bremer. The veteran British military man, a committed Christian, told me he had listened to every word I had just spoken about the true path to peace and security. Then he looked me straight in the eye and said, "Micah is right; Rumsfeld is wrong."

After September 11, but before the United States had began to sound the drumbeat for war with Iraq, I attended a meeting at the White House. A new director for the White House Office on Faith-Based and Community Initiatives was to be announced, and about thirty religious leaders were invited to a ceremony in the Roosevelt Room. President Bush was there to introduce his new director, Jim Towey, somebody I had known and respected for many years. When he was finished speaking, and in his typical fashion, the president came down from behind the lectern to mingle and talk to the diverse group of faith-based leaders.

The president walked over to me, grabbed one of my cheeks in each hand, and asked, "Jim, how ya doin', how ya doin'?!" I confess, it was a little startling for me and the other invited guests who were observing this presidential interaction in the middle of the room. "I'm doing just fine," I answered the president, after he released my cheeks from his warm Texan embrace. "But," I went on, "I was concerned about your speech the other night" (the 2002 State of the Union address). He looked at me intently as I continued, "You said that unless we focus all our energy, resources, and resolve on stopping the terrorists, we will never be secure. But unless we also focus our attention, energy, and will on defeating poverty, we will lose both the war on poverty and the fight against terrorism." He nodded his head, as if in agreement, and said, "That's why we need the leadership of all of you in the religious community." I responded, "Well, we'll do our best, but this battle to overcome poverty and violence will take all of us—and our government's commitment too." For a few more minutes we went back and forth in what was a good discussion, observed with great interest by my colleagues. Mr. Bush was right about the critical role of the religious community, but he was not addressing the vital responsibilities his own administration also had.

The enormous task of securing peace and security requires qualities that are specifically religious, I would suggest, especially the energy of hope. The prophets begin with critique and judgment, but they always end in hope.

And without that spiritual energy and the moral authority of the religious community, we may never be able to break the hold of poverty over the world's poorest people or free ourselves from the ever present threats of violence. But that's our job, our vocation, and Micah is crying out for us to pay attention. Micah says we have an alternative—work hard, fashion another response, show a better way. Don't just say your governments are wrong. They may be wrong, but you've got to show a better way. Hear the voice of the prophet Micah saying, Don't be tempted by your military might and power, but instead "do justice, love kindness, and walk humbly with your God."

I saw the potential power of Micah's vision in Queretaro, Mexico during the early fall of 2003. Several hundred evangelical leaders were gathered in this Mexican industrial city, from Christian relief and development organizations around the world. They had a big and bold idea. From more than 250 agencies in 50 countries, these evangelical poverty fighters from mostly Africa, Asia, and Latin America (with allies from the U.K., Europe, and a few from the U.S.) were calling themselves the "Micah Network." Inspired by the ancient Hebrew prophet to "do justice, love kindness, and walk humbly with your God," they were ready to issue a challenge to their own churches and to the government leaders in each of their countries, right out of the prophetic biblical tradition. They are calling it, appropriately, "The Micah Challenge," and are directing it straight to the heart of globalization and its impact on the poor. The backdrop of failed trade talks at a recent World Trade Organization meeting in nearby Cancun, Mexico, was clearly on people's minds. I was there to speak, to the "prophetic call" of Micah, and to strategize with these brothers and sisters about what a global campaign might look like.

No longer willing to just "pull the bodies out of the river," these evangelical Christians, mostly from the southern hemisphere, were ready to "go upstream and find out what or who was throwing them in!" Having worked in poor communities for many years (and won great credibility in doing so) these community development agencies had decided to now turn to advocacy as well—prophetic advocacy on behalf of the poor. And they had entered into a clear partnership with the World Evangelical Alliance (comprised of church associations in 120 countries). That partnership would unite evangelical churches around the world (now comprising 200–400 million Christians) with their evangelical relief and development organizations in the common cause of biblical justice.

The Micah Challenge mission statement begins with a clear declaration that will warm the hearts of people across the world who long for justice. It reads simply, "The World Evangelical Alliance (WEA) and the Micah Network are creating a global evangelical campaign to mobilize Christians against poverty."[6] Their strategy was to promote "integral mission," where the proclamation and demonstration of the gospel are deeply connected, so that evangelism and social justice both have clear consequences for the other. Then they would prophetically call upon the political leaders of the world and seek to influence them to seek justice for the poor and rescue the needy as the Bible instructs.

The Micah Challenge is taking direct aim at the implementation of the Millennium Development Goals, agreed to by 147 nations, to cut extreme global poverty in half by 2015. The Micah Network believes that achieving those goals will indeed require a "spiritual engine" that provides both moral energy and political accountability. They intend to raise a strong "evangelical voice" to political decision-makers in their own countries, in the wealthy nations, at the United Nations, World Bank, International Monetary Fund, World Trade Organization, and other international bodies. Their advocacy will be local, national, and global, holding the nations accountable to what they have already said and agreed to. The Micah Network is ready to collaborate with others, whenever possible, but their strong appeal will be to and from evangelical Christians. Given the amazing growth of evangelical Christianity around the world, especially in the global south, the emergence of the Micah Challenge could be of great significance. As one delegate from a developing nation remarked quietly and prayerfully after the morning session on the vision of Micah for today's world, "We could be starting history in this room." Indeed.

PART IV

Spiritual Values and Economic Justice

When Did Jesus Become Pro-Rich?

CHAPTER 13

The Poor You Will Always Have with You?

What Does the Bible Say about Poverty?

I OFTEN DO A LITTLE Bible quiz for audiences I'm speaking to. I ask this question: "What is the most famous biblical text in America about the poor?" Every time, I mean every single time, I receive the same answer. "The poor you will always have with you!" they shout out. It's the Scripture from Mark's gospel, chapter 14, verse 7: "For you always have the poor with you, and you can show kindness to them whenever you wish; but you will not always have me." But it's only the first part of the verse they remember, not the part about being kind to the poor, and they miss a few other things—like the whole context and meaning of the text. This highly American commentary on a famous biblical text is worth some further reflection.

First of all there is the context. Jesus was at Bethany "in the house of Simon the leper," the Scripture says. There's a clue to the meaning of the story right away. Jesus and his disciples are "at the table" with a leper—the worst outcast of his society. This was not a dinner with business executives from the chamber of commerce, nor a prayer breakfast with the president and members of Congress. Rather it was like having supper with somebody that everyone knows has AIDS.

While they were eating, "a woman came with an alabaster jar of very costly ointment of nard, and she broke open the jar and poured the ointment on his head." Now this was not a prank or an act of social terrorism; it was a cultural sign of high regard and religious devotion toward a person due great respect. But instead of being pleased that their master was so honored in this way, Jesus's disciples began to complain among themselves and criticize the woman's lavish act of worship. "Why was the ointment wasted in this way?" they murmured. "For this ointment could have been sold for more than three hundred denarii, and the money given to the poor." Why waste money on worship, they were saying, when there are so many needs out there? (Another rendition of the story suggests they were not so pure in motivation and might have wanted to keep some of the money for themselves, but let's give them the benefit of the doubt and say they were honest left-wingers who wanted to spend more on the poor.) The poor woman had probably expended her whole savings account to afford this special gift to Jesus, and his followers just "scolded her."

But Jesus says to them, "Leave her alone; why do you trouble her? She has performed a good service for me. For you always have the poor with you, and you can show kindness to them whenever you wish; but you will not always have me. She has done what she could; she has anointed my body before my burial. Truly I tell you, wherever the good news is proclaimed in the whole world, what she has done will be told in remembrance of her." So what is Jesus really saying here?

Remember the context. They are at the dinner table with a leper, and Jesus is making an assumption about his disciples' continuing *proximity* to the poor. He is saying, in effect, "Look, you will always have the poor with you" *because* you are my disciples. You know who we spend our time with, who we share meals with, who listens to our message, who we focus our attention on. You've been watching me, and you know what my priorities are. You know who comes first in the kingdom of God. So, you will always be near the poor, you'll always be with them, and you will always have the opportunity to share with them.

Indeed, biblical scholars trace Jesus's teaching here directly back to the Hebrew Scriptures in Deuteronomy, chapter 15. "Give liberally and be ungrudging when you do so, for on this account the Lord your God will bless you in all your work and in all that you undertake. Since there will never cease to be some in need on the earth, I therefore command you, 'Open your hand to the poor and needy neighbor in your land.'" Jesus is assuming the *social location* of his followers will always put them in close proximity to the poor and easily able to reach out to them.

And he's telling his disciples not to be so "politically correct," as it were. Don't be grudging and cheap when it comes to your worship and devotion. You can be as extravagant in your worship as this poor woman has been and still be very generous toward the poor, who will always be at your side. I remember how Dorothy Day, the founder of the Catholic Worker Movement and an American saint who lived her life with the poor, always loved beautiful cathedrals and churches. She too, like the woman in the gospel story, wanted to express her devotion to Jesus in beautiful extravagance. After attending mass every morning, she went back to her house of hospitality to take care of the poor and protest the injustice of their condition.

But how do modern Americans interpret this text? We simply use it as an excuse. "The poor you will always have with you" gets translated into "There is nothing we can do about poverty, and the poor will always be there, so why bother?" Yet that's not what the text is saying at all. The critical difference between Jesus's disciples and a middle-class church is precisely this: our lack of proximity to the poor. The continuing relationship to the poor that Jesus assumes will be natural for his disciples is unnatural to an affluent church. The "social location" of the affluent Christians has changed; we are no longer "with" the poor, and they are no longer with us. The middle-class church doesn't know the poor and they don't know us. Wealthy Christians talk *about* the poor but have no friends who are poor. So they merely speculate on the reasons for their condition, often placing the blame on the poor themselves.

So because of our isolation from the poor, American Christians *get the text wrong!* We misuse it to justify ourselves and don't realize how this story offers a deep biblical challenge to how we live. Social location often determines biblical interpretation, and that truth goes a long way toward understanding why Christians from the United States and many other wealthy countries simply miss some of the most central themes of the Scriptures.

All that came home to me dramatically during my first year of seminary. I was freshly converted out of the student movements of the 1960s and I wanted to go to a theological school where they took the Bible seriously. So I chose Trinity Evangelical Divinity School, outside of Chicago, instead of one of the more "liberal" seminaries in the country. Almost immediately upon our arrival, a small group of activist evangelical seminarians began to form, and we quickly turned our attention to the Bible.

I've told the story many times about how we discovered a "Bible full of holes," when it came to the question of the poor. Here's what we did. Our band of eager young first-year seminary students did a thorough study to find *every* verse in the Bible that dealt with the poor. We scoured the Old and New Testaments for every single reference to poor people, to wealth and poverty, to injustice and oppression, and to what the response to all those subjects was to be for the people of God.

We found *several thousand* verses in the Bible on the poor and God's response to injustice. We found it to be the second most prominent theme in the Hebrew Scriptures (Old Testament)—the first was idolatry, and the two often were related. One of every sixteen verses in the New Testament is about the poor or the subject of money (Mammon, as the gospels call it). In the first three (Synoptic) gospels it is one out of ten verses, and in the book of Luke, it is one in seven!

After we completed our study, we all sat in a circle to discuss how the subject had been treated in the various churches in which we had grown up. Astoundingly, but also tellingly, not one of us could remember even one sermon on the poor from the pulpit of our home churches. In the Bible, the poor were everywhere; yet the subject was not to be found in our churches.

Then we decided to try what became a famous experiment. One member of our group took an old Bible and a new pair of scissors and began the long process of literally cutting out every single biblical text about the poor. It took him a very long time.

The prophets were simply decimated. When he got to the resounding command of Amos to "let justice roll down like waters, and righteousness like an ever-flowing stream," he just cut it out. When he found God speaking through Isaiah to say, "Is not this the fast that I choose: to loose the bonds of injustice, to undo the thongs of the yoke, and let the oppressed go free?" he just cut it out. When he discovered the summation of God's call in Micah to "do justice, love kindness, and walk humbly with your God," he just cut it out.

He cut out almost everything that the Hebrew prophets had to say about how nations, rulers, and all of us are instructed to treat the poor. Much of the Psalms also disappeared, where God is seen as the defender and deliverer of the oppressed. And all references to the Hebrew tradition of Jubilee had to be cut where, from Leviticus onward, the practice of a periodic "leveling" was lifted up as crucial to the health of a society—slaves were to be set free, debts canceled, and land redistributed to its rightful owners. It was all too dangerous to remain in the Bible.

When he got to the New Testament, the seminarian with the scissors had a lot of work to do. He began with the thankful prayer of a simple peasant woman who would bear the new messiah. Mary's famous Magnificat prophesied the meaning of the coming of Jesus: "He has brought down the powerful from their thrones, and lifted up the lowly; he has filled the hungry with good things, and sent the rich empty away." Because Mary didn't sound like a religious service provider with a faith-based federal grant, but instead like a social revolutionary; her prayer had to be cut. Then there was Jesus's first sermon at Nazareth, his "Nazareth manifesto," where he announced his messianic vocation. Hearkening back to Isaiah, Jesus proclaimed his own mission statement by saying, "The Spirit of the Lord is upon me, because he has anointed me to bring good news to the poor. He has sent me to proclaim release to the captives and recovery of sight to the blind, to let the oppressed go free, to proclaim the year of the Lord's favor." Because all the biblical scholars agree Jesus is talking about that Jubilee thing again, this was a mission statement that had to be cut before it reached committee. His Sermon on the Mount, and especially the Beatitudes, threatened to turn the world (as we know it) upside down by saying, in his kingdom, the blessed ones are the poor, the meek, the merciful, the peacemakers, the persecuted, and the ones who are hungry and thirsty for justice. It clearly had to go.

That account of how the early church began to practice economic sharing, after the Spirit landed on them, would be pretty incredulous to churches today. And so would the totally unrealistic assertion that "there was not a needy person among them," even if Paul was encouraging economic redistribution as a sign of fellowship wherever he went. Snip, snip, snip. All the stuff from John about not having the love of God in you unless you open your heart to the needy just doesn't apply to some of our most important and pious church leaders, not to mention our television evangelists. And the idea from James that "faith without works is dead" was dangerously close to the "social gospel." So some more cuts were in order.

When the zealous seminarian was done with all his editorial cuts, that old Bible would hardly hold together, it was so sliced up. It was literally falling apart in our hands. What we had done was to create a Bible full of holes.

I began taking that damaged and fragile Bible out with me when I preached. I'd hold it up high above American congregations and say, "Brothers and sisters, this *is* our American Bible; it is full of holes." Each one of us might as well take our Bibles, a pair of scissors, and begin cutting out all the Scriptures we pay no attention to, all the biblical texts that we just ignore.

We still have that old Bible full of holes. It serves as a constant reminder to me of how you can miss so much, even when it is right in front of your eyes. I learned in my little home church that people can really love the Bible, believe they are basing their lives upon it, and yet completely miss some of its most central themes. We don't see what would most challenge us and perhaps change our lives.

Yet, down deep in our souls, we do know the poor are there: in the heart of God, in the compassion of Christ, and in our own communities—if we would just open our eyes. Revealing the poor in the Scriptures and in our own world is always the prophetic task of faith. To discover the forgotten poor is more than the work of "social action," as some would call it. It is rather to put our Bibles back together again. Indeed, it is nothing less than to restore the integrity of the Word of God—in our lives, our congregations, our communities, and our world. What could be more important?

That prophetic vocation has been taken up by servants of the poor across history and around the world. Some have been famous and others virtually unknown. It was absolutely central to the heart of Dr. Martin Luther King Jr.'s message and mission. I am writing this chapter on the day after the official U.S. holiday commemorating the birth of King. And yesterday, I took another group of evangelical seminarians to the ancestral family home of Dr. Howard Thurman in Daytona Beach, Florida. We went to his old house in the black section of that southern city because Thurman was a key spiritual and political mentor to Dr. King. Prolific author of more than twenty books on theology, philosophy, and spirituality, this Christian mystic wrote a little book called *Jesus and the Disinherited,* about the crucial place of the poor in the ministry of Jesus. Reportedly, King often carried it with him as he traveled the country trying to bring hope to the disinherited.

In a recent book about King by historian Stewart Burns, the religious commitment of the great civil rights leader is the central focus. Often treated as a mostly cultural Southern phenomenon by previous biographers, King's religion has often taken a backseat to his social and political accomplishments. The new volume is called *To the Mountaintop: Martin Luther King Jr.'s Sacred Journey to Save America, 1955–1968.*[1] In it, King's deep biblical faith is described as the driving force of his mission and movement by historian Burns who himself has no religious motivations. According to the publisher's description, the book "argues that King saw himself as a reluctant, unworthy, and sinful messiah, mandated by God to free his people from the slavery of racial oppression and to rescue America from the cancer of racism and discrimination that was destroying its soul. His models were Moses and Jesus." Many close to King have testified that he became more and more religious in the last years of his life, more committed as a Christian, and more conscious of his call to follow in the steps of Jesus's suffering and even his death.

We stood on the front porch of Howard Thurman's home as I shared with the seminarians how central Thurman was to King, Jesus was to them both, and the poor were to both men's understanding of Jesus. I could tell that these Fuller Seminary students understood, and their questions indicated their own hopes to follow in the steps of these great prophets.

My son Luke was also learning about Dr. King in his kindergarten class at our Washington, D.C. public school. Every day for a week, he would come home with more stories and ask me new questions about the civil rights leader, what he did, and why. Together, we read children's books about King and watched some of the movies and documentaries airing about the freedom movement. Luke was very puzzled and upset with the footage of the angry white mobs hurling abuse and attacking civil rights activists. Dr. King's mission to "set the people free" made total sense to him and was completely consistent with his five-year-old understanding of the mission of Jesus. Just like the seminarians I met in Florida, kids like Luke seem to get it, and that is a very hopeful sign for the future.

I also recall a group of seminarians in South Africa, whom I invited to come with me on a visit to see Archbishop Desmond Tutu in Cape Town. Tutu, like King, had won the Nobel Peace Prize by helping to lead his own people to freedom, by putting his faith in action. They were very excited about meeting the great man and sat nervously in the archbishop's parlor awaiting his entrance. Suddenly, the door opened, and in walked Desmond Tutu, wearing a sweatshirt and slippers and carrying a tray full of tea, cups, and milk. "Oh, Bishop!" the awed seminarians exclaimed as they knelt down awkwardly trying to kiss his ring. But holding the tea tray, Bishop Tutu just kept smiling and asking, "Milk, sugar, or black?" Finally we all sat down, and the students were educated by a genuine church leader, who quietly explained Jesus's call to liberate the poor and the captives.

On another occasion Tutu was to speak to a group of several thousand Australian young people at a large church-sponsored festival. The news of his winning the Nobel Prize had just been announced, and a huge throng of students was waiting outside as his car pulled up to the conference center. In that crowd was a teenage girl who had been terribly scarred by a childhood accident. Standing with her friends, she watched as the speaker for the conference seemed to be heading in her direction. She moved to one side, but he still appeared to be coming toward her! She kept moving and he kept coming until Desmond Tutu stopped directly in front of the awestruck teenager. Looking right into her disfigured face, the champion of the South African poor smiled and said, "It is so good to see you!" Tutu has an intuition for those left out, and it has taken him to jail, like his kindred spirit Martin Luther King Jr., as well as to the bishop's palace.

But it is not just the famous leaders like King, Tutu, and Dorothy Day who have responded to the call of Christ to serve the poor. I remember once being interrogated by the dreaded and notorious South African Security Police many years before the end of apartheid. As an American, I suspected I would be all right, but I was worried about my companion, Jam Jam, a twenty-year-old activist from the black township where we had been picked up. He had already been to prison where, like many young South African youth, he had been tortured. Automatic weapons were trained on us during the whole interrogation. Then the huge and imposing Security Police captain raised an enormous arm toward my young friend and threatened him. "We know who you are, we know what happened in prison, and if you persist in making this trouble, we'll send you back there for more of the same!"

Without any hesitation, the young man reached into his pocket, pulled out his New Testament, and placed it on the table. "Sir," he said respectfully but confidently, "I am a Christian, and I am not afraid." His Bible had no holes.

One of my lesser-known mentors was an old Pentecostal woman in our neighborhood named Mary Glover. She taught me more about the call of Jesus to the poor than any seminary professor I ever had. Mary was like a self-appointed missionary in our poor community, and she was a regular volunteer in our weekly food line. So poor that she too needed the bag of groceries passed out each week, Mary often said the prayer before we opened the doors each Saturday morning—simply because she was our best pray-er. Mary was one of those people who pray like they know to whom they're talking. You got the sense that she'd been carrying on a running conversation with her Lord for a very long time. It was always worth getting up and heading down to our neighborhood center just to hear Mary Glover pray. The prayer was more or less the same each week, but it made a profound impression on me.

She'd usually start by saying something like, "Thank you, Lord, for waking us up this morning! Thank you, Lord, that our walls were not our grave and our bed was not our cooling board! Thank you, Lord!" Then she would always pray the same words, as a long line of people waited outside in the rain, cold, and heat for a simple bag of groceries, a mere twenty blocks from the White House—by itself a striking metaphor in the capital city of the wealthiest and most powerful nation on earth. Here's what Mary Glover always prayed, "Lord, we know that you'll be comin' through this line today, so Lord, help us to treat you well."

Her prayer comes right out of the twenty-fifth chapter of Matthew's gospel, which was the passage that brought me back to Christian faith. I've read most every commentary on the text, and no biblical scholar gets it better than Mary Glover. It's a scene of judgment, and it says,

> When the Son of Man comes in his glory, and all the angels with him, then he will sit on the throne of his glory. All the nations will be gathered before him, and he will separate people one from another as a shepherd separates the sheep from the goats, and he will put the sheep at his right hand and the goats at the left. Then the king will say to those at his right hand, "Come, you that are blessed by my Father, inherit the kingdom prepared for you from the foundation of the world; for I was hungry and you gave me food, I was thirsty and you gave me something to drink, I was a stranger and you welcomed me, I was naked and you gave me clothing, I was sick and you took care of me, I was in prison and you visited me." Then the righteous will answer him, "Lord, when was it that we saw you hungry and gave you food, or thirsty and gave you something to drink? And when was it that we saw you a stranger and welcomed you, or naked and gave you clothing? And when was it that we saw you sick or in prison and visited you?" And the king will answer them, "Truly I tell you, just as you did it to one of the least of these who are members of my family, you did it to me." Then he will say to those at his left hand, "You that are accursed, depart from me into the eternal fire prepared for the devil and his angels; for I was hungry and you gave me no food, I was thirsty and you gave me nothing to drink, I was a stranger and you did not welcome me, naked and you did not give me clothing, sick and in prison and you did not visit me." Then they also will answer, "Lord, when was it that we saw you hungry or thirsty or a stranger or naked or sick or in prison, and did not take care of you?" Then he will answer them, "Truly I tell you, just as you did not do it to one of the least of these, you did not do it to me." And these will go away into eternal punishment, but the righteous into eternal life. (Matthew 25:31–46)

What's always been most striking to me is that the people gathered in front of the throne of Christ in this story all really believe they are among his followers. And they must be completely stunned to learn that they will be separated and judged by how they have treated the poor—the poor! This judgment is not about right doctrine or good theology, not about personal piety or sexual ethics, not about church leadership or about success in ministry. It's about how we treated the most vulnerable people in our society, whom Jesus calls "the least of these." Jesus is, in effect, saying, I'll know how much you love me by how you treat them. Whatever you do for them, it's like you've done it for me. And, conversely, ignoring them is like ignoring me. Jesus is casting his lot with the poor, almost taking up residence among them. Mother Teresa once said about this Scripture that Jesus appears in "the distressing disguise of the poor." As a young student and activist, I had never encountered anything like this passage before and had never heard about it in the church. None of the radical writers I was reading in the late 1960s were as radical as this. It was enough to make me sign up and decide to try and be a follower of this radical Jesus.

Mary Glover understands how radical Jesus is, though she would have never said it that way. But her weekly prayer in the shadows of Washington's symbols of power served to keep us all straight about who was really important in the kingdom of God. I quoted Mary Glover's prayer so much on the road that it eventually was added to the official prayer book of the World Council of Churches. She was amazed, then asked me what the World Council of Churches was. A song was even written about Mary's prayer by Ken Medema, an extraordinary improvisational musician and my good friend. I remember when he composed it—in a van on the way from O'Hare Airport to an event we were doing together in Peoria, Illinois. The song is called "Coal Black Jesus," and the words go like this:

I'm just a coal black Jesus with a hole in his shoes,
On a D.C. street with no more to lose,
Get into the line and there you'll stand
And sing, "Sweet Mother Mary, put some food in my hand."
A coal black Jesus with a hole in his shoes,
On a D.C. street we got no more to lose,
Get into the line and there you'll stand
Saying, "Sweet Mother Mary, put some food in my hand."[2]

In the east end of London, in a large black church, I once repeated Mary's prayer from the food line. The mostly low-income and Pentecostal audience rose in response and took a special offering for Mary and the food line. We presented it to her at our annual food line recognition dinner, and then played Ken's song. She loved it and commented that Ken "didn't sing like a white man." I reminded her that Ken had been blind from birth and probably didn't really know he was a white man! All the volunteers and the recipients of the food line rose to pay tribute to Mary Glover, our not so famous (but getting there) champion of the poor. Mary was at a loss for words (the first time I think) and had tears streaming down her face. "I'm just a humble Christian woman whose prayer has gone around the world," she said. And that's what she is—a Christian who gets it, whose Bible is not full of holes, who knows what God requires, and understands what her Jesus would have her do. Mary Glover gets it, like Martin and Desmond and a South African kid named Jam Jam. And maybe the rest of us are starting to get it as well.

CHAPTER 14

Poor People Are Trapped—in the Debate about Poverty

Breaking the Left/Right Impasse

SHE WAS WORKING the drive-through window at 4:00 in the afternoon. But whenever there was a lull between orders, the young woman kept returning to a table in the corner of the restaurant. Three kids were sitting there, with schoolbooks, papers, and pencils all spread out, doing their homework. And Mom was helping as best she could while keeping straight the orders for Whoppers, fries, and chicken nuggets. Given her low wages, this single mother was no doubt balancing more than fast food and homework—but also deciding between paying the rent, going to the doctor and affording prescriptions when somebody gets sick, or buying winter boots for her kids. She has become an icon for me. I call her Burger King Mom.

In election years, the pundits talk often about Soccer Mom and how she will vote. Both the Democrats and the Republicans court her. Since the president went to Daytona, there is a new electoral icon; he's called NASCAR Dad, and his support is crucial, especially for Republicans. Also, in the 2004 election, attention focused on Security Mom. But who will speak to or for Burger King Mom? She exists in both the red and blue states, but neither party is much interested in her or her family's issues. She is part of the low-income demographic that is most unrepresented in American politics, with the lowest levels of both voter registration and turnout, and includes a high percentage of immigrants. Many low-income people have a hard time connecting to voting: too complicated, too many other things to worry about, too little confidence that the outcome makes much difference for them.

The Republicans look after their wealthy constituents, and the Democrats want to be the champions of the middle class, but neither prioritizes the needs of the poor. Is that because the problems of poverty are disappearing in America? On the contrary, the poverty rate (including for children) has risen over the last three years, more people than ever are without health insurance, increasing numbers of people can't find affordable housing, and the minimum wage hasn't been raised for eight years.

Most Americans believe that if you work hard and full-time, you should not be poor. But the truth is that many working families are, and many low-income breadwinners must hold down *multiple* jobs just to survive. With stagnant wages in an economy that is growing for some but clearly not for others, more and more people and their children are simply being left out and left behind. When work no longer works to support a family, the existence of a genuine opportunity society and the ethic of work itself are at risk.

The media have yet to report on the condition of low-income American families who have also become the casualties of war. The ongoing costs of the war with Iraq, along with the cost of tax cuts for the wealthy, are leading to a crisis for America's poorest children, as U.S. domestic needs have been literally pushed off the political agenda. The consequences of this silent war are being felt most severely in the poorest parts of the United States, where low-income families are desperately clutching onto the bottom rungs of the failing economy.

The truth is that hungry people are going without food stamps, poor children are going without health care, elderly are going without medicine, and schoolchildren are going without textbooks because of war, tax cuts, and a lack of both attention and compassion from our political leaders. The moral contradictions are too great to ignore. The deepening injustice of America's domestic priorities is increasingly impossible to justify. It's becoming a religious issue.

That lack of attention means that poor people are trapped, and not just in poverty. They are trapped in the debate about poverty. They have become pawns in a political chess game, props for media performances, and, worst of all, victims and cannon fodder for the ideological war between liberals and conservatives, Left and Right. Poverty will not be overcome, or even significantly reduced, until the debate over poverty is set free from its ideological captivity.

We have astounding statistics like these: One of every six children in America is poor (13 million children in America!), 36 million people live below the poverty line (more people than live in any state, including California), 4 million families are hungry to the point where people in the household are skipping meals (and three times that number are "food insecure," meaning at some time during the year they have difficulty providing enough food for members of their family), 45 million Americans are without health insurance—including 8.4 million children, and 14 million families have critical housing needs as affordable housing disappears, rents and housing prices soar, and homelessness is on the rise—especially for families in urban areas.[1]

If anything, these statistics underestimate the problem, since what constitutes the "poverty line" hasn't been adjusted for forty years. In 2003, it stood at $18,810 for a family of four. But try to imagine living on even $35,000 as a family of four in today's economy. Consider the actual (and growing) costs for basic items like food, housing, transportation, and health care combined with job losses and stagnating wages for many lower-income working people, and as many as 40 percent of American families have incomes low enough to be "poor" in our modern economic environment.

Perhaps the greatest scandal of all is the absolutely inferior education that poor children in America are subject to. Because education is so key in changing all the above statistics, to rob low-income families of the one thing that would most liberate their children is especially cruel and evil. The truth that nobody wants to really say is that affluent American parents would simply never tolerate the disastrous public schools to which so many poor families are forced to send their children. And it is simply not a coincidence that the vast majority of the children in those schools are children of color—mostly black or Hispanic.

But instead of politicians taking aim at those scandalous facts in the world's richest country, they spend their time just blaming each other for the problem. This is a political debate that is not evaluated by results and outcomes; rather it is only used to score points against the opposition.

It's become standard practice in Washington, D.C. First you name the problem, then you make the public afraid of it, and finally you blame it on your opponents. Armies of spin masters, media consultants, and pollsters are employed for this ideological warfare, and you know who ultimately wins by taking a poll—an election is simply the final poll. What the politicians never get back to is solving the problem. The problem is simply used in the battle for political power; it is not something to be really addressed.

When poverty becomes one of the issues in the political battleground, poor people themselves become superfluous. They are just not the issue; winning the political war is the issue. Low-income Americans become an ideological sacrifice on the altars of both conservative and liberal fights for power.

The media play their part in the political wars by preferring stories with conflict—as almost any reporter will admit. For almost all the media "debates" on issues, there are generally only two sides, usually conservative and liberal. "Preinterviews" screen the prospective talking head guests for the discussion to see if they have enough conflict to make the show interesting to viewers. Sometimes, if it seems the guests might have too much common ground in their views, or if they don't fall into neat categories, or if they aren't combative enough, or if they suggest strategies based on cooperation rather than conflict, producers will turn to more entertainingly conflictual and predictably ideological spokespeople.

Think about it. Do most serious, and often complicated, social issues have just two sides? Isn't it more likely that a variety of perspectives and approaches might be necessary to first understand and then solve a problem? But actually solving problems like poverty would be far less interesting to the media than just staging a fight over it.

The answer to this captive and political self-serving debate over poverty may be twofold. The first answer is to insist that the debate over poverty be disciplined by results. Political leaders must be forced to commit to outcomes, to actually and concretely change the above statistics. In the United

Kingdom, the Labour government has made a specific commitment to reducing child poverty: They have aimed to cut the child poverty rate in half by 2010. That is a concrete and public goal to which political leaders can be held accountable. The same could be done in the areas of health care, housing, hunger, education, jobs, and more.

To be disciplined by results requires us to be less concerned about ideological presuppositions and more focused on what actually works. What is liberal or conservative would be replaced by what's right and what works. It is a solution-based approach to overcoming poverty, not a blame-based debate. Instead of just pointing fingers and blaming the other side, we focus together on finding solutions that might really work.

I've done literally hundreds of town meetings around the country, and when I get to the subject of poverty, the blaming begins. In the room are usually civic and religious leaders—mayors and city-council members, business executives, educational and law-enforcement officials, pastors and lay church leaders. I often ask who is responsible for the poor children who are falling through the cracks in their community. Immediately, one side says it's the Democrats whose programs have failed, and the other quickly counters that it's the Republicans whose policies have abandoned the poor.

It's very interesting. I ask them who is responsible, and they instead tell me who is to blame. When that's pointed out and I suggest there is more than enough blame for child poverty to go around, I ask who the leaders are in their community. "We are," they finally say. "Then who is responsible?" I ask again. That's when they look at each other and admit that *they* should be the ones responsible. And that's when we begin to talk about a strategy that might actually work to reduce child poverty, address real community issues like drugs and youth violence, and create safe and stable communities of opportunity and hope, instead of stagnation and despair.

The second answer is to discover that poverty is a spiritual and religious issue and not a "left-wing" issue. On my last book tour, perhaps the greatest surprise and satisfaction came in many interviews on "Christian radio." The talk shows on National Public Radio stations, network affiliates, and various community radio outlets were more typical for me. But I was greatly encouraged by the interest from local and national Christian radio, and even more heartened by the response.

What I found on conservative Christian radio shows was a deepening concern, on the part of both interviewers and callers, for people who are poor. Most significant was the breaking out of old ideological categories. In an interview on the Salem Network (the largest chain of Christian radio stations in the country), the host said to me and to his audience, "You know, poverty is not a left-wing issue; it's a Christian issue, and it's time for us all to recognize that." Another show's host acknowledged, "You know, Jim, most of us wouldn't have had anything to do with you just a few years ago. We thought talking about poverty was left-wing. But many of us are coming around and want to be with you now." Comment after comment, and caller after caller, expressed similar views.

Christian radio may be changing, but maybe I am too. I now believe that if poverty is to be overcome, it will take the insights and energies of both conservatives and liberals. As long as poverty fighting is seen as merely a left-wing issue, we will never succeed. And it's not just a matter of perception; it's also a question of content.

America needs a new and balanced vision for how poverty might be overcome. Instead of just rehashing old ideas, we must seek a comprehensive plan for change, involving every sector of society—not just the government, nor just the "market," not just churches and charities, as the various competing ideological options often suggest. Rather, we should focus on the stories on the ground from the most successful and inspiring projects around the country that are truly making a difference, and listen to the new approaches they suggest.

Those stories all suggest that it will take the best values and efforts from both conservatives and liberals if we are going to really make a difference in people's lives. Liberals must no longer be content to just "service poverty" instead of overcoming it, and conservatives must stop merely blaming poor people for their poverty instead of taking some responsibility themselves. Real solutions to poverty will require both liberals and conservatives to take new responsibility and lead us all to new approaches that transcend the old political options of Left and Right.

Conservatives have been right in saying that the hold of poverty over people's lives will not be broken until we confront the problem of broken families. Family breakdown *is* a cause of poverty, and it further traps single parents and their kids in a continuing cycle of impoverishment, even when other social and economic factors are involved. My neighborhood has had 80 percent single-parent families during most of the three decades we have lived there. In a situation of such massive family breakdown, no merely economic initiatives to overcome poverty can possibly succeed unless we are simultaneously reweaving the web of family and community—a cultural solution. To promote and support marriage and stable two-parent families is an anti-poverty measure, as virtually all the social data show. To ridicule traditional family patterns as essentially bourgeois or patriarchal, as too many left-wing intellectuals have done, has devastating consequences for poor people. Sure, many liberals have good family values, but most are often hesitant to talk about them, fearing they will sound like the religious Right.

For many years at our Sojourners Neighborhood Center, and at similar programs around the country, personal motivation and responsibility have been stressed to young people at risk. Bad personal and moral choices do land people in poverty or keep them there. Out in the suburbs, affluence buffers the many bad choices kids make, giving them second, third, and many more chances. But to an inner-city kid, living in a poor and violent neighborhood, a bad choice could cost you your life.

Similarly, sexual promiscuity is often covered over by money and lots of abortions in wealthy communities. But in poor neighborhoods, kids having kids is killing people's chances of ever escaping poverty. One can be committed to the conservative bedrock values of personal and moral responsibility, marriage, and family values without resorting to the kind of mean-spirited scapegoating of women, single mothers, and gays and lesbians that some on the religious Right have engaged in. To blame all of them for the breakdown of the family is not only mean, but also, frankly, stupid. Sound conservative personal and family values should not be simply conceded to the political right wing.

But liberals are right in saying that personal behavior is not the only issue in poverty—not by a long shot. Structural issues are also involved, as the Bible itself points out in holding kings and rulers, employers, landlords, and judges responsible for injustice. Good family values don't assure you of a job that pays a living family wage, and the stagnating wages that American workers now experience are a major cause of family instability. Some companies used to talk about treating their workers as "family" in important ways, but that concept seems utterly outmoded in a cutthroat global economy that makes profit maximization the only bottom line. Today it is large corporations that push down wages, cut health benefits, lay off workers, and export good jobs overseas; they are the biggest violators of "family values" and the principal force destabilizing family life in America.

Nor are sound family values enough to make you a homeowner, as my next-door neighbor, Thelma, found out. She could never afford the middle-class resource of down-payment money to buy her house, and she ended up paying for it four times over in rent through many years. She ultimately lost her family home and all the middle-class tax deductions and equity that could have helped her children with college and more. Overcoming poverty also takes some responsibility on the part of private business to see to the common good and not just the bottom line; it entails better corporate and banking policies *and* effective government action where the market has failed to address fundamental issues of fairness and justice.

In Denver, on a speaking trip, I learned that you had to work 144 hours per week at minimum wage in order to find any sort of affordable housing in that city.[2] Other cities are even worse. It will take more than good personal motivation and responsibility to make enough affordable housing available to those working families who have no place to live; it will take focused government action to ensure a sufficient stock of available affordable housing. It may also require changing planning and zoning laws, which keep low- and moderate-income housing out of many wealthy areas. And how will marriage and good moral choices provide your family the health care it needs? We have all heard many stories of how exorbitant medical bills can virtually wipe out a family's income and future. Can we include forty-five million Americans into the health-care system without some government intervention? I doubt it.

Indeed, what can destroy family life and values is losing a job that provides the capacity to support your family or being unable to find affordable housing, quality health care, or educational opportunity for your children. Conversely, steady employment at a livable family income, access to health care, a path to homeownership, and the chance to send your kids to good schools can lay the best foundation for solid and successful family values.

Both conservatives and liberals are coming to see the crucial leadership role of nongovernmental organizations in the so-called civil society for resolving the toughest issues of poverty. The role of faith-based organizations, in particular, is being more and more widely viewed as making a crucial contribution. But anybody who is serious about the problems of poverty knows that the resources to solve the problem simply don't sufficiently exist in the civil society. Government, on all levels, must be involved. How, when, and where is the most important question now. The real choices will not be defined as big versus small government but, rather, how government can be effective in helping to mobilize new multisector partnerships and target its resources in the most strategic ways. Rather than creating new government programs for every problem, government must see what's already working and then figure out how to "scale up" those solutions. And, most of our values suggest that a good society should provide a "safety net" for those who really need it, enabling them to live in dignity and security.

There is also growing agreement across the political spectrum that racism is still very real and that we won't succeed in overcoming poverty without also dismantling the structures of racial prejudice that still work to maintain economic injustice. The Justice Department and the FBI have released devastating studies demonstrating the stark differences in arrest, conviction, and sentencing between white offenders and black and Hispanic offenders—for the same crimes and even in the same cities and neighborhoods. In the criminal justice system, as in many other social systems in America, race still makes a real difference in how people are treated. And this is now, in the new millennium, not just twenty, fifty, or one hundred years ago. It is crucial that both conservatives and liberals work to overcome the continuing impact of race as a cause of poverty.

That kind of approach is neither Left nor Right, in the traditional ways. The truth is that overcoming poverty will take both liberals and conservatives and those who are neither. It will take all of our best values and insights, then require us to find solutions that might move each of us to new places of commitment and responsibility. And that's what would make the greatest difference of all.

Is the country ready for a new debate on poverty?

What's clear is that there are Republicans who really do care about the poor, Democrats who really don't, and of course vice versa. What's also clear is that the political debate over poverty in Washington is still stuck in old language, historical baggage, and partisan warfare. What's "liberal" or "conservative," or what might tip upcoming elections one way or the other, weighs very heavily in the political decision making that will so affect the lives of our poorest citizens.

Despite that, Call to Renewal, a faith-based network of churches overcoming poverty, has been carrying on a regular dialogue with top Senate staff members on both the Democratic and the Republican sides of the aisle who are longing for new criteria in the poverty debate: what's right and what works. What brings the dialogue participants together is the conviction that overcoming poverty must become a bipartisan issue and a nonpartisan cause. Many are veterans of the Senate and say the reasons they keep coming despite very busy schedules are threefold. First, it is one of the few genuinely bipartisan discussions left on Capitol Hill. Second, it is a civil conversation, also quite rare these days in the halls of Congress. (The participants say they don't want to misbehave in front of religious leaders!) And third, these key Senate staffers often long for what they call the "seminal moral questions" that always focus our discussions. Many of them came to public life because of serious ideals and values, and our Senate dialogue reminds them of why they came to the Senate in the first place.

But in Congress, an ideological cleavage still separates those who see policy and funding issues at the heart of reducing poverty and those who point to the deep cultural roots in family breakdown, sexual behavior, or personal responsibility. I'm always amazed at how politicians can make such false choices between these very real causalities while practitioners who actually live and work with poor people just shake their heads as they hear such an impoverished and futile debate. For example, in the welfare-reform debate we always faced a battle between those who are "pro-funding" in welfare reauthorization and those who are "pro-family." Of course, Call to Renewal, and all the faith-based leaders around our table, are decidedly both.

When debates are framed wrongly, they almost inevitably turn out badly. That happens all the time on Capitol Hill. In the welfare debate, focusing on simply reducing welfare rolls, instead of reducing poverty, is still the major problem. Most people involved in anti-poverty efforts would agree now that helping low-income people find "self-sufficiency" is far preferable to a system of endless subsidy. But what are the best ways to support people in moving from subsidy to sustenance? And if work is the best way out of poverty (as most of us now agree), how do we make work really work in America? What do people need to support successful work in terms of child care, real education and training, health care, and affordable housing?

The welfare-reform debate should become a national discussion about how to overcome poverty in America. In fact, the debate doesn't really make any sense apart from the goal of poverty reduction. Let's state our goal clearly and unanimously: that welfare reform should be judged by how well we are actually reducing poverty. Then let's have the most honest debate we've ever had about just how to do that.

As I travel around the country and listen to people across the political spectrum, I often sense that the country may be ready for a new debate on poverty that puts the old liberal and conservative labels aside. But I also live in Washington, where the political elites in both parties are clearly not ready for a new discussion. Maybe it's time to help them.

The best role for faith-based initiatives in America is not only in the provision of social services, but also in the shaping of public policy to secure social justice. We learn that lesson often in Washington, D.C.

Call to Renewal's Pentecost 2002 Mobilization was named Speaking the Truth About Poverty. It drew more than three hundred faith-based leaders from forty-two states to press their senators toward a compassionate and just reauthorization of welfare reform. Out of eighty-four potential Senate visits, we had eighty-three—a remarkable accomplishment in this town. Twenty national church and organizational leaders had a very positive dialogue with a bipartisan group of senators and the key Senate staff members who were crafting a bipartisan welfare bill. They also met with our whole group of grassroots church-based anti-poverty workers to discuss their progress and to ask what the most important issues ought to be.

Over and over again, our state delegations heard this response from lawmakers: "We can't do this without you." They wanted to hear stories of what was working in local communities and on the street. They wanted our facts, research, and experience. And they were told about the human face of poverty.

Those who came were pastors and laypeople, executive directors of faith-based organizations and heads of denominations, community organizers and service providers, and former welfare recipients who came with moving testimonies of how they have escaped poverty. The faith-based leaders run shelters and food banks, do job training and economic development, provide health care and education, and coordinate councils of churches and interfaith coalitions that address the most basic problems in their communities.

We told the senators that we too believed in work over welfare but that if single moms are to be successful in work, they need real help with child-care support. And unless dead-end jobs are to be the end result of welfare reform, single parents' desire for more education and training must also be supported and generously counted toward "work requirements." We said that legal immigrants who work and pay taxes should be eligible for the assistance they need too. We said that successful programs to support healthy marriages and families will help overcome poverty, as long as we protect against domestic violence and adequately fund other programs—that we must stop making false choices between being pro-family or pro-funding. We testified about how faith-based initiatives are finding real solutions to poverty but that churches and congregations can't succeed without good public policy.

Perhaps because of that hope for partnership with faith-based initiatives, most of our delegates felt that the senators listened to their words about what low-income people really need to succeed. Just to make the point even more credible, many low-income people themselves were among our delegations—people who could say what had worked for them.

For the three hundred national and local church leaders who journeyed to the nation's capital, it was an inspiring and invigorating time. But for me and many others, the most moving moment in the three days of intense advocacy and witness came at our Tuesday morning prayer breakfast. It was the only prayer breakfast in Washington I can remember that focused on poverty and the biblical imperative to overcome it. Two people were recognized for their work for justice.

Call to Renewal gave its first annual Joseph Award, given to a person who faithfully used a position of influence to benefit those in poverty, to Representative Tony Hall from Ohio. (In the book of Genesis, Joseph was a captured slave who rose to be second only to the Pharaoh. His vision moved Pharaoh to exclaim, "Can we find anyone else like this—one in whom is the spirit of God?" And when famine struck and the people cried to Pharaoh for bread, Pharaoh said, "Go to Joseph.") Tony Hall gave a very personal testimony of how he was changed by his faith in Christ and a trip to Ethiopia, where he saw hundreds of people dying every hour. Since then Tony has traveled all over the world to be with hungry people and has committed his life to relieving their suffering. He spoke compellingly about passages in Proverbs where one who serves the poor "honors God" and is said to be actually "lending to God." Hall said, "Can you imagine that?" We honored Tony Hall as a good friend of Call to Renewal, as one who has "lent" so much of his energy, time, and career to God in his service for justice.

The Call's first annual Amos Award, was given to honor a person who came from a humble background to serve God and community. (The prophet Amos was a humble shepherd who became one of the strongest voices for God's judgment against poverty and injustice.) It was presented to Reverend Darren Ferguson. I recalled how I first met Darren at Sing Sing prison, where I spoke to the prisoners several years ago. When I wrote to ask when they wanted me to come, they wrote back and said, "Well, we're free most nights. . . . We're a pretty captive audience here!" I'll never forget what one of the inmates said the night I arrived at the prison: "We are all from only about five neighborhoods in New York City. It's like a train that begins in my neighborhood. You get on the train when you're nine or ten years old, and the train ends up here at Sing Sing." Then he said, "When I get out, I want to go back and stop that train."

Two years later, I met Darren again. He was leading a Call to Renewal town meeting in New York City, back in his old neighborhood, trying to stop that train. Darren Ferguson is now the youth minister at historic Abyssinian Baptist Church in Harlem, where the former Sing Sing inmate does amazing work with urban young people and ex-offenders who are usually forgotten and invisible to officials in Washington. There wasn't a dry eye at the prayer breakfast as Darren warned us to "be careful" with the lives of our youngest and most vulnerable citizens, for in their future lies our success or failure. Tony and Darren reminded us all of why we do this work and of who has called us to it.

In my opening remarks to our 2002 Mobilization, I reminded the faith-based leaders that our vocation is not only to "pull people out of the river, but to go upstream to find out what or who is pushing them in." In Washington, the faith-based providers came upstream. In the midst of a debate on historic social welfare legislation and on the occasion of the church's season of Pentecost, the timing seemed right.

The result of the coming of the Spirit in Jerusalem two thousand years ago, says the book of Acts, was an economic sharing so transformational that "there was not a needy person among them." For another generation of Christian disciples in Washington, D.C., that became not only a prayer, but a commitment. As the quiet voices of prayer were mingled on the west lawn of the Capitol on Monday night, a participant was heard to comment, "This is what Pentecost must have sounded like."

For those who care about poverty in America, the coming years are a critical time, a turning point similar to the New Deal of the 1930s or the War on Poverty of the 1960s. Now, as then, we can make a difference in the lives of millions of people. It is a time for people of faith to speak, act, and pray on behalf of those still trapped in poverty—and in the debate about poverty.

Too many working people remain poor. Their jobs are mostly entry-level and minimum wage, with few or no benefits. We must change the political debate to measure our success by reducing the number of people in poverty, rather than just by reducing welfare rolls. That's a political and moral message that most people agree with. Both Democratic and Republican politicians need to change their criteria. The image of a poor working mother is different than the old welfare stereotypes. And most of us want to support her. This is no time to cut our concern, care, and commitment to poor women and their children. Low-income people need work supports that help develop self-sufficiency, and that costs money. But in the long term, it's the best investment we can make.

We must have a clear moral message. Budgets with billions of dollars of increases for the military and massive tax cuts for the wealthiest—while cutting funding for overcoming poverty—should be named as *morally* unacceptable. Rather, funding for real solutions to poverty needs to be increased. Let's reward people's efforts to improve their lives.

In introducing his welfare plan, President Bush spoke movingly about the barriers faced by single mothers. "Across America, no doubt about it, single mothers do heroic work," Bush said. "They have the toughest job in our country. Raising children by themselves is an incredibly hard job."[3] He's right. And the single largest problem facing those single mothers who work is the limited availability and high cost of child care. Along with more assistance to help meet the costs of child care, a variety of additional programs need funding—from improved facilities to better training for child-care workers. And as the Children's Defense Fund points out, only one in seven children who are eligible for federal child-care assistance receives it.[4] Any poverty-reduction package must include substantial increases in funding for child care.

Proposals to fund marriage-promotion programs have generated the greatest controversy. Studies bear out that children with a single parent are far more likely to be poor. Conservatives insist that promoting marriage should be central to reducing poverty, while liberals insist that funding antipoverty programs will promote marriage. It's a false choice. Of course healthy marriages are good for economic stability, and economic stability is good for healthy marriages. Why can't we do both—adequately fund necessary programs and develop initiatives that help establish and maintain good marriages? The faith community can play an important role on this issue.

Many legal immigrants today work hard and pay taxes, and they should be entitled to benefits when in need. The biblical book of Leviticus speaks to immigration policies: "The alien who resides with you shall be to you as the citizen among you" (Leviticus 19:33).

Many of those still receiving welfare assistance are the hardest to employ—people with disabilities or substance abuse or victims of domestic violence. They are the people who need the most support in order to remove those barriers and find jobs. We must develop, strengthen, and fund programs in those areas.

Churches and nonprofit organizations could be the catalysts and conveners of new partnerships for real solutions to poverty. We need new approaches beyond either relying on government programs alone or hoping that churches and charities can, by themselves, take care of the problem. Churches and faith-based organizations have led by example in creating innovative new programs and partnerships in communities around the country. Call to

Renewal has been working with those grassroots organizations in virtually every state, and I've been very impressed at the work they are doing. We must find ways to strengthen and replicate those programs.

But effective local community service and development are not substitutes for advocacy, for supporting good public policy in Washington, D.C., and in the states. Now we must bring to the policy debate the experiences and lessons of faith-based organizations. If political leaders would lift up our role, they must also listen to our voice. And we have a simple message that must be carried to Washington: People who work full time should not be poor. Despite our incessant rhetoric of prosperity, too many vulnerable people and families simply don't make enough to successfully piece life together for themselves or their children. That view from the streets is far different from the worldviews of the multimillion dollar, corporate-sponsored, made-for-TV political campaign ads that now pass for public discourse in America. From the national political podiums, we hear both parties chanting the mantra, "Leave no child behind." It's time for that oft-repeated promise to be kept.

When my son Luke was two years old, he would sometimes put his hand on my cheek and turn my head toward him when he thought I wasn't paying close enough attention to what he wanted. It was his way of getting me to focus. That's what we need to do now—focus on the relationships, values, and commitments that give our lives coherence and purpose.

But first we must face the fact that what's happening in the streets of America is connected to what's happening in the hearts of Americans. The greatest moral question in American politics today is, What is our prosperity for? Will it serve as an excuse to forget those left behind? Or will it include those who have fallen through the cracks in our society, including almost thirteen million children? The biblical prophets say that a society's integrity is judged, not by its wealth and power, but by how it treats its most vulnerable members.

When our unprecedented economic growth exists alongside the embarrassing fact that one in six American children is still poor, and one in three children of color, the moral underpinnings of our prosperity are in great disarray. The post–World War II prosperity benefited most American families. But the prosperity of the 1990s sent a financial windfall straight to the very top, while leaving those at the bottom even farther behind and creating a new middle-class anxiety. We now have record prosperity and rising inequality at the same time. The rising tide has lifted all the yachts, but not yet all the boats.

It is time to assemble the ideas and practices that are already working to address and solve the problems of poverty. That will require the insights and energies of both liberals and conservatives and the formation of new partnerships between government, the private sector, and the civil society—including faith-based organizations—if we are to overcome the material poverty that shames our prosperity.

More and more we see how our anxious striving after affluence has also created a spiritual poverty. It is the great myth of modern advertising that mere prosperity can give us happy, fulfilled, and purposeful lives. Nobody wants to say out loud that shopping doesn't satisfy the deepest longings of the human heart.

We underestimate our spiritual impoverishment. The consumer economy is putting enormous pressure on all of us. The market fuels a never-ending and relentless cycle of working and buying, which not only undermines our personal integrity but destroys our sense of moral balance. We simply spend too much time and energy worrying about "things."

The economy may be booming, but how happy is life inside the gated communities? The poor aren't the only ones worried about their kids; the affluent are also concerned about their children's values or, worse, the consequences of their own values showing up in their kids. It is no longer just the kids of inner-city poverty who are erupting in societal violence. Many of the shooting sprees in schools that have shocked the nation were carried out by middle-class white kids from two-parent homes. Listening to them talk, you quickly discover that something has gone terribly wrong in their value system. Affluence often helps to mask moral and spiritual poverty.

When I ask students why they spend far more time in volunteer service projects than would be necessary for a balanced résumé, they tell me they are looking for "meaning" and "connection." That search requires us all to ask how much is enough and whether more time is more important than more money or things—questions that strike terror into modern advertisers. And our youth are right in sensing that service for social justice is key to overcoming our spiritual poverty. It's not just about some people doing for others; it's about all of us getting healed.

Only a values-based politics can overcome our material and spiritual poverty. It will require a transition from the rhetoric that blames to the solutions that heal—implementing real answers, rather than just finding more ways to argue over the questions.

But all that will require some soul-searching. The problems and solutions to poverty in the midst of prosperity are spiritual and not just political. Many voices now are calling for a renewed commitment to the common good over the bottom line and for an ethic of both personal and social responsibility. And as more and more people are finding, engagement in our own communities provides practical solutions toward overcoming poverty, leads to personal satisfaction, and creates the rebirth of a vibrant citizen politics. These are the ingredients for reawakening genuine democracy.

The poor have been virtually missing in action from American religion and politics. The shame of American poverty can be overcome, but not without first ending the liberal/conservative warfare and transforming the fruitless poverty debate with good theology and spiritual commitment. Ignoring the poor has fundamentally distorted the theology and practice in the churches of the affluent nations and thus made any prophetic role impossible. In contrast, the priority of the poor among the fast-growing churches in the developing world has been key to their success and their prophetic energy. The Bible suggests that the poor can literally save the faith of the churches, which could then help to "heal the nations"—perhaps the most critical spiritual and social task in a world so deeply and dangerously divided.

The good news is that religious leaders and communities from across the theological and political spectrum are responding to the vacuum of political leadership on poverty. In fact, poverty is becoming the defining moral issue for many in the faith community, including evangelicals and Pentecostals, Catholics and mainline Protestants, and, of course, the black churches. While divided on other issues like gay marriage and abortion, church leaders are declaring a determined "unity" to make poverty a *religious issue*.

At Call to Renewal's Pentecost 2004 mobilization, Making Poverty a Religious and Electoral Issue, there was a moving and historic demonstration of "unity" on grounds of faith—a faith that does prioritize the needs of the poorest among us, a faith that includes compassion, a faith that does justice. In a liturgical act of worship as part of a service in the Washington National Cathedral, nearly forty Christian leaders signed a declaration of unity, a covenant to act together in that critical election year to make poverty both a religious and electoral issue.

We were evangelical and mainline, Catholic and Protestant, black, Hispanic, Asian, and white, making a common declaration across the theological and political spectrum of the church's life. We said that what united us was our common faith in Christ and his commandment to stand with his poorest children. All of us were there in the National Cathedral to tell the nation that in election year 2004, America's poorest citizens must not be forgotten! And that is good news indeed.

Unity Statement on Overcoming Poverty

How good and pleasant it is when the people of God live together in unity!

As Christian leaders in the United States, we recognize that we live in a time when political and social issues threaten to divide the church. Although there are issues on which we do not agree, we come together to affirm that justice for those in our society who live in poverty is, for all of us, a deeply held religious belief on which we are firmly united. We affirm God's vision of a good society offered to us by the prophet Isaiah. His words are as relevant today as they were three thousand years ago, and show us the way forward. Isaiah envisions a society where:

"No more shall there be in it an infant that lives but a few days, or an old person who does not live a lifetime. . . . They shall build houses and inhabit them: they shall plant vineyards and eat their fruit. They shall not build and another inhabit: they shall not plant and another eat; for like the days of a tree shall the days of my people be, and my chosen will long enjoy the work of their hands. They shall not labor in vain, or bear their children for calamity; for they shall be offspring blessed by the Lord. . . ." (Isaiah 65:20–25)

In America, people who work should not be poor, but today many are. We must ensure that all people who are able to work have jobs where they do not labor in vain, but have access to quality health care, decent housing, and a living income to support their families. The future of our country depends upon strong and stable families that can successfully raise their children. We must also ensure that those who are unable to work are cared for by our society.

We therefore covenant with each other that in this election year, we will pray together and work together for policies that can achieve these goals. We will ensure that overcoming poverty becomes a bipartisan commitment and a non-partisan cause, one that links religious values with economic justice, moral behavior with political commitment. We will raise this conviction in the public dialogue, and we will seek to hold all our political leaders accountable to its achievement. In our work together, we will "make every effort to maintain the unity of the Spirit in the bond of peace." (Ephesians 4:3)

May 24, 2004

CHAPTER 15

Isaiah's Platform

Budgets Are Moral Documents

NEARLY THREE THOUSAND YEARS AGO, the biblical prophet Isaiah offered us God's vision of a good society. His vision is set forth in the Unity Statement on Overcoming Poverty (see p. 239). It includes fair and good wages, housing and health, safety and security. In America, people who work should not be poor, but today many are. We must ensure that all people who are able to work have jobs where they do not labor in vain, have access to good health care and decent housing, and are able to support their families. The future of our country depends on strong and stable families that can successfully raise their children. We must also ensure that those who are unable to work are cared for by our society. Economic security for all our people is vital to our national security. Isaiah's platform links religious values with economic justice, moral behavior with political commitment. And it reflects the conviction that overcoming poverty must become a bipartisan commitment and a nonpartisan cause.

The starting point to check how our society measures up to Isaiah's platform is by examining our federal budget. Budgets are moral documents. They clearly reveal the priorities of a family, a church, an organization, a city, or a nation. A budget shows what we most care about and how that compares to other things we care about. So when politicians present their budgets, they are really presenting their priorities. It is worth paying close attention.

For some time now, the U.S. government's annual budgets have showed record deficits. They reflect billions of dollars of tax cuts that provide most of their benefits to the wealthiest Americans. They include huge increases for the costs of war, and they slash domestic spending. Core government programs that create affordable housing, curb juvenile delinquency, hire police officers, aid schools, make child care available to low-income working mothers, and guarantee children's health insurance have been drastically cut. These budgets reflect the federal government's priorities.

The deficits increase each year and may rise up to $1 trillion over the next five years. Military spending has ballooned to $420 billion (at this writing), well beyond the biggest military budgets at the height of the Cold War. However, most of the increases are not directed toward counteracting new threats from terrorism, but for weapons systems that are guaranteed to leave no defense contractor behind.

There is no money in the federal budget for the states, which are confronting huge deficits and the prospect of draconian cuts in education and social services, mostly to the poor. In fact, the federal government suggests that states could meet their budget challenges with the "flexibility" to cut programs like health insurance for the nation's poorest children.

Such budgets are justified by the "war against terrorism." In a time of war, the logic suggests, there are sacrifices that just have to be made. But these budgets don't really reflect the choice between "guns and butter," as the traditional language goes, but are full of both "caviar and missiles," as commentator Mark Shields so aptly put it.[1] The military gets the largest increases, the rich get enormous tax cuts, and the poor get left behind.

Clearly, the big money in the federal budget is now for war and tax cuts. The lists of new domestic initiatives proposed in each annual State of the Union message, while often good, are relatively low-cost items and ultimately more symbolic than substantial. Without the crucial funding for programs that directly and effectively reduce poverty, "compassionate conservatism," the mantra of the Bush administration, is in grave danger of becoming compassionless conservatism. The federal budget now seems to have a "compassionate deficit."

George Bush's faith-based initiative (which I supported) promised equal access to government funding for effective programs run by religious communities, which would no longer face discrimination just for having a cross, Star of David, or crescent on the wall. But the faith-based initiative has now been reduced to equal access for religious organizations to the budget crumbs falling from the federal table. What a tragic outcome to the promise and rhetoric of the early days of the Bush administration! But budgets are moral documents, and they reveal a government's true priorities.

I often do conservative talk shows on television. Whenever you mention poverty in a venue like that, they scream that you're engaging in class warfare and promptly declare war on you. I've decided that the right wing is correct on this: There is a class war, but they and their political allies are the ones who have declared it. As Episcopal Bishop John Bryson Chane said at a recent Sojourners chapel service, "We've gone from a war on poverty to a war on the poor."

Susan Pace Hamill, a University of Alabama tax law professor, took a sabbatical to earn a master of theological studies degree. She wrote her thesis on "An Argument for Tax Reform Based on Judeo-Christian Ethics." In it she applied "the moral principles of Judeo-Christian ethics" to Alabama's tax system, seeing reform as "a critically important step toward ensuring that Alabama's children, especially children from low-income families, enjoy an opportunity to build a positive future."[2]

Those "moral principles" came into sharp focus in three events over the summer of 2003. Each event became a modern "parable" about the spiritual connections between budgets, priorities, and faith.

The first parable related directly to Alabama and its Republican governor's proposal to reform the state tax system. The other two showed the same principles on the national scale. One was about the exclusion of seven million low-income working families—and their twelve million children—from the child tax credit that other families were receiving. The other was the annual IRS annual report, showing huge increases in the wealth of America's four hundred richest taxpayers.

In all three, the moral contradictions are too great to ignore. The deepening injustice of America's growing wealth chasm is increasingly impossible to justify. It's becoming a moral, and even a religious, issue.

Alabama has long had one of the most regressive tax systems in the country. A family of four earning $4,600 a year has to pay income taxes—a lower threshold than any other state. Property taxes are the lowest in the nation, which primarily benefits the timber industry in a state where 71 percent of the land is timber. The state sales tax is 4 percent, but local governments are free to add to it. Many do; in some counties it's as high as 11 percent, even on groceries. People with incomes below $13,000 pay 10.9 percent of their income in taxes, while those who make more than $229,000 pay only 4 percent. How's that for fair?

Professor Hamill's thesis was published in the *Alabama Law Review* and came to the attention of the new governor, Bob Riley, a conservative Republican and former member of Congress. Alabama, like most of the fifty states, faced a severe budget crisis with a deficit of $700 million. Yet it is obligated by its constitution to have a balanced budget. So on May 19, 2003, Governor Riley addressed a special session of the state legislature.[3] "We cannot balance our budget with cuts alone," Riley said, "not unless we are willing to lay off thousands of teachers and cancel all extra-curricular activities, open prison doors and put convicted felons back on the streets, and force thousands of seniors out of nursing homes and take away their prescription drugs."

The governor then went on to propose a tax-reform package that included higher property taxes, higher income taxes on the wealthy, and no income taxes on the poorest people. The plan raised the threshold to pay income tax for families of four to $17,000—paying for it in part by raising corporate taxes on the timber industry.

Governor Riley said, "I have spent most of my life fighting higher taxes. While in Congress, I always voted against tax increases because I know the hardships they place on a family and on a business. No one wants to raise taxes—especially me. And I don't like being forced to do it now—but I believe we have no other choice."

The plan was approved by the state legislature and then went to a statewide public referendum, where it went down to defeat—due in large part to a huge advertising campaign by the state's wealthy business and special interests. But Alabama's churches—including the Methodists, Presbyterians, Southern Baptists, Episcopalians, and Catholics, along with Jewish leaders—supported the changes. Though the tax-reform plan ultimately failed, I believe it was one of the most important political stories in many years and, just perhaps, has planted a seed that will only grow in the future.

The key question is, What caused the governor's change of heart? It turns out that he is deeply Christian and realized that his faith had something to say about the budget and tax situation. "According to our Christian ethic, we're supposed to love God, love each other, and help take care of our poor," he was quoted by CBS News. "And this is a step in the right direction."[4] Here is a conservative Republican governor who has been reading his Bible and decided to put his Christian faith first.

While the White House and many congressional Republicans seem to believe that the solution to every problem is to cut taxes, there was a breath of fresh air coming from Alabama. Governor Bob Riley was proposing to raise taxes on the rich, while cutting them for poor working families.

I often hear people say that the Bible talks about individual charity and has nothing to say about government policies on budgets and tax cuts. Here's one Christian politician whose ideology has been altered by his faith and who was trying to do the right thing. Maybe some of his former colleagues in Washington will get the message.

The second parable of the summer of 2003 was the exclusion of low-income working families from the child tax credit. That debacle became a dramatic parable revealing a spiritual lesson about what happens to poor families and their children again and again in the halls of power in Washington, D.C., and in statehouses across the country: They are simply left out.

Most of the country understood that the $350 billion tax cut passed by the Congress in 2003 primarily benefited the wealthiest of Americans. Estimates were that each millionaire would receive $93,000. Yet 1 percent of the total tax cut—$3.5 billion—could not be found for families who struggle mightily just to get by. Part of the bill was a child tax credit accelerated for middle- and upper-income families, and checks of $400 were to be sent out. The Senate added an amendment to ensure the *refundability* of the child credit, so that working families who earn between $10,500 and $26,625 would also benefit.

But, the *New York Times* reported, House and Senate Republicans removed the child tax credit from most families who make under $26,625 in a last-minute revision of the tax cut bill that President Bush signed into law. That effectively prevented almost twelve million children, one in every six in America, from receiving any tax benefit at all. Middle- and upper-middle-income families would see an increase in their child tax credits from $600 to $1,000, but low-income families and their children would be systematically excluded. The inclusion of these families in child tax credit benefits was in the original Senate package but was stripped out in a late-night conference committee, reportedly to make room for more dividend and capital gains tax cuts for wealthier Americans.[5]

When the deed was revealed and the storm broke, two senators, Arkansas' Blanche Lincoln and Maine's Olympia Snowe, one a Democrat and one a Republican, demanded that the injustice be corrected. Not surprisingly, both were women, and more attuned to family needs. The Senate quickly fixed the omission in a 94–2 vote, in a way that cost the Treasury nothing.[6] But the Republican leadership of the House, seemingly oblivious to the political damage feared by the White House, brazenly tacked the low-income family child tax credit onto another tax cut for wealthier families—in other words, using the restoration of the measure for poor families to increase tax cuts for the rich.

The issue then deadlocked, and some Republicans actually admitted their tactic was an attempt to kill the child tax credit restoration altogether. As checks went in the mail for middle-class families, low-income working parents wondered why they got left out in the cold. The House of Representatives majority leader Tom DeLay's answer: "There are a lot of other things that are more important than that."[7] Unfortunately, Governor Riley's former Republican colleagues in Washington, led by DeLay, hadn't been reading their Bibles the way Riley had. And Tom DeLay was in danger of making himself into the Bull Connor of the modern anti-poverty movement; consistently blocking almost everything that would benefit poor people.

The third parable dramatically showed what is happening to the distribution of income in America. In its annual tax analysis for 2000, the IRS reported that the top four hundred taxpayers—only 0.00014 percent of the population—now take in more than 1 percent of the total income of all taxpayers. Meantime, their tax payments plummeted, mostly due to substantial reductions in capital gains tax rates. Over the nine years from 1992 to 2000, the average annual income of the top four hundred increased to $174 million, while the average income for the bottom 90 percent was $27,000, according to the *New York Times*. Even the *Wall Street Journal* called it "so much money in so few hands . . . a startling accumulation of wealth at the very top of the income pyramid." The "income gap," wrote the *Journal,* is becoming a "vast chasm." For many religious people across the theological spectrum, that inequality is also becoming a moral issue.[8]

I've been reading *The Message,* a paraphrase of the Bible by evangelical pastor Eugene Peterson that has sold millions of copies. Its renditions are vividly contemporary and include some truth telling from the prophets Amos and Isaiah about our current situation. "People hate this kind of talk. Raw truth is never popular" (Amos 5:10). "Doom to you who legislate evil, who make laws that make victims—laws that make misery for the poor, that rob my destitute people of dignity, exploiting defenseless widows, taking advantage of homeless children" (Isaiah 10:1–2).

It's the kind of talk we don't hear much these days in America. But we need it. If the Hebrew prophets were around today, they would surely be preaching about our tax and budget policies that enrich the wealthy and "make misery for the poor." And I don't think they would have worried much when accused of class warfare.

If biblical prophets like Amos and Isaiah had read the news about what happened to child tax credits for low-income families, for example, they surely would be out screaming on the White House lawn about the justice of God—and be quickly led away by the Secret Service.

Government spending programs sometimes provoke legitimate concerns about effectiveness. Nobody wants to waste money on things that don't work, and we all prefer public investment in proven and pragmatic programs. But this was not even a government spending program. It was a child tax credit that would have put money directly into the hands of the poorest mothers and fathers who are trying desperately to raise their children. "These are the people who need it the most and who will spend it the most," said Senator Blanche Lincoln, whose provision to include low-income families was dropped from the final bill. Senator Olympia Snowe finally voted against the tax cut, calling the omission "ill-founded" and "unfair." What does such a clear and revealing decision tell us about the priorities of many lawmakers?

Apparently, what is good for middle- and upper-income families and children is too good for the poor. Apparently, stimulating the economy with middle-class mall shopping is a good thing, but increasing the grocery budget for low-income single moms is not. Apparently, reducing taxes on stock dividends and capital gains for our wealthiest citizens was the highest priority for the congressional leaders, and there was simply no room left, under the tax cut ceiling, to do anything for poor families. Apparently, the Republican preference of putting money back into people's hands, rather than spending it on government programs, doesn't apply to the poor. We do have our priorities after all.

Let's tell it as the prophets might have: The decision to drop child tax credits for America's poorest families and children in favor of further tax cuts for the rich is morally offensive. It is blatant disregard for the poor and an outrageous bias toward the rich. In religious terms, the exclusion of any benefits for poor children in a new tax bill should have been named as a political sin. And those politicians who utter the words of religion and faith, yet who supported this exclusion of the poor, deserve to be called hypocrites. The White House, which approves all these choices, engages in moral double-talk when it espouses faith-based initiatives, then allows the abandonment of poor families. The Republican House and Senate leaders who made these choices against the poor should be ashamed of themselves. And the president too ought to be ashamed for allowing something to happen that is so conspicuously wrong.

Perhaps it's time for our religious leaders to head for the Capitol building and the White House lawn. *Outrageous, shameful,* and *intolerable* are all appropriate words in response to these three news stories—and modern parables—that hit the front pages. The connections between the three stories were impossible to miss, yet most in the political and media world did. But the moral contradictions are becoming too great to ignore, and the deepening injustice of America's growing wealth chasm is impossible to justify. It is indeed becoming a moral, and even religious, issue.

Call to Renewal convened a press conference on Capitol Hill to urge President Bush to restore the child tax credit for low-income families and unequivocally lay down the law to the Republican House leadership. Senator Lincoln and Senator Snowe joined the religious leaders in sponsoring the press conference. I spoke, along with John Carr of the U.S. Conference of Catholic Bishops, Episcopal Bishop of Washington John Bryson Chane, and Dave Donaldson of the evangelical We Care America. At the press conference, we released a letter to the president, signed by nearly thirty religious leaders across the theological and political spectrum, which vowed that faith-based leaders were not going away on this very "biblical" issue.

Yet whenever the growing economic disparity is raised, the political and media elites immediately call it class warfare. There is indeed class warfare raging in this country, but not from those who speak for the poorest Americans. It is the class warfare of tax cuts and budget priorities that make the rich richer while further decimating low- and middle-income families. For many in the religious community across the theological spectrum, that inequality is becoming intolerable.

The ongoing costs of the war, combined with tax cuts for the wealthy, have led to a crisis for America's poorest children. Indeed, America's poor were the first casualties of the Iraq war, as U.S. domestic needs were pushed off the political agenda. Leaders from both political parties agreed to budget resolutions containing billions of dollars in increased spending for the military *and* tax cuts, while resources for important domestic programs fell below the amount needed even to maintain current services in a deteriorating situation for the poor.

The consequences of these actions have become a silent war, felt most severely in the poorest parts of the United States, where low-income families are desperately clutching onto the bottom rungs of the failing economy. Virtually every state in America is suffering terrible budget deficits. But the federal budgets offer no relief for states, and no solutions to the deficits except further cuts to critically needed domestic poverty programs, child health care, and education.

Clearly, the sacrifices for the war in Iraq will be borne by those in most need—who will bear the brunt of inevitable spending cuts to vital social programs—and by future generations who will ultimately pay for the record-setting deficits. David Firestone, writing in the *New York Times*, put it well: "When President Bush informed the nation . . . that remaining in Iraq next year will cost another $87 billion, many of those who will actually pay that bill were unable to watch. They had already been put to bed by their parents."[9]

Even before the stunning announcement of the costs of the war in Iraq, the news for the poor had been getting worse and worse. According to the Census Bureau, the number of people in poverty increased by 3 million from 2001 to 2003 and is now 35.9 million people in 7.6 million families, including 12.9 million children.[10] A National Low Income Housing Coalition study found no state where a low-income worker can reasonably afford a one- to two-bedroom apartment, and forty states where such housing would require an income of more than twice the minimum wage.[11] Despite this, a bill passed the U.S. House that would reduce the number of the low-income families who receive housing vouchers by more than 150,000.

And then there are the continuing casualties, both killed and wounded. As of November 2004, the U.S. death toll stands at more than twelve-hundred fifty men and women—with thousands more wounded[12]—since George W. Bush landed on the USS *Abraham Lincoln* aircraft carrier wearing a military flight suit to proclaim that the primary conflict was over and that "the United States and our allies have prevailed." Deadly attacks still occur every day. And of course we don't hear much about Iraqi casualties.

The faith-based initiative, which many of us have supported, is in danger of becoming a hollow program without consistent domestic tax and spending policies that effectively reduce poverty. And the drastic federal and state budget cuts will be acutely felt by faith-based service providers who often bear the brunt of increased poverty in their communities. I talk to many people who run these programs and who have been supportive of the faith-based initiative. Frankly, many are angry and feel betrayed.

The government's budgets are a disaster for the poor, a windfall for the wealthiest, and thus directly conflict with biblical priorities. Budgets are moral documents. It may be controversial, but it is not inappropriate to name the federal budgets now being passed as "unbiblical." And it is time for religious people to clearly and prophetically respond. We need a "faith-based initiative" against budget priorities that neglect poor people.

In June of 2003, a wide cross-section of church and faith-based leaders, convened by Call to Renewal, undertook just such an effort. They sent a letter to the White House calling their further support for faith-based initiatives "increasingly untenable" without domestic policies and budget priorities that benefit poor families and children. The letter was followed by a meeting at the White House with the top domestic policy officials in the Bush administration. In the room were the heads of both evangelical and mainline Protestant denominations, representatives from the U.S. Conference of Catholic Bishops and both men's and women's Catholic orders, and key faith-based social service and advocacy organizations like the Salvation Army, World Vision, Bread for the World, and Evangelicals for Social Action.

At a dramatic point in the meeting, one evangelical leader confronted the White House officials, saying, "I voted for President Bush, I supported you all because I liked the language of 'compassionate conservatism.' But I am just a hair's breadth away from concluding the whole thing was a trick, just to focus our attention away from poverty policy and put the whole burden of reducing poverty back on the shoulders of the faith community." You could have heard a pin drop in the room, as the White House leaders fell silent and finally said they would agree to meet with us again. The impressive and diverse group of religious leaders then took their message to Capitol Hill, where they spoke to leaders of both Democratic and Republican parties and conducted a rare prayer service to remember the nation's poorest children with dozens of members of the House of Representatives.

In his column in the *Washington Post* that week, E. J. Dionne commented on the religious leaders' letter:

"... it's precisely because I don't want the faith-based approach to be a cover for the wholesale abandonment of government's responsibilities that I share the dismay of the religious leaders who wrote to Bush. Millions of working people are poor—and lack health insurance and adequate child care—even though they do all the things that society and our religious traditions say they should. Religious groups will never have the money to transform the material conditions of these families. But relatively modest government outlays could make their lives much better.

There is a religious mandate for such an approach. "Jewish prophets and Catholic teaching both speak of God's special concern for the poor. This is perhaps the most radical teaching of faith, that the value of life is not contingent on wealth or strength or skill, that value is a reflection of God's image."

Those thoughtful words are George W. Bush's. Is it too much to ask him to explain how his policies live up to that vision?

© the *Washington Post,* June 10, 2003

The text of the letter to the White House is below.

June 9, 2003

Dear Mr. President,

We are all leaders in the faith community, whose churches and faith-based organizations are on the front lines of fighting poverty. Many of us have supported your faith-based initiative from the beginning of the administration. Several of us have met with you to discuss the churches' role in overcoming poverty and have offered solid support to our friends John DiIulio and Jim Towey, who have led your Office of Faith Based and Community Initiatives. But while we have consistently backed faith-based approaches to poverty reduction, we have also insisted they must be accompanied by policies that really do assist low-income families and children as they seek self-sufficiency.

Mr. President, it is a critical time for poor people in America. Poor people are suffering; and our faith-based service providers see it every day in communities across the country. The poor are suffering because of a weakening economy. The poor are suffering because of resources being diverted to war and homeland security. And the poor are suffering because of a lack of attention in national public policy.

We are writing because of our deep moral concern about consistency in your administration's support for effective policies that help alleviate poverty. We believe a lack of focus on the poor in the critical areas of budget priorities and tax policy is creating a crisis for low-income people. We believe the budget your administration has put forward fails to protect and promote the well being of our poorest and most vulnerable citizens. The tax cut just passed by the Congress with your support provides virtually no help for those at the bottom of the economic ladder, while those at the top reap windfalls. The resulting spending cuts, at both federal and state levels, in the critical areas of health care, education, and social services, will fall heaviest on the poor. Budgets are moral documents.

You have taken many positive steps with regard to international aid and development, such as the HIV/AIDS initiative, and we would like to see that compassion manifest here at home. In significant social programs, like welfare reform, we have supported the proposals of your administration to strengthen marriage and family as effective antipoverty measures; but the companion pro-family commitments to invest in adequate child care, education, and training for our poorest families have fallen short in your administration's proposals. The most effective and bipartisan public policies for reducing poverty have not been adequately supported by your administration.

Over the past several years, we have advocated several policy initiatives in addition to the "faith-based initiative" that would help low-income people in this country. These include TANF reauthorization that makes poverty reduction a priority, targeted tax relief for low-income families, and funding for proven programs that would effectively reduce poverty. We believe administration support for such policies would be consistent with your stated commitment of being compassionate toward the poor, especially since you have spoken more about issues of poverty than many of your predecessors.

We recall your Notre Dame address two years ago, where you pointed out: "Government has an important role. It will never be replaced by charities. . . . Yet, government must also do more to take the side of charities and community healers, and support their work. . . . Government must be active enough to fund services for the poor—and humble enough to let good people in local communities provide those services."

Mr. President, "the good people" who provide such services are feeling overwhelmed by increasing need and diminishing resources. And many are feeling betrayed. The lack of a consistent, coherent, and integrated domestic policy that benefits low-income people makes our continued support for your faith-based initiative increasingly untenable. Mr. President, the poor are suffering, and without serious changes in the policies of your administration, they will suffer even more.

When you announced the faith-based initiative, you pledged that: "I want to ensure that faith-based and community groups will always have a place at the table in our deliberations." Mr. President, it's time to bring faith-based organizations to the table where policy decisions are being made. We are concerned that the needs of poor people in America seem to have little influence in the critical policy decisions your administration is making. The faith-based initiative seems to be the only place in your administration where poverty is prioritized, yet we know that faith-based initiatives alone will never be sufficient to solve the problems of poverty. As we have discussed with you the faith-based initiative, we now want to engage your administration in a serious conversation about domestic social policy. Mr. President, it's time to talk.

Rolling to Overcome Poverty

The children from St. Luke's School in St. Paul, Minnesota, came out to give us water bottles and display their handmade signs about Catholic social teaching on poverty. It was the first stop on Call to Renewal's "Rolling to Overcome Poverty" fifteen-city, twelve-day bus tour in mid-October 2004, just before the election, and we were marching with hundreds of people from Minnesota's diverse faith communities on a four-mile "people's pilgrimage" to the state capitol.

I was immediately drawn to the kids who were about the same size as my son Luke, so I found myself among the first- and second-graders. A seven-year-old girl held up the brightly colored poster she had made to illustrate Catholic social teaching on poverty, a sign made by many of the kids. It said "Every Life is Sacred. Stop Poverty." When I smiled at her, she looked up at me and said, "Thank you for doing this." That kept me going for days.

She had given us the theme for the bus tour, and the next stop on the march gave us the national context. We stopped at the House of Hope Presbyterian Church, where an overnight shelter took the overflow from a city shelter; the night before, three families, with eleven children under the age of six, had stayed there, and they would again that night. "Overflow" became the watchword for the tour as we encountered countless shelters, food programs, and faith-based ministries seriously overextended in a country where the poverty rate has risen every year for the last three years.

The next day we stood under the dome of the Wisconsin state capitol in Madison as those in prison ministry, within earshot of the state's political leaders, told stories of life at the bottom. Biblical wisdom suggests that the truth about a society is best known from the bottom. In the afternoon we heard the success stories of formerly homeless children in Baptist-run, affordable family apartments, children whose grade point averages went from 1.2 to 3.2. None of them dropped out of school and all went on to college. Stability, security, and opportunity help poor kids to succeed. Imagine that.

In Milwaukee, the director of the Hunger Task Force seemed very tired as she spoke of all the food pantries and food lines trying to meet the growing needs of hungry families, and we wondered together why these programs put into place a generation ago as temporary measures have become our permanent way of dealing with poverty. Shall poor families and children forever shop at food banks and stand in line to eat at soup kitchens? I said repeatedly on the tour that we can't just keep pulling bodies out of the river and not send somebody upstream to see what or who is throwing them in. The very last event of the tour, many days and cities later, was breakfast at a women's and children's shelter run by the Salvation Army. I was watching my own two sons playing with the homeless children as I ate with the

Salvationist "commander" who oversees the shelter. We talked about William Booth, the founder of the Salvation Army in nineteenth-century England, who also used to also say that we can't just keep picking people up at the bottom of the cliff and not climb the mountain to see who is throwing them off the edge.

In Chicago, we learned the average age of a homeless person is now nine years. A broad cross section of church people and leaders came together at North Park University to do something about unacceptable poverty in the Windy City—that night, eight hundred people signed a covenant. I preached again that night, along with James Forbes from Riverside Church, but the real sermon was the congregation and the diverse group of religious leaders sitting side by side on the stage. Many remarked that they have never seen such a broad group of religious leaders or such a cross section of the whole religious community in Chicago. Poverty had brought us all together.

In Grand Rapids, Michigan, the state's Republican member of Congress said the crisis of poverty had to break into the busy lives of his colleagues, and the independent mayor of the city called for nothing less than a "national agenda to eliminate poverty." But the prophetic word of the evening to the packed church came from an immigrant mother who apologized that English was her second language before saying, "We are children of God—that should count for something." She got a standing ovation. Almost every evening service or rally featured the testimonies of people struggling with poverty, and their stories changed the nature of all the events.

In my hometown of Detroit, I visited a walk-in center with hundreds of homeless and hopeless people, who made Third Street look like a Third World city. Two weeks before, a poor, confused woman at the center had plunged a knife into the back of a staff member she had never met, and killed him. His name was Fernando Garcia; he had been homeless himself, and now he leaves behind a wife and kids who will miss their father. I wondered how we have allowed so many people to simply get lost, and I remembered the words of Jesus, "As you have done to the least of these, you have done to me."

A midweek gathering in a church basement in Toledo again drew a large and diverse group of religious leaders determined to do something about poverty. The service that evening in an activist black church in Cincinnati's Over-the-Rhine neighborhood felt like a justice revival. A young man from

Justice for Janitors spoke of his children and their families' life on the edge, but then spoke about how faith both sustains and motivates him to work for change. Cleveland is now ground zero for poverty in America—the poorest big city in America, with an appalling 47 percent of its children below the poverty line. One woman told us how her husband had left her and their children and how "sometimes these situations happen so fast they just spiral you down to the bottom." But the city's religious communities are seeing that bottom ranking as an opportunity to unite, and have created Greater Clevelanders Together Overcoming Poverty. That night they united in a moving cathedral service sending a signal to their city that they will no longer tolerate such poverty in their community. In nearby Akron, the stories of those who had lost jobs were devastating, a loss causing people to "live in fear." And, said the director of a jobs center, "This fear is not from a foreign threat." We heard similar stories in Pittsburgh, where evangelical college students organized an event to help put their faith into action against poverty. Indeed.

While we were on the tour, an astounding new study by the Annie E. Casey, Ford, and Rockefeller foundations—"Working Hard, Falling Short"—revealed that 9.2 million families, including twenty million children, now make such low wages that they are barely able to survive. Married couples are members of half of those households, debunking yet another myth about poverty being automatically solved by marriage. At every stop we passed out the Isaiah Platform, drawn from Isaiah 65, which speaks of God's vision for a good society with good wages, good health, good houses, and safety and security for all God's children.

The tour ended the way it had started, with a march—this time through the streets of Philadelphia—and a service at Philadelphia Cathedral. The excitement and energy of the procession, led by young African American drummers, galvanized our tired bodies after twelve days on the road.

I told tales from the road and reflected on what we had seen and heard. We saw amazing *ministries* without which, I believe, the nation would fall apart in about 48 hours. We also saw *models* of concrete answers to the problems of poverty that offer islands of hope in a sea of despair and show the way forward if we have the political will to invest in real solutions. But all along the way, there was talk of *movement* in the air. It's time, I preached, to shift our thinking "from ministry to movement."

"Poverty is a religious issue!" we proclaimed across the Midwest. The cry of the poor rings from cover to cover in the Bible; God hears the cry of the poor—do we? In most tour cities, newspaper stories, radio reports, and even some television coverage helped put poverty on the agenda as we had hoped. But this year, the presidential candidates weren't talking about poverty. They wanted to present themselves as champions of the middle class, even while many were slipping down from the middle. As the election drew nearer, there was no more talk of "Two Americas" that Democratic candidate John Edwards had raised in the primary season. Yet two Americas is the reality we found on the ground in every city we visited.

We thought the election would be the focus of the Rolling to Overcome Poverty bus tour. But after a few cities, we realized the possibilities were much deeper. The faith communities we met were more interested in November 3 and beyond, and they made it clear that no matter who won the election we must be at their door the next day with a faith-based movement that finally demands real action and solutions to overcome poverty in America.

CHAPTER 16

Amos and Enron

What Scandalizes God?

WHEN AMERICA'S CORPORATE scandals were making news (remember, before the war with Iraq?), I wrote a column entitled "The Sin of Enron." Sounds like a great Old Testament saga, doesn't it? Well, this may be a more biblical tale than we think. Before going to church on a Sunday morning during the crisis, I heard then Treasury Secretary Paul O'Neill on *Fox News Sunday*. "Part of the genius of capitalism," he said, is that "people get to make good decisions or bad decisions. And they get to pay the consequences or to enjoy the fruits of their decisions."[1] O'Neill was doing the Sunday morning news circuit to talk about Enron, the huge energy company that had just gone bankrupt, destroying both the jobs and the life savings of thousands of its employees, yet enriching the corporation's top executives.

O'Neill got it wrong. In fact, the Enron scandal teaches a different lesson than O'Neill's: The people on top of the American economy get rich no matter whether they make good or bad decisions, while workers and consumers are the ones who suffer from all their bad ones. In the Enron case, the company executives overestimated the company value, ran it into the ground, lied to their employees about the company's stability, encouraged Enron's workers to invest their pension funds in company stock, and then imposed rules against selling that stock, while at the same time, arranging an executive bailout for themselves worth a fortune. Shortly before the collapse, the now-indicted Enron CEO Kenneth Lay quietly sold his company stock for $101 million. Timing is everything.

We now know the basics. Enron stock went from $85 per share to 26 cents a share, and the company went from number seven on the Fortune 500 list to declaring bankruptcy. As the value was plummeting, twenty-nine top executives cashed in their stock options for $1.1 billion while blocking employees from doing the same. While the top few made out like bandits, thousands of employees saw their entire life savings, their pension funds, and their jobs disappear. The audit firm that reviewed Enron's books, Arthur Andersen, shredded the company's documents. While Lay was busy reassuring employees and investors that the stock would recover, a senior financial employee sent him a letter that warned, "I am incredibly nervous that we will implode in a wave of accounting scandals," and described a "veil of secrecy" inside the company.

Enron was one of the best-connected corporations in the country. The Houston company had been longtime contributors and old friends to the Bush family, both father and son, and had extensive access to Washington, D.C., politics. Attorney General John Ashcroft recused himself from the investigation due to the nearly $60,000 in campaign contributions he had received from Enron. Enron executives met six times with Vice President Dick Cheney and his staff on the administration's Energy Task Force, and the oil giant helped shape (some say virtually dictated) a policy based on government deregulation and the marginalizing of both conservation and alternative energy sources. Of course, such influence was downplayed because, it was argued, George Bush and Cheney already agreed with the oil company's view of America's energy future. What a surprise.

A big Washington political topic during the Enron debate was a couple of urgent phone calls made from Ken Lay to O'Neill at Treasury and Secretary Donald Evans at the Commerce Department, perhaps hoping for some last-minute administration help for old friend Enron. The Bush administration points to the fact that no help was offered, another testimony to its belief in capitalism's survival of the fittest. But the episode again demonstrated the access and the survival of the richest, while all the ordinary employees lost their livelihoods and life savings.

No one seemed to worry about the fact that Ken Lay's calls instantly got through to cabinet secretaries. The relationship between money and access is a given that nobody in Washington even questions anymore. Democrats were careful about critiquing too strongly since Enron was so bipartisan in its buying of influence—three-fourths of the Senate and half of the House benefited from Enron cash. A Center for Responsive Politics/*Washington Post* study shows that from 1992 to 2001, Enron's corporate "soft money" contributions totaled $1.4 million to Republicans and $850,000 to Democrats. As for those Cabinet calls? Rest assured that faith-based organizations and advocacy groups fighting child poverty don't get through nearly as easily.

The Enron scandal once again made painfully obvious the extent to which money controls our political process; it's not hyperbole to say that our political system is for sale to the highest bidder. Events like these show how the very nature of democracy in America is under threat. We now have a system of big corporations, big accounting firms, big law firms, and big politicians scratching each other's backs while millions of Americans struggle just to get by.

Every week the news brings more reports of corporations revealed to have been "cooking" their books—using accounting tricks to mask expenses, hide losses, and inflate profits. When the truth is revealed and investigators finally step in, thousands of people are left jobless, or without their retirement funds, while top executives walk away with millions.

The list includes Adelphia Communications, the sixth largest cable provider in the country, which filed for bankruptcy after having inflated its revenue and earning statements, with $3 billion in off-the-books personal borrowing by the founding family. Xerox, which was fined $10 million to settle fraud charges by the SEC after having improperly recorded $6.4 billion in revenue. WorldCom, which filed for bankruptcy and faced SEC fraud charges after "hiding" nearly $4 billion in expenses. Five top officials pleaded guilty to criminal charges, and the former CEO was indicted. Employees of WorldCom who owned shares of the company in their retirement portfolios collectively lost at least $1.1 billion in three years. So far, nearly twenty thousand people have been laid off. And, of course, there's the continuing saga of Enron.

In the midst of the corporate scandal story, questions arose about some of George W. Bush's activities as a director of Harken Energy. Insider trading and low-interest loans were alleged. And a lawsuit was filed against Vice President Cheney alleging fraudulent accounting practices when he was CEO of Halliburton, which since then has, of course, received the lion's share of the most lucrative contracts in postwar Iraq.

In 2001 alone, 270 corporations "restated" the numbers in their financial statements. From 1997 to 2001, a total of 1,089 companies apparently did so. All told, these transactions have cost investors billions of dollars.

Americans have a love-hate relationship with both government and business. The climate seems to shift like a pendulum between eras of an "anything goes" mentality and periods of stricter government regulation. The excesses of the 1920s, leading to the Great Depression, were followed by the reforms of Franklin Roosevelt, including the creation of the Securities and Exchange Commission and tighter regulations.

Beginning with the Reagan administration and culminating in the Gingrich "revolution" of 1994, many of these regulations were relaxed. At the urging of their biggest corporate contributors, Congress passed legislation shielding accounting and law firms from liability for false reporting and made shareholder suits more difficult. The trend has continued through the Bush administration's close relationship with the corporate world—in relaxing environmental regulations on clean air and water, diminishing worker protections, and appointing a new fraternity of former corporate executives to oversee the businesses and industries they used to run.

It increasingly appears that the economic "boom" of the 1990s may have been a house of cards built on fraud. The pied piper of the bull market and the illusive dream of endless profits put the economy and the culture into an addictive state of financial irresponsibility.

A 2002 Gallup poll on public attitudes showed a sharp increase in people who view big business as a greater threat to the nation than big labor or big government. When asked which they view as "the biggest threat to the country in the future—big business, big labor, or big government," 38 percent named big business, up sharply from 22 percent when the question was last asked in October 2000, and the highest it has been since the question was first asked in 1985. A greater percentage of Americans, 47 percent, still views big government as a larger threat. But the percentage of the population believing government to be the worst threat is now the lowest since

1981, and down significantly from 65 percent in October 2000.[2] Another poll in June 2003 found that nearly one-third of the American people have "very little" confidence in big business—also considerably higher than those who expressed "very little" confidence in Congress or organized labor, both at 20 percent.[3]

President Bush's July 9, 2002, speech, at the height of the corporate scandal crisis, was delivered with a high moral tone: "There's no capitalism without conscience. There is no wealth without character." But he had very little to offer in the way of concrete solutions. The president's words were strong, but his proposals lacked the power of clear specificity or real reform. Contrary to reformer Theodore Roosevelt, George Bush speaks loudly but carries a small stick when it comes to the behavior of large corporations. As in many things, the president's rhetoric evidences fine speechwriters, but his proposals demonstrate cautious policy advisers.

The president says the problem is a few bad apples. He fails to see that a tree whose growth is all at the top, with bare branches at the bottom, is in real danger of falling over. And at a deeper level, George Bush doesn't seem to grasp that the tree of the American economy is rooted in the toxic soil of unbridled materialism, a culture that extols greed, a false standard of values that puts short-term profits over societal health, and a distorted calculus that measures human worth by personal income instead of character, integrity, and generosity. Unhealthy roots and soil tend to produce stunted growth and bad fruit. Can we really expect an administration to clean up the mess when its own house may be dirty—or at least tainted?

The Senate finally passed unanimously a series of accounting and corporate regulatory measures considerably tougher than what the president had suggested. They included, by a 97-to-0 vote, a new chapter in the criminal code that makes any "scheme or artifice" to defraud stockholders a criminal offense. "If you steal a $500 television set, you can go to jail. Apparently if you steal $500 million from your corporation and your pension holders and everyone else, then nothing happens. This makes sure something will happen," said Senator Patrick Leahy.[4]

Thankfully, the laws are being belatedly enforced. In the Enron case alone, twenty-eight former executives have been indicted, including the former chief executive officer and chief financial officer.

Corporate CEOs, no less than everyone else, have a responsibility to the common good, not just to the bottom line. The entrepreneurial spirit and social innovation fostered by a market economy has benefited many and should not be overly encumbered by unnecessary or stifling regulations. But left to its own devices and human weakness (okay, let's call it sin), the market will too often disintegrate into greed and corruption, as each week's reports of corporate wrongdoing painfully reveal. Capitalism needs rules, or it easily becomes destructive. A healthy, balanced relationship between free enterprise, on the one hand, and public accountability and regulation, on the other, is morally and practically essential, as many are now coming to see.

Senator Leahy is right: But let's call theft *theft*. And when there is theft, let us hope that "something happens"—for as folksinger Woody Guthrie reminded us a long time ago, "Some will rob you with a six gun, and some with a fountain pen."

The great crisis of American democracy today is the division of wealth. The Republicans hate it when people talk this way and, frankly, so do many Democrats, but the political issues now are increasingly *populist* ones—which is another way to speak of "class warfare"—the rich against the rest of us.

But the real battles are seldom discussed in a media that really hates the class warfare subject (perhaps, in part, because so many media reporters and pundits have gone through a dramatic conversion from middle to upper incomes themselves). Nobody comments on the new pattern. What invariably happens now when a company announces huge lay-offs for workers? First their stock goes up, and then their executive compensation goes up! It's become a battle between Main Street and Wall Street: Main Street bears the brunt of economic change, while Wall Street gets the reward. Only a few politicians are willing to call that "immoral," and they get charged with class warfare. Each night, CNN announces a new list of American companies exporting jobs overseas.

Although the economy grew at a 6 percent annual rate in the last half of 2003, raising profits and the stock market, the unemployment rate was still at 5.6 percent, and the average length of unemployment was the highest since 1984. Many people have simply given up looking for work, resulting in the percentage of the population officially in the labor force being the lowest in years.

Columnist Richard Reeves, one of the few journalists who write about this subject, suggests that politics is "evolving as class warfare, a term despised by Republicans in power who think, quite correctly, that they have won that war, fair, square and profitably." Reeves quotes a breathtakingly honest editorial phrase from *Newsday*: "Though there are still skirmishes and pockets of resistance in outlying areas of the country, it is now probably safe to say that major combat operations in the war between the classes have ended. The rich have prevailed."

Reeves quotes *Business Week,* hardly a populist broadsheet, reporting in November 2003 that "a better measure of voter sentiment is personal income. While incomes rose in the Bush years, most of the gain has been limited to the wealthy. After adjusting for inflation, median after tax family income dropped 3.3 percent from 2000 to 2002. . . . The roughly 2.5 million families who will make more than $225,000 in 2004 will enjoy more than 40-percent of the benefits of the three Bush-era cuts, while 70 million families making up to $60,000 will get only about 20-percent."[5]

But I want to get back to where I was headed before listening to Treasury Secretary O'Neill's Sunday morning homily. I wonder if he or the other executives at Enron made it to church or synagogue that weekend. And if they made it, did they hear anything about their business and political dealings?

"Truth be told, the corporate crisis is as much spiritual as it is financial," says David Batstone, the executive editor of *Sojourners* and professor at the University of San Francisco, in his new book, *Saving the Corporate Soul.*[6]

Let me be blunt. The behavior of Enron and other guilty corporate executives is a direct violation of biblical ethics; the teachings of both Christian and Jewish faith excoriate the greed, selfishness, and cheating that have been revealed and condemn, in the harshest terms, their callous and cruel mistreatment of employees. Read your Bible.

The strongest media critics of Enron call it putting self-interest above the public interest; the Bible would just call it a sin. I don't know about the church- or synagogue-going habits of Enron's top executives, but if they do attend services, I wonder if they ever hear a religious word about the practices of arranging huge personal bonuses and escape hatches while destroying the lives of people who work for you. Where do Enron executives go to church?

It's time for the pulpit to speak—to bring the Word of God to bear on the moral issues of the American economy. The Bible speaks of such things from beginning to end, so why not our pastors and preachers? In addition to doing the weekend news shows, O'Neill and the others should have to hear about all this in their congregations.

> Hear this, you that trample on the poor and take from them their jobs and retirement funds. You say, "When will the Sabbath be over so we can make the measure small and the payment great and practice deceit with false accounting?" Therefore, you who afflict the righteous, who take a bribe, who push aside the needy: You have built huge estates of hewn stone, but you shall not live in them. You have your stock options, but you shall not cash them in. For I know how many are your transgressions and how great are your sins—you who afflict the righteous, who take a bribe, and push aside the needy in the gate.

No, Amos didn't quite say that (but it's only a slight paraphrase). I suspect, though, that if the eighth-century goat-herding prophet were in America today, he probably would say something quite close to this. Does anyone really doubt that? The Hebrew prophet's condemnation of the corrupt wealth of his era applies quite dramatically to our current situation. Amos lived in a time not unlike our own—one of great prosperity, but prosperity built upon corruption and oppression, leading to a great and growing inequality.

Not all wealthy people favor the stacking up of more and more privileges for the rich. I remember a rally on the Capitol grounds just outside Congress, where a political debate was raging over the "estate tax." This is a tax that affects only the top 2 percent of the American people and is levied on the estates of America's wealthiest people to moderate the enormous passing on of wealth from one generation to another. Originally proposed by a Republican, President Theodore Roosevelt, to counter the institutionalizing of wealth aristocracies, like Europe had, the tax was first passed in 1916 and supported by both parties for the eight decades since. But a vigorous campaign was begun by conservative forces and the Bush administration against what they call the "death tax." The 2001 tax cut gradually phases out the estate tax through 2009, repeals it entirely in 2010, and then restores it in 2011 as part of the Bush administration's plan to sell the overall package. Even this gradual repeal robs the federal budget of $982 billion in the next twenty years. Efforts are now being made in Congress to make the repeal permanent; legislation passed the House in 2003 but failed in the Senate.[7]

Gathered outside the Capitol were those who supported the estate tax as a healthy social tool to temper inherited fortunes and give something back to society from those who have benefited most. Among them were senators, social justice advocates, and leaders of philanthropic organizations (among them faith-based groups) that have benefited greatly from the estate tax by providing the opportunity for many wealthy people to give to charity, rather than seeing their inheritances heavily taxed. If you take away the tax burden on the largest inheritances, many might not opt to give their money to such worthy causes. Also on hand were a few very wealthy individuals who favored the estate tax, including Bill Gates Sr., the father of the richest person in the world.

I spoke on behalf of the estate tax and quoted from the prophets Amos, Isaiah, Jeremiah, and Micah, suggesting that some regular leveling of riches in society was a consistent theme of the biblical sages. The common good was their abiding concern, and the wealthy were regularly chastised for their selfishness and greed. I quoted all four prophets:

From Isaiah:

> Ah, you who join house to house, who add field to field,
> until there is room for no one but you,
> and you are left to live alone in the midst of the land. (5:8)

From Micah:

> The powerful dictate what they desire,
> thus they pervert justice. (7:3)

From Jeremiah:

> They do not judge with justice the cause of the orphan, to make it prosper,
> and they do not defend the rights of the needy. (5:28)

And, from Amos, the passages I just mentioned.

The powerful words of the biblical prophets really energized the crowd, and I concluded by suggesting that "Amos, Isaiah, Jeremiah, and Micah can be counted on to support the estate tax!" The next speaker was Bill Gates Sr. He is a very tall and dignified-looking man, who stood quietly at the podium for a few moments and seemed to be gathering his thoughts. Then he introduced himself: "I am Amos Isaiah Jeremiah Micah Gates!" The delighted audience shouted its approval to a wealthy man with a social conscience.

I've had the opportunity since to speak with Bill Gates Sr. on several occasions and discovered that he is an active Episcopal layman. Together with Chuck Collins, Gates has coauthored a book called *Wealth and Commonwealth: Why America Should Tax Accumulated Fortunes,* which lays out a comprehensive and compelling argument for tax justice. In an article in *Sojourners* based on the book, they wrote, "Society has an enormous claim upon the fortunes of the wealthy. This is rooted not only in most religious traditions, but also in an honest accounting of society's substantial investment in creating the fertile ground for wealth-creation. . . . Judaism, Christianity, and Islam all affirm the right of individual ownership and private property, but there are moral limits imposed on absolute private ownership of wealth and property. Each tradition affirms that we are not individuals alone but exist in community—a community that makes claims upon us. The notion that 'it is all mine' is a violation of these teachings and traditions. . . . Society's claim on individual accumulated wealth is a fundamentally American notion, rooted in recognition of society's direct and indirect investment in an individual's success. In other words, we didn't get here on our own."[8]

I spoke at the launching of the book in Washington, D.C., and was struck by the religious and moral self-reflection of a very wealthy Christian. Bill Gates Sr. presented both the practical and ethical case for the wealthy paying significant taxes. "I believe that one's obligations to society grow in proportion to how much one has benefited," said Gates. That's good theology.

With political leaders so beholden to the money of wealthy individuals and corporations, and with the media elites joining the ranks of the rich and famous, who will sound the call to justice? Where will a new populist energy and constituency come from? Perhaps, just perhaps, from voices in the religious community—inspired from reading their prophets and Jesus, motivated by more than material pleasures, and linked, by the gospel, to the fate of those marginalized by the greed of corporate elites.

What if the new populism was more biblical than just political? What if the calls for economic justice were made in the name of Jesus—or Amos, Isaiah, Jeremiah, and Micah—instead of from more ideological sources and causes? What if a more "religious populism" began to emerge? What if behavior in the economic spheres of our lives became the substance of adult Sunday school curriculums and Bible study groups? And what if the hard political questions about corporate responsibility, tax benefits, trade policies, budget priorities, and campaign financing were coming from religious congregations that political leaders couldn't afford to ignore? Nothing could do more to bring about a change of fortunes in the battles of class warfare.

CHAPTER 17

The Tipping Point

Faith and Global Poverty

FOR THE FIRST TIME in history we have the information, knowledge, technology, and resources to bring the worst of global poverty to an end. What we don't have is the moral and political will to do so. And it is becoming clear that it will take a new moral energy to create that political will. I believe the religious communities of the world could provide the "tipping point" in the struggle to eliminate the world's most extreme poverty. Faith communities could provide the crucial social leadership the world desperately needs, and I don't see where else that prophetic leadership might come from.

Malcolm Gladwell, in his best-selling book, *The Tipping Point,* talks of how an idea, product, or behavior moves from the edges of a society to broad acceptance, consumption, or practice, often suddenly and unexpectedly. Along the way there is a "tipping point" that transforms a minority perception to a majority embrace. Today, a sizable and growing number of individuals and institutions have identified the deep chasm of global poverty as their central moral concern and have made significant commitments to overcome the global apathy that leads to massive suffering and death. But we have not yet reached the tipping point—when the world demands solutions.

The most astute observers of the issue now realize that only a new moral, spiritual, and even religious sensibility, in relation to the problems of global poverty, will enable us to reach that critical tipping point. Even some of the world's political leaders who are focused on this question (whether they themselves are religious or not) are coming to realize the need for a moral imperative. And it is clear that the moral imperative must now focus on the interrelated issues of debt, aid, trade, and AIDS. This will be the "template" for real solutions. None of these elements will be enough in themselves; only all four essential ingredients in combination will produce lasting results in overcoming global poverty.

In a 2004 speech to a conference of mostly faith-based development agencies in the United Kingdom, Chancellor of the Exchequer Gordon Brown gave a sobering report on how the world was failing to keep the promises of the Millennium Development Goals in the crucial areas of education, health, and targeted poverty reduction.[1] Despite the commitments made by 147 nations to cut extreme poverty in half by the year 2015, global progress was significantly behind schedule. As to the causes of the thirty thousand infant deaths that still occur each day in the poorest parts of the world, Brown pointed to our moral apathy: "And let us be clear: it is not that the knowledge to avoid these infant deaths does not exist; it is not that the drugs to avoid infant deaths do not exist; it is not that the expertise does not exist; it is not that the means to achieve our goals do not exist. It is that the political will does not exist. In the nineteenth century you could say that it was inadequate science, technology and knowledge that prevented us saving lives. Now, with the science, technology and knowledge available, we must face the truth that the real barrier is indifference."

Point by point, the chancellor went through the alarming facts and statistics showing how these once hopeful goals had already fallen far short. "If we let things slip," he predicted, "the Millennium Goals will become just another dream we once had, and we will indeed be sitting back on our sofas and switching on our TVs and—I am afraid—watching people die on our screens for the rest of our lives. We will be the generation that betrayed its own heart." He ended his speech with a passionate appeal to the nongovernmental organizations (NGOs) and the faith communities in particular, quoting the prophet Isaiah: "I appeal to NGOs and faith groups: to hold us accountable, to be the conscience of the world, to be the voice that guides us at this crucial crossroads, to work together with no one ever subordinating their own objectives but recognizing that each of our objectives can be better realized if we can agree on the financing to underpin them. In 2015 we cannot look back and say: 'It was not us who acted, it had to be left to the next generation. It was not now, but some other distant time in the future.' That is not good enough. When the need is urgent and our responsibilities clear; and even when the path ahead is difficult, hard, and long, let us not lose hope but have the courage in our shared resolve to find the will to act. And let us say to each other in the words of Isaiah 'though you were wearied by the length of your way, you did not say it was hopeless—you found new life in your strength.' The strength together to fight poverty, remove destitution, end illiteracy, cure disease. The challenge for our time and for our generation. And let us achieve it together."

On an earlier occasion, Gordon Brown said to me, "The most important social movement in Britain since Wilberforce was Jubilee 2000. Without that campaign, led by your church people, our government simply would not have cancelled the debts of the poorest countries." William Wilberforce, the eighteenth-century British parliamentarian who was converted in the Wesleyan revival, became the political leader of the historic anti-slavery campaign, which was sparked by spiritual renewal. Brown led the Labour government of Tony Blair in their decision in 2000 to cancel the bilateral debts to Britain of the world's most impoverished nations, and Jubilee 2000 was the church-initiated movement for debt cancellation. "It's obviously only a start to completing this process of debt relief and poverty reduction," Brown said, "but it is the important start that I think everyone is looking for." And it serves as perhaps the best modern case study of what a faith-based initiative can do.

Jubilee: Cancel the Debt

On a trip to London in 1999, I was amazed when I saw the words *Jubilee 2000* emblazoned high atop the millennial countdown clock at Piccadilly Circus—the closest thing central London has to Times Square. I knew the grassroots campaign begun by religious and secular poverty activists in Britain to cancel the debt of the world's poorest nations had been having remarkable success, but I didn't expect to see its name in London's lights on the way to a theater in early January.

Over an Indian meal with our friends Peter and Dee Price, Joy and I discussed the extraordinary success of Jubilee 2000. Peter had just become a bishop in the Church of England and was a staunch supporter of the campaign. He told us about the humble beginnings of the effort. Too radical a pipe dream, said most. Sure, forgiving the crushing debt of the world's poorest nations would, more than anything else, begin to make poverty reduction possible. But how could you ever convince the world's wealthiest countries, their banks, and the World Bank and International Monetary Fund (IMF) to ever forgive those debts? Even the biggest international aid and development groups were skeptical of the idea at first.

But at the beginning of the new millennium, Jubilee 2000 reported enormous progress in moving the world toward a cancellation of the poorest nation's debts. In Britain, where the movement began, Chancellor of the Exchequer Gordon Brown and Prime Minister Tony Blair had already announced their goal to end the debt of the world's poorest countries by the end of the year and called upon other nations to follow suit. President Bill Clinton also announced his desire to cancel the debt owed to the United States, and bipartisan support was growing in Congress. The G–7 countries had begun to take some positive steps toward debt relief, and even the World Bank and the IMF were exploring how the crushing debt of the world's most impoverished nations might indeed be relieved, using the savings to reduce poverty in those countries.

It was extraordinary that the biblical principles of Jubilee had become part of the international economic discussion. World Bank president James Wolfensohn and the leaders of the IMF now knew that Leviticus 25 proclaims the biblical Jubilee—a periodic economic redistribution in which slaves are set free, land is returned, and debts are forgiven. Jubilee is a call for a regular "leveling" of things, necessary because of the human tendency toward overaccumulation by some while others lose ground. The Bible doesn't propose any blueprint for an economic system, but rather insists that all human economic arrangements be subject to the demands of God's justice, that great gaps be avoided or rectified, and that the poor are not left behind. The World Bank and the IMF have been discussing the implications of such biblical texts with religious leaders and must cope with an international grassroots campaign that has enlisted a variety of supporters, from U2's Bono to the pope. Did anybody really think that Bill Clinton and Tony Blair would have been calling for debt cancellation without such international pressure from a movement that began with religious imperatives?

While there is much left to do to definitively cancel all that debt (especially the multilateral debts held by international bodies like the World Bank and the IMF), an enormous amount of progress has already been made. Jubilee 2000 was up in lights! Just shows you what a grassroots campaign can do.

On the same day that all the U.S. newspapers carried stories on the third presidential debate in the election of 2000, another front-page article appeared in the *New York Times*.[2] That story started with the headline "Congressional Leadership Agrees to Debt Relief for Poor Nations." The *Times* reported an agreement to fund $435 million for the U.S. part of the Cologne initiative for debt cancellation (where the G7 nations decided to act). The amount would now be included in the final appropriations bills moving through Congress.

Representative Sonny Callahan, an Alabama Republican and the ranking member of the House committee that controls the foreign aid budget, was quoted in the story: "The debt relief issue is now a speeding train. We've got the Pope and every missionary in the world involved in this thing, and they persuaded just about everyone here that this is the noble thing to do." How did debt relief come to be such a "speeding train"? The *Times* reporter noted this "is a sign that street protests and parish activism about the problems of globalization have had an impact on Congress." And President Clinton proclaimed, "It's not often we have a chance to do something that economists tell us is a financial imperative and religious leaders say is a moral imperative."

In an impassioned debate in the House of Representatives, Republican Congressman Spencer Bachus of Alabama, a leader in the struggle for debt reduction, said, "Debt relief is not an end in itself; it is a means to an end. It is not a total solution to poverty, to hunger, to disease; but it is the first step. It is where the journey should begin to free these countries of the burden of debt, the chains of poverty, and the shackles of despair, to enable them to minister to the economic and social needs of their people, of their children."[3]

David Beckmann, president of Bread for the World, tells the story of how Representative Bachus came to this belief. Two laywomen in Alabama got their church's hunger committee involved in Bread for the World. They invited Bachus, their member of Congress, to a dinner at which Beckmann spoke. During the dinner, they urged Bachus to cosponsor Bread's antihunger legislation. He had never before sponsored such legislation, but he called the next evening and said, "I doubt that this will win me many votes, but I don't want to be responsible for even one child going hungry."

At the time the Jubilee 2000 network was getting organized, Bachus had become chair of the international subcommittee of the House Banking Committee, where congressional action on debt relief would have to begin. When the committee held a hearing on poor-country debt, Bachus declared, "If we don't write off some of this debt, poor people in these countries will be suffering for the rest of their lives. And we'll be suffering a lot longer than that." He held up a statement from Pope John Paul II and said, "I haven't read much by Catholics before, but I don't know how any Christian could read what the pope is saying here and not agree that we need to do something about the debt of these countries."[4]

According to Beckmann, "Bachus lobbied his conservative colleagues, including the Republican leadership of the House, for U.S. participation in international debt relief. Bachus says that he had come to see the world differently because of the church people back home who had approached him about Jubilee." Beckmann concludes the story by reminding us that it was the grassroots activism of those two women in Alabama, along with thousands of others across the country, that led to the passage of debt relief.

The diversity of the debt-relief coalition was on display at a White House meeting in 2000 as televangelist Pat Robertson and U2 lead singer Bono appeared at a press briefing with the president to urge passage of the appropriation. Following a meeting with Bono, even archconservative Senator Jesse Helms has gotten on board the train. In a *New York Times* interview, Bono said, "When I met with Senator Jesse Helms, he wept. I talked to him about the biblical origin of the idea of Jubilee Year, the idea that every 49 years, you were supposed to release people from their debt and slaves were supposed to be set free. It's very punk rock for God, but I think it's in Leviticus. He was genuinely moved by the story of the continent of Africa, and he said to me, 'America needs to do more.' I think he felt it as a burden on a spiritual level."[5] The *New York Times* story concluded by calling this "a victory for a coalition of rock stars, religious figures, and charity groups that have made debt forgiveness a moral touchstone for wealthy nations." Jubilee 2000 stands as an example of how a movement of concerned and active people, grounded in moral and religious beliefs, can "change the wind" to accomplish what only a few short years ago seemed impossible.

At the annual meetings of the IMF and World Bank, supporters of debt relief are now always present—both in the suites and in the streets. Partly because of that pressure, the World Bank now argues that economic growth alone is not sufficient to overcome poverty as long as existing political and economic systems favor the rich over the poor. The director of one World Bank study noted, "In order to increase poor people's share of this growth, we're going to have to address inequalities." Jubilee 2000 is now the Jubilee Network, and its supporters still call on the World Bank and IMF to cancel 100 percent of the debt owed to them. Along with street demonstrations, there are also meetings with top officials of the institutions and even public dialogues between critics from a variety of NGOs and financial leaders. Anne Pettifor, former U.K. director of Jubilee 2000, has said, "It's very seldom you get powerful people listening to civil society."

At one World Bank/IMF meeting, Wolfensohn noted the demonstrations in his speech: "Outside these walls, young people are demonstrating against globalization. I believe deeply that many of them are asking legitimate questions, and I embrace the commitment of a new generation to fight poverty. I share their passion and their questioning."[6] That continual questioning must go on because the full debt cancellation needed has yet to be accomplished at the World Bank and the IMF. Congresswoman Maxine Waters, a supporter of the effort, noted especially the role played by Jubilee 2000: "This [debt cancellation] is really a spiritual movement. And I want you to know it could not have been, had it not been for Jubilee 2000. They have been able to organize religious organizations all over the world to come together with nongovernmental organizations and to really move this issue forward. It would not have happened without them."

Since the historic protests in Seattle against unjust globalization, and since the Jubilee 2000 movement, no G-8 meeting of the world's leading industrial nations will escape moral scrutiny and public accountability. It's been only six years since seventy thousand people formed a human chain around the G-8 meeting in Birmingham, England, marking the first major appearance of the Jubilee movement. Much has been accomplished in the years since then. The network of faith communities, nongovernmental organizations, celebrities, and millions of people around the world has brought a moral spotlight to the unsustainable indebtedness and the systemic poverty of the world's poorest countries. And the world has begun to address the problem.

The limited debt relief provided to these poor countries has made some significant changes in living conditions. Jubilee USA reports, for example, that in Uganda, debt savings were used to double elementary school enrollment; in Mozambique, half a million people were vaccinated against deadly diseases. Tanzania used debt savings to eliminate school fees, and 1.5 million children will be able to return to school this year, while in Honduras, the savings went toward access to junior high school for all young people.

Yet major problems remain with the Heavily Indebted Poor Countries (HIPC) Initiative—as the current World Bank program is called. At the end of 2003, after seven years of the initiative, a report by the British New Economics Foundation found that of the forty-two countries considered to be "heavily indebted poor," only eight had passed the "completion point" where debts were cancelled, rather than the twenty-one the original timetable called for. Only nineteen others had qualified for some debt relief.[7]

It also provides too little relief. The HIPC Initiative defines "debt sustainability" as countries whose outstanding debt is 150 percent more than its annual exports. Only those countries are eligible for relief. But that definition is already outdated. As prices drop for exports, the relief will end up being too small to reduce the debt for many countries to even that "sustainable" level of 150 percent of exports. Uganda, for example, has completed the program but still has an "unsustainable" debt projected by the World Bank at 250 percent of exports. Jubilee notes that half of the countries in Africa pay more on debt service than on health care while six thousand people a day are dying from AIDS. Estimates are that stopping the epidemic would cost $7–10 billion annually, while Africa pays $13.5 billion annually on debt service. It is past time to label such policies for what they are: illogical and unconscionable.

Fair Trade

More equitable trade practices are also crucial for seriously reducing poverty. Countries that rely on the export of raw materials to the industrialized world are at the mercy of market forces that lead to further indebtedness. Simply providing more debt relief and aid without changing the rules of the game is not the solution to global poverty. Justice and sustainability are better long-term solutions than benevolence. A worldwide movement of protest against unfair "free trade" administered by the World Trade Organization (WTO) is now under way. The current wave of protests aimed at unjust globalization policies began in Seattle in December 1999.

On a Sunday night, just before the week of scheduled protests that would rock the WTO meeting and the world, I preached in Seattle's St. James Cathedral. From the pulpit, I looked out over the standing-room-only crowd and could feel the electric excitement. We were all gathered for a religious service organized by Jubilee 2000.

Just before I preached, the text was read from Leviticus 25. As I listened to the prophetic Scripture being read, I marveled at how it was being used that night—as a relevant contribution to a public discussion on the rules for global economics!

The official World Trade Organization meetings planned for Seattle were never meant to be public. Quiet and private WTO proceedings of a very elite group had been scheduled to determine the rules of the global economy. But the events of the next several days would shout a message heard around the globe: that discussions about how to conduct international trade would no longer be private conversations. Instead of a small, behind-the-scenes meeting to determine the rules of global trade, a very noisy public debate ensued, asking who makes those rules, who benefits, and who suffers.

The issue in Seattle was not whether there should be global trade. There is, and there will be. The question is, What will the rules of trade be? There are and will be rules, and somebody will make them. But who will profit from those rules, and who will be left behind? The street prophets in Seattle said that the rules should protect the lives of workers, the environment, and human rights.

Until now, the definition of *free trade* preferred by the world's largest corporations have paid little or no attention to workers' rights, environmental threats, and political oppression. Within developed countries, there are rules that companies must abide by. In the United States, for example, you can't legally pour sewage into rivers, foul the air, or produce your goods in exploitative and unsafe sweatshops. There are rules. How do we construct fair rules for a global economy?

The existing conditions of international trade that are spurring the new movement are highlighted by the British organization Christian Aid in its Trade Justice campaign:[8]

Top-Ten Facts

1. International trade is worth $10 million a minute.
2. Poor countries account for only 0.4 per cent of world trade. Since 1980 their share has halved.
3. The United Nations estimates that unfair trade rules deny poor countries $700 billion every year. Less than 0.01 per cent of this could save the sight of 30 million people.
4. Income per person in the poorest countries in Africa has fallen by a quarter in the last 20 years.
5. The three richest people in the world control more wealth than all 600 million in the world's poorest countries.
6. Nearly half the world's population (2.8 billion people) lives on less than US$2 per day.
7. The prices of many poor countries' key exports are at a 150-year low.
8. Global trade is regulated through policies and priorities set by international institutions, including rules made at the World Trade Organization (WTO) and conditions attached to loans provided by the World Bank and International Monetary Fund. These bodies are controlled by governments.

9. At the last full meeting of the WTO, the European Union had 500 negotiators and Haiti had none.
10. After the last round of trade negotiations rich countries estimated that they would gain by $141.8 billion per year and Africa would be $2.6 billion per year worse off.

In the past decade, a series of regional free trade agreements have exacerbated these realities—from NAFTA (North American Free Trade Association) to FTAA (Free Trade Association of the Americas) to CAFTA (Central America Free Trade Association), agreements that benefit the largest corporations and wealthiest countries without adequately protecting workers, human rights, and the environment. Such agreements have helped to fuel the growing anti-globalization movement, which protests these realities.

But as we noted earlier, while it is good to protest, having an alternative is better. And there are alternatives. The anti-globalization movement sums it up with its slogan, "Another world is possible." *Fair* trade or trade justice, rather than mere *free* trade, is the objective. Achieving this requires fundamentally changing the underlying principles of the world trading system so that they also benefit poor countries and people, not just the wealthiest.

One of the keys is who gets to sell what in whose market. Agreements that force poor countries to open their markets to the wealthy countries, while their protectionist policies block exports from developing nations, simply perpetuate and exacerbate the rich-poor gap. And the failure to require policies from corporations that protect labor rights and the environment means that poor workers and farmers are further harmed.

In developing new policies, poor countries must become active participants—able to choose and help shape the policies that will work best to reduce poverty. And poverty reduction, rather than simply profit maximization, must be a primary criterion by which to judge any trade policy. Only with that clear focus will the lives of the majority of people in the poor countries genuinely improve.

Trade policies should be determined with the active involvement of the people who will be affected—poor people themselves, nongovernmental organizations that work with them, and, increasingly, faith-based organizations. Rather than the closed-door meetings in resort hotels that still characterize trade summits today, they should be open debates with high public participation and reporting. Civil society groups must then be involved in monitoring and evaluating trade agreements.

For poverty reduction to succeed, however, the development of democratic governance and "transparency" in poor countries must also continue. Aid and trade that simply line the Swiss bank accounts of already wealthy and oppressive despots in poor nations will change nothing. Endemic corruption in the developing world is also an important cause of poverty, along with the oppressive structures and rules of the global economy that keep some poor and others rich. And, of course, the expansion of fair and just trade improves living conditions, which will also provide an incentive for greater democracy. People who are educated and no longer hungry are also more able to freely participate in their society. And, as I noted in chapter 12, the biblical prophet Micah proclaims that this will make us more secure.

Along with advocacy for fair trade policies, we all have the opportunity to *buy* fair trade products through a growing network of nonprofit organizations that deal directly with producers, eliminating the middlemen and paying a fair, above-market price. Such fair trade products now being marketed in the United States, the United Kingdom, and throughout the developed world include coffee, tea, cocoa, chocolate, clothing, and crafts.

One of the major products in the growing fair trade movement is, of course, coffee. Although a very heavily traded commodity, most coffee farmers living in the poorest countries of the world receive little. Coffee prices in the market fluctuate widely, and there are many steps between the farmer and the supermarket. Fair trade coffee is giving these farmers the alternative to deal directly with organizations that distribute the coffee in the United States and elsewhere. The prices they receive are above the market rate, enabling them to better support their families.

Equal Exchange, an active fair trade coffee distributor in the United States., was founded in 1986 and is now the largest fair-trade-certified coffee company in North America, with seventeen trading partners in ten countries in Latin America, Africa, and Asia. Although not faith-based, Equal Exchange has partnered with national church organizations in a growing "Interfaith Coffee Program" to market coffee for fellowship hours after worship. In 2004, these partnerships included Lutheran World Relief, Catholic

Relief Services, the American Friends Service Committee, the Presbyterian Church (USA), the Unitarian Universalist Service Committee, the United Church of Christ, the United Methodist Committee on Relief, and Brethren Witness.

The organization notes, "Equal Exchange's fair trade practices help build pride, independence and community empowerment for small farmers and their families. A coffee processing plant in El Salvador, community stores in Colombia, the training of doctors and nurses in Mexico, reforestation programs in Costa Rica, new schools in Peru—these are all examples of the initiatives that co-ops have taken in their own communities with the income from fair trade."[9]

Another fair trade coffee distributor, Pura Vida,[10] was founded in the late 1980s by two Christian business students, Chris Dearnley and John Sage. Dearnley and Sage openly seek to root their work in their faith, which they believe calls them to serve those in poverty. The organization is charitably owned, so that all of its resources go to help at-risk children and families in coffee-growing countries. The programs they have helped to fund include computer centers, soup kitchens, Saturday kids' clubs, and soccer teams. Pura Vida is also developing marketing partnerships with other organizations, including Sojourners—we now market our own "SojoBlend" coffee.

Pura Vida's mission statement says, We believe . . .

- In a different approach to business. One driven by good rather than greed. One that sees capitalism as an agent for compassion, faith as an engine for action.
- That the earth itself is our greatest production line and that the better we take care of it, the more it will provide.
- That every business is composed, first and foremost, of human beings and that the fairer they are treated the more fulfilling and productive their lives will be.
- That the most gratifying form of corporate profit is the kind returned to the people and communities that made it possible.
- That if Pura Vida can successfully operate according to this philosophy, then maybe every company in the world can too.

Rigid criteria must be met to receive certification as fair trade coffee. In the U.S., TransFair USA issues a "Fair Trade Certified™" label that guarantees these standards have been met.[11] Those include:

- Paying a fair price so farmers are ensured a living wage, and offering employees opportunities for advancement;
- Buying coffee directly so the proceeds go to the farmers and their communities, rather than middlemen;
- Engaging in environmentally friendly practices, including ecologically sustainable farming;
- Buying from democratic cooperatives or farms with independent democratic unions
- Providing financial and technical assistance, as well as advance credit to help farmers until the coffee crop is harvested.

Celeste Kennel-Shank, a student journalist from Goshen College in Indiana, visited El Salvador in January 2004 as part of a group looking into how fair trade coffee has affected the lives of small farmers. She wrote in *Sojourners* about her conversation with Luis Castillo, a member of the Las Colinas cooperative: "He works every day during the harvest from before dawn until after nightfall, and the reward for his labor is that his seven sons are educated, he said with quiet pride evident in his face. With their increased income, Castillo and his wife, Reina, were able to buy vitamin-enriched milk when their youngest son lost his hair from malnutrition last year. Though his family has no luxuries, Castillo is committed to helping less-advantaged neighbors. After the earthquakes in El Salvador in 2001, pre-harvest financing received from Equal Exchange supported families in the cooperative. Castillo shared his income with indigenous communities living more remotely, to help fend off starvation."[12]

There is also a growing market for handicrafts. Ten Thousand Villages was founded in 1946 as a nonprofit program of the Mennonite Central Committee (MCC), the relief and development agency of Mennonite and Brethren in Christ churches in North America. It now operates over 180 stores across North America staffed by thousands of volunteers. Products from more than thirty countries are sold, including everything from textiles to ceramics to art, toys, and games.[13]

MCC operates under the guidelines of the International Fair Trade Association, which notes that "fair trade is better than aid—it builds a sustainable future on artisans' own abilities. Artisans receive a fair price for their goods, enabling them to improve their quality of life. And, cultural exchange and understanding is facilitated as consumers are told about the people who made the handicraft they bought."[14]

The fair trade consumer movement is much further along in the United Kingdom and other European countries than in the United States, but it is growing steadily here too. In London supermarkets, one can find a whole array of fair trade products now. People would rather buy something they know was traded fairly, where the proceeds go to the producers themselves and where the environment is also protected. That kind of shopping just feels better and concretely contributes to a more just world.

A New Marshall Plan

Definitive debt cancellation, wise aid programs, and trade justice are all crucial for addressing global economic inequality and improving the lives of millions of the poorest of the poor around the world. Debt, aid, and trade are the pillars of global poverty reduction, and they are becoming the moral imperatives of a growing popular movement that has strong support in the faith community.

Diverse Christian groups are now mobilizing around the issues of international debt, aid, and trade. At one such event in London, I preached a sermon on this movement in the historic Wesley Chapel and from the revivalist John Wesley's pulpit. I couldn't escape the feelings of excitement as I wondered if a new movement to overcome global poverty might be rising up, born of spiritual renewal just like the anti-slavery campaign more than two hundred years ago.

After the service, I met again with some of the British government leaders who are most involved in the issues of debt, aid, and trade. America's leading ally in the world today is utterly convinced that terrorism will not be defeated without a new initiative for the economic development of the world's poorest nations. And the United Nations' stated goals to dramatically reduce global poverty are now helping to shape British foreign policy. The United Nations' Millennium Summit, held in the fall of 2000, made a commitment to cut global poverty in half, reduce infant mortality by two-thirds, and make primary education available to all children—all by 2015. These Millennium Development Goals (MDGs)[15] and their 2015 commitments are a part of every discussion of international affairs I have with the U.K. leaders of both church and state.

The MDGs are indeed an important and urgent commitment. Today, some eight hundred million people around the world are malnourished. According to UNICEF and the World Health Organization, 30,500 children die every day in the developing world from hunger and preventable diseases. Almost three billion people, nearly half the world's population, live on less than $2 a day, 1.2 billion of them on less than $1 a day.

In 1970, the United Nations estimated that commitments to significantly reduce global poverty could be met by increasing development aid to poor countries by $50 billion per year, and it set a goal to accomplish this with developed countries spending 0.7 percent of their GNP. They haven't even come close—the current average is between 0.2 and 0.3 percent.

In several speeches over the past few years, Chancellor of the Exchequer Gordon Brown has called for a new "Marshall Plan" of aid to developing countries in order to accomplish the ambitious Millennium Development Goals. In a speech on December 17, 2001, at the National Press Club in Washington, Chancellor Brown recalled how U.S. Secretary of State George Marshall had committed the resources needed to rebuild Europe after World War II, believing that a true victory in that war would require a global fight against "hunger, poverty, desperation, and chaos." Today, said Brown, "there cannot be a solution to the urgent problems of poverty the poorest countries face without a . . . substantial increase in development funds for investment in the very least developed countries. . . . We must move from providing short-term aid just to compensate for poverty to a higher and more sustainable purpose, that of aid as long-term investment to tackle the causes of poverty by promoting growth."

Brown is right—an international war against terrorism that doesn't target global poverty is doomed to failure. The U.S. government's 44 percent increase in the military budget since September 11 (from $306 billion in FY

2001 to $441 billion in FY 2004) will do little to eradicate the conditions of poverty, injustice, and lack of democracy, which breed terrorism.

Clare Short, former British secretary of state for international development, put it this way when I interviewed her in the fall of 2003: "I do think one of the greatest dangers to the future security and safety of the world is the level of poverty and inequality in a world of great riches, of technology, capital, and knowledge, and now with the speed of communications that means we both have the capacity to spread basic development and the basic decencies of life across the globe—that could be easily done, it's a matter of will. It's not a matter of affordability, and it's a grave injustice when people can see each other and some fifth of humanity are living in abject poverty, the sort of conditions that were in Britain at the time of the Industrial Revolution—child labor, illiteracy, the lack of clean water, curable diseases killing people, women dying in childbirth—all that is the condition of a fifth of humanity. With a bit of effort, those people could be given the chance to improve their lives, get their kids to school, have some basic healthcare, be able to work—fairer trade rules so they could export their products, etc. The world wouldn't be equal, but it would start to be decent and there would be no one living in those abject conditions. So, I think this is the biggest moral issue we face."[16]

I've often met with Brown, Short, and several other members of the British Parliament. All of them are convinced that churches and faith-based organizations could play a decisive role in convincing the people and governments in the West of the political and moral imperatives of dramatically reducing global poverty. The British leaders believe that unless the United States and the United Kingdom can be persuaded to lead in this effort, it cannot really succeed, and that the churches must help their governments to act.

Together the churches and the faith-based development agencies must create transatlantic and international alliances aimed at mobilizing our own people and pushing our governments toward effective moral and political leadership in seriously reducing global poverty. The Millennium Development Goals are not at all out of reach, but their achievement requires the creation of a new political will. They are indeed possible, but not without a spiritual engine to drive them forward. That is indeed what the religious community can best provide. Together, we can commit to work to make the possible . . . possible.

The Moral Crisis of HIV/AIDS

Today we must aim our moral energy at the developing world in places like AIDS-infected sub-Saharan Africa and in response to the urgent moral call to dramatically reduce global poverty. Our churches must convince our political leaders that such a moral and political initiative aimed at the root causes of global injustice will enable the war against terrorism to succeed far better than dropping more bombs on more countries. This may not be the kind of faith-based initiative that our political leaders had in mind, but it may be the kind of witness that many churches may now be ready to offer.

There are many signs that the churches may be ready to take just such an initiative. One very powerful sign is how a growing number of church-based organizations and leaders are being drawn to respond to the crisis of HIV/AIDS. After many years of reluctance to engage the issue, many surprising people and groups (including many conservative evangelical leaders and organizations) are joining the fight against AIDS.

Today, nine thousand people will die of HIV/AIDS. And almost all of them will be poor—they simply can't afford the drug treatments that prolong life. World Vision says it well: "For the majority of patients in poor countries, AIDS is a death sentence—not a chronic, manageable condition."[17]

Today, fourteen thousand new people will be infected with the disease, most of them in poor countries. Forty-two million already have it, twenty-two million have already died, and the World Health Organization puts the number of new infections at five million. The world has never seen a public health crisis like this. Given the numbers, one could say it is a crisis of biblical proportions.

U.N. Secretary General Kofi Annan has said prophetically, "For me it's not just statistics. . . . I've seen the human suffering and the pain, and what is even more difficult is when you see somebody lying there dying who knows that there is medication and medicine somewhere else in the world that can save her. But she can't have it, because she is poor and lives in a poor country. Where is our common humanity? How do you explain it to her, that in certain parts of the world AIDS is a disease that can be treated and one can live with and function? But in her particular situation, it's a death sentence."[18] People dying of AIDS, who know that elsewhere medication is saving the lives of others, know that they are dying because they are poor. That, indeed, is a biblical matter.

The only good news is that the churches are changing. We must admit that the churches have been slow to respond. AIDS has carried a sexual stigma for the churches, and we didn't want to deal with it. The early perceptions of the disease were mostly associated with homosexuality—and the church didn't want to deal with that either.

But most victims of HIV/AIDS today are women and children—infected by the promiscuity of men and exacerbated by their poverty. An entire generation of children—thirteen million worldwide—have been orphaned by AIDS and face a bleak future without our immediate support.

We are finally seeing new church leadership and, perhaps most significantly, new evangelical leadership. We can now point to strong and unequivocal statements of commitment from groups like World Vision, World Relief, and the National Association of Evangelicals, along with the Catholic bishops and the mainline Protestant Church World Service. HIV/AIDS is awakening the conscience of the churches.

Some have likened AIDS to a modern-day leprosy—the terrible scourge of Jesus's day. The gospels note that Jesus went out of his way to embrace the lepers who were isolated and abandoned by the society and the religious people of his day. He instructed his followers to heed his example. Today, Christians are starting to follow Jesus, who said, "I was sick and you took care of me." Against a health crisis unlike anything the world has ever seen, the faith community is beginning to raise up a prophetic voice and undertake a new faith-based initiative on HIV/AIDS. To take such an initiative that will help in both the prevention and treatment of AIDS is now quite simply a matter of good faith—and of making good on faith. More than twelve million orphaned children in Africa alone demand our response. There can be no excuses. It is a moral imperative, as well as a political necessity. For all of us, *it is a matter of faith*.

In his State of the Union address in 2003, President Bush outlined a bold role for the United States by pledging $15 billion over five years to fight HIV/AIDS in Africa and the Caribbean. The initiative promises to provide life-prolonging treatment to two million people and to prevent seven million new infections. In addition to this desperately needed new money, the president stated a principle that activists and people of faith have embraced in the fight against HIV/AIDS. "In an age of miraculous medicines," Bush said, "no person should have to hear 'You've got AIDS. We can't help you. Go home and die.'" But that bold plan has yet to be fully funded or implemented.

Adam Taylor of Global Justice explains, "With Bush's plan the devil lies in the details. The political landscape around HIV/AIDS will be defined by how this new commitment is implemented. The advocacy battle must turn to getting the details right. While the details may not generate the same degree of outrage, they will determine how many lives are lost or saved. In the AIDS initiative, Bush exacerbates an ongoing streak of the United States going it alone in addressing global crises. In the first year, only 10 percent—$200 million—of the new money will go through the Global Fund for HIV/AIDS, Malaria, and Tuberculosis. The Global Fund represents a new multilateral mechanism designed to pool money from wealthy nations and deliver it to scientifically sound and fiscally accountable programs in countries most heavily impacted by HIV/AIDS. The fund represents one of our best hopes, because it significantly depoliticizes aid and gives both the donor and recipient country a say in how money is used. A pledge of only $200 million would mean a significant step backward in U.S. leadership toward the fund and could cripple the fund in the future. And the World Health Organization conservatively estimates that a fair U.S. share of AIDS assistance would be at least $3.5 billion a year, while the Bush initiative includes less than $2 billion in its first year."[19]

Taylor is the executive director of a very hopeful network of students and young people in the United States and around the world who are coming together over the AIDS crisis, much as student movements have done in the past around great social issues. He likens their activism to the role of students in the civil rights movement, which was led by groups such as the Student Nonviolent Coordinating Committee, which trained, educated, and mobilized a generation of students to fight for freedom. Taylor and the young activists who make up the Global Justice network make the moral and political connections of HIV/AIDS to global poverty.

Says Taylor, "The longer our nation waits to address the crisis with the urgency and priority it deserves, the more lives will be lost and the more costly it becomes to turn back the course of the epidemic. The war against HIV/AIDS must be wrapped around underlying issues of poverty, inequality, and marginalization. HIV disproportionately impacts 'the least of these' in relation to both color and class. And along with aid, the administration should support full debt cancellation for poor nations in order to free up desperately needed resources for health."[20]

I stood outside the U.S. Treasury Building on World AIDS Day 2004 at a religious vigil that the young people had organized. Adam Taylor said that we stand at a "crossroads" in the fight against HIV/AIDS, and he likened the solutions to the battle against global poverty: "We have the tools of prevention, treatment, and care to stop this deadly epidemic. What is needed most is the personal, societal, and political will. To paraphrase Rabbi Hillel: If not now, when? If not people of faith, who?"

Adam was one of my students at Harvard and is now a leader of a very hopeful new generation of Christian activists. His words at the U.S. Treasury bring us all the way back around to the beginning of this chapter. Success in the fight to overcome global poverty is more a battle for the necessary moral and political will, rather than a problem of resources, information, or technology. It is a battle of the spirit—and therefore a task for the community of faith.

The Millennium Development Goals

In September 2000, at the United Nations headquarters in New York City, a Millennium Summit was held with the largest gathering ever of heads of state. These world leaders agreed on an ambitious agenda for reducing global poverty: the Millennium Development Goals.[21] For each goal, one or more targets were set, most for 2015, using 1990 as a benchmark:

1. Eradicate extreme poverty and hunger

Target for 2015: Halve the proportion of people living on less than a dollar a day and those who suffer from hunger. More than a billion people still live on less than US$1 a day: sub-Saharan Africa, Latin America and the Caribbean, and parts of Europe and Central Asia are falling short of the poverty target.

2. Achieve universal primary education

Target for 2015: Ensure that all boys and girls complete primary school. As many as 113 million children do not attend school, but the target is within reach. India, for example, should have 95 percent of its children in school by 2005.

3. Promote gender equality and empower women

Targets for 2005 and 2015: Eliminate gender disparities in primary and secondary education preferably by 2005, and at all levels by 2015. Two-thirds of illiterates are women, and the rate of employment among women is two-thirds that of men. The proportion of seats in parliaments held by women is increasing, reaching about one third in Argentina, Mozambique and South Africa.

4. Reduce child mortality

Target for 2015: Reduce by two thirds the mortality rate among children under five. Every year nearly 11 million young children die before their fifth birthday, mainly from preventable illnesses, but that number is down from 15 million in 1980.

5. Improve maternal health

Target for 2015: Reduce by three-quarters the ratio of women dying in childbirth. In the developing world, the risk of dying in childbirth is one in 48, but virtually all countries now have safe motherhood programs.

6. Combat HIV/AIDS, malaria and other diseases

Target for 2015: Halt and begin to reverse the spread of HIV/AIDS and the incidence of malaria and other major diseases. Forty million people are living with HIV, including five million newly infected in 2001. Countries like Brazil, Senegal, Thailand and Uganda have shown that the spread of HIV can be stemmed.

7. Ensure environmental sustainability

Targets:

• *By 2015, reduce by half the proportion of people without access to safe drinking water.*

• *By 2020 achieve significant improvement in the lives of at least 100 million slum dwellers.*

More than one billion people lack access to safe drinking water and more than two billion lack sanitation. During the 1990s, however, nearly one billion people gained access to safe water and the same number to sanitation.

8. Develop a global partnership for development

Targets:

• *Develop further an open trading and financial system that includes a commitment to good governance, development and poverty reduction—nationally and internationally*

• *Address the least developed countries' special needs, and the special needs of landlocked and small island developing States*

• *Deal comprehensively with developing countries' debt problems*

• *Develop decent and productive work for youth*

• *In cooperation with pharmaceutical companies, provide access to affordable essential drugs in developing countries*

• *In cooperation with the private sector, make available the benefits of new technologies—especially information and communications technologies.*

Indicators:

Many developing countries spend more on debt service than on social services. New aid commitments made in the first half of 2002 could mean an additional $12 billion per year by 2006.

UNDP, in collaboration with national governments, is coordinating reporting by countries on progress towards the UN Millennium Development Goals. The framework for reporting includes eight goals—based on the UN Millennium Declaration. For each goal there is one or more specific targets, along with specific social, economic and environmental indicators used to track progress toward the goals.

The eight goals represent a partnership between the developed countries and the developing countries determined, as the Millennium Declaration states, "to create an environment—at the national and global levels alike—that is conducive to development and the elimination of poverty."

PART V

Spiritual Values and Social Issues

When Did Jesus Become a Selective Moralist?

CHAPTER 18

A Consistent Ethic of Life

Abortion and Capital Punishment

THE MEETING ROOM was packed to overflowing with students sitting on the floor, standing along the back wall, and spilling out the doorways to the corridor. The topic was social justice, and I was about to engage a new generation of Christian activists. Our discussion ranged from growing economic inequality in the United States, to the failure of the American war in Iraq, to the world's pandemic of HIV/AIDS, to the most effective means for global poverty reduction, to protection of the environment, to . . . abortion. This was the University of Notre Dame, and these were young *Catholic* student activists.

Political liberals generally fail to comprehend how deep and fundamental the conviction on "the sacredness of human life" is for millions of Christians, especially Catholics and evangelicals, in forming their view of abortion. They include those who are quite committed and even radical on other issues of justice and peace. Left-wing political correctness, which now includes a rigid litmus test of being pro-choice on abortion, really breaks down here. And the conventional liberal political wisdom that conservatives on abortion are conservative on everything else too is just plain wrong. Christians who are economic populists, peacemaking internationalists, and committed feminists can also be pro-life on the issue of abortion, especially if they are also Catholics or evangelicals. The roots of this conviction are deeply biblical, perhaps best expressed in Psalm 139:13: "For it was you who formed my inward parts; you knit me together in my mother's womb." For many, it is also consistent with a commitment to nonviolence as a gospel way of life.

And there are literally millions of votes at stake in this liberal miscalculation. Virtually everywhere I go, especially during an election year, I encounter many moderate and progressive Christians who find it painfully difficult or even impossible to vote "Democrat" given the party's highly ideological and very rigid stance on this critical moral issue, a stance that they regard as "pro-abortion." Except for this major and, in some cases, insurmountable obstacle, these voters would be casting Democratic ballots.

Ironically, the Republicans, who actively and successfully court the votes of many Christians on abortion, are much more ecumenical in their own toleration of a variety of views within their own party. For example, fellow Republicans have not enforced anti-abortion orthodoxies on their rising new star, California Governor Arnold Schwarzenegger, whose "pro-choice" views seem not to be a problem (not even, apparently, for the religious Right, whose high-profile leaders have been virtually silent about Schwarzenegger's pro-choice abortion stance and even his admitted history of "womanizing"). Indeed, there is now a long list of pro-choice Republicans who appeal to the more culturally liberal but affluent suburbanites whose support the party seems to regard as crucial to its success. Former New York mayor Rudolph Giuliani is another prominent pro-choice Republican with a bright future in the party. The Republican Party does indeed take a strong anti-abortion stance in its party platforms but then allows for a wide variety of opinions among its party members based on either conscience or pragmatic political calculations with regional demographic differences and targeted constituencies in mind.

But to be a pro-life Democrat is to be a very lonely political creature in America. Former Pennsylvania governor Robert Casey, a Catholic, was denied the opportunity to speak at the 1992 and 1996 Democratic conventions, in part because of his pro-life views. It didn't matter that Casey was very progressive on both economic and foreign policy questions and was a particularly outspoken supporter of women's rights; he just didn't have the right position on abortion. Former Ohio congressman and evangelical Christian Tony Hall experienced similar discrimination as a pro-life Democrat, despite being perhaps the most courageous congressional champion on issues of hunger and poverty. Jesse Jackson was virtually forced to change his pro-life views, which were typical for a black church minister, when he decided to run for the presidency, as have many other Democrats as they seek higher offices. Nobody gets to speak from the Democratic Convention rostrum whose abortion views don't toe the party line. And the Democratic National Committee refuses even to allow a link from their website to the principal pro-life Democratic group, which now includes many officeholders. High-profile pro-life Democrats, like the new Senate minority leader, Harry Reid, and former House whip, David Bonier, provide some hope for a broader dialog.

As I told the Notre Dame students, the Democrats are being quite rigid on the issue of abortion. On pragmatic grounds alone, not to mention the issue's importance as a matter of conscience for many Christians and others, the Democratic Party could take a much more respectful and even dialogic approach. Democrats, like Republicans, could still take a strong party stance (their official position being pro-choice) yet offer some space for different positions along a spectrum of practical options. Such a respect of conscience on abortion and a less dismissive approach to conscientious dissenters to Democratic orthodoxy would allow many pro-life and progressive Christians the "permission" they need to vote Democratic. Again, there are millions of votes at stake here.

But if the Democrats were really smart, they would do something more. The Democrats could affirm that they are still the pro-choice party but then also say what most Americans instinctively believe: that the abortion rate in America is much too high for a good and healthy society that respects both women and children. They could make a serious public commitment to actually do something about significantly reducing the abortion rate. Abortion is historically used as a symbolic issue in campaigns and then forgotten when the election is over. From year to year, the abortion rate doesn't change much, even when it has been a serious campaign issue. Republicans literally win elections on the basis of their anti-abortion position and then proceed to ignore the issue (and the nation's highest abortion rate in the world) by doing nothing to reduce the number of abortions.

The Democrats could vow to change that symbolic debate and political inaction by uniting both pro-choice and pro-life constituencies around goals that could actually become the basis for some new common ground—that is, really targeting the problems of teen pregnancy and adoption reform, which are so critical to reducing abortion, while offering real support for women, especially low-income women, at greater risk for unwanted pregnancies. Providing meaningful alternatives to women caught in difficult decisions about unexpected pregnancies could also be a common project. They could follow the lead of a new alliance of both pro-choice and pro-life women who call themselves the Common Ground Network and work together to find real solutions to the abortion problem, rather than just engage in endless debates. Pro-choice feminist leaders like Naomi Wolfe are taking the "moral issue" of abortion seriously, while female pro-life leaders often take women's rights and responsibilities more seriously than do many of their conservative male colleagues.

Indeed, the rhetoric of the Democratic Party on abortion would seem to support such a plan of action. The 2000 Democratic Party Platform says, "Our goal is to make abortion less necessary and more rare. . . . We must continue to support efforts to reduce unintended pregnancies, and we call upon all Americans to take personal responsibility to meet this important goal." Indeed, candidate Bill Clinton ran on the promise to make abortion "safe, legal, and rare" and then did nothing more than to continue to keep abortion legal. The 2000 platform even said that the party recognizes "different views on issues of personal conscience like abortion and capital punishment. We view this diversity of views as a source of strength, not as signs of weakness." These words were removed from the 2004 platform. Restoring them would at least provide a basis for a better practice. Just providing a website link to Democrats for Life, allowing pro-life Democrats to speak at party conventions, and, most important, taking a proactive initiative to dramatically reduce abortions would be a real turnaround. Given the bitter partisan division on the issue of abortion, it may be that the Democrats are the only ones who could initiate a common project to make abortion truly "rare" in America.

But underneath the strong convictions felt by the Notre Dame students on abortion is something deeper than politics. The most thoughtful ones speak of "a consistent ethic of life," which derives from the heart of Catholic social teaching. It was Chicago Cardinal Joseph Bernardin who coined the phrase "a seamless garment of life," which clearly links the "life issues" of abortion, euthanasia, capital punishment, nuclear weapons, poverty, and racism as critical components of a consistent ethic of life.[1] The Catholic Bishops themselves teach against single-issue voting, which focuses on only one issue, like abortion, to the neglect of all the rest. In their 2004 statement on faithful citizenship, the U.S Conference of Catholic Bishops noted, "We hope that voters will examine the position of candidates on the full range of issues, as well as on their personal integrity, philosophy, and performance. We are convinced that a consistent ethic of life should be the moral framework from which to address issues in the political arena."[2]

One young student at Notre Dame passionately reminded the group that a legal practice that kills four thousand unborn children every day is an urgent moral imperative. But she was then reminded that nine thousand people each day now die of AIDS, thirty thousand children perish every day because of hunger and diseases mostly due to poverty, and as many as half a million are lost each year in international conflicts and wars. All agreed that a more consistent ethic of human life is sorely needed.

The tragedy is that in America today, one can't vote for a consistent ethic of life. Republicans stress some of the life issues, Democrats some of the others, while both violate the seamless garment of life on several vital matters. But the consistent life ethic still serves as an invaluable plumb line by which to evaluate all political candidates and parties.

The other most controversial "life issue" is, of course, the death penalty. On June 11, 2001, Timothy McVeigh was executed for the bombing of the Oklahoma City Murrah Federal Building, which killed 168 people, including nineteen children. There was overwhelming public approval: Gallup polling showed 80 percent support. In fact, Gallup found 23 percent of those who supported the execution were generally opposed to the death penalty. But McVeigh was an aberration—a confessed, unremorseful mass murderer. And, unlike many, he had competent legal counsel every step of the way.

I was working late one night when I heard the news that McVeigh had instructed his lawyers to discontinue his appeals. He would be executed within days, and the reporter jumped right to how McVeigh would spend his last four nights on earth. The overwhelming feeling I had was deep sadness. I was first sad, yet again, for all those who lost their loved ones in the Oklahoma City bombing. Some of those crime victims were expressing gratitude for the upcoming execution of McVeigh, and others felt the way that Bud Welch did. "To me the death penalty is vengeance, and vengeance doesn't really help anyone in the healing process. Of course, our first reaction is to strike back. But if we permit ourselves to think through our feelings, we might get to a different place. I was taught that even the souls of dastardly criminals should be saved. I think it is necessary, even for the soul of Timothy McVeigh. I think my daughter's position on this would be the same as mine." So said Welch in a 1997 *Time* article; his daughter, Julie-Marie, was killed in the bombing.[3] Welch is on the board of directors of Crime Victims for a Just Society.

Like the rest of the country, the victims' families were divided on the question of capital punishment. The *Washington Post* published a front-page story on the contradictory reactions of families.[4] Some couldn't wait to see McVeigh die; others had become eloquent opponents of the death penalty. One of the most poignant stories was that of Marsha and Tom Kight. Marsha's twenty-three-year-old daughter was killed in the bombing, yet Mrs. Kight was not interested in watching McVeigh die. "Viewing the execution does not seem mentally healthy to me," she said. "It won't bring my daughter back." But Mr. Kight has chosen to witness the execution. "I hope it will give me peace knowing he will be eliminated."

But this execution was not likely to provide any of them with what they really need to heal the pain and loss they feel. Honestly, nor will time or anything else we can really offer. Even in the loving arms of a God who also grieves such terrible human losses, the pain never really does go away.

I was also sad for the spectacle that this public execution promised to become. Some fifteen hundred media people applied for press credentials, T-shirt and button sellers applied for vendor permits, and thousands of pro- and anti–death penalty advocates planned demonstrations. I was also sad that McVeigh would get his wish: to die with the hope of becoming a martyr and spreading his hateful message. And I was sad that a child of God, created with such possibilities, became such a hateful and remorseless killer. I was sad for Timothy McVeigh's immortal soul.

Attorney General Ashcroft made the decision to have the execution transmitted via closed-circuit TV to a federal facility in Oklahoma for family members who wished to view the execution. Almost three hundred survivors and family members (of nearly three thousand) had expressed a desire to watch, and the official witness room at the federal prison in Indiana, where the execution was carried out, only held ten. In many ways, the whole nation was watching. It was on all of our minds, no matter what our views on capital punishment.

After a visit to Oklahoma City, I wrote of the emotional impact of the memorial field of empty chairs symbolizing the victims of the bombing. Some were little chairs that reminded us of the children who were killed in this unspeakable violence. On the first anniversary of the bombing, a simple service was held at the memorial. The families and friends of those who had died in the blast stood silent for 168 seconds, one for every person who was killed. The service ended with the ringing of church bells and the song "Let There Be Peace on Earth." At the same time, the final preparations were being made in Terre Haute, Indiana, for another death, Timothy McVeigh's.

I am against the death penalty in principle. We simply should not kill to show we are against killing. It's also easy to make a fatal mistake, as alarming DNA testing has demonstrated. The death penalty is clearly biased against the poor, who cannot afford adequate legal representation, and is outrageously disproportionate along racial lines. Few white-collar killers sit on death row, and fewer are ever executed. And there is no real evidence that it deters murder; it just satisfies revenge.

But that's a human impulse that is easy to understand in the wake of a crime as heinous as the bombing in Oklahoma City. If anybody ever deserved the death penalty, it is McVeigh. Like many Americans, I was angry with him. I reacted like a father when he described the children in the daycare center he murdered as "collateral damage" in his political war. He's not just remorseless and cold, as many have accused; he's proud of what he did. But I still didn't think we should kill him.

McVeigh was a mass murderer, but he was no longer a threat in prison. McVeigh wanted to be executed. Why give him what he wanted? He waived all appeals in order to speed the process. In his mind, he hoped to become a martyr to a small but dangerous group of twisted souls like himself. Why stoop to his level and kill him as he so ruthlessly killed others? Instead of giving McVeigh the dramatic end he desired, why didn't we really punish him for what he had done?

In response to a column I wrote on McVeigh, some criticized me as too harsh when I said he should be sentenced to a life of solitary confinement and hard labor, that he be allowed no public voice, and that his life should end in obscurity instead of the celebrity he so craved. Maybe I *was* angry that day and thinking about my two-and-a-half-year-old son who also attended a preschool at the time. But I believe it is incumbent upon opponents of the death penalty to offer alternatives commensurate with the crime. To be against capital punishment does not require us to be against punishment for such a heinous crime. It was not only a moral contradiction to kill McVeigh for his killings; it was also not punishment enough.

But an assessment of the overall state of the death penalty after McVeigh shows that the tide may be turning. A spring 2003 Gallup poll showed 74 percent favoring the death penalty and only 24 percent opposing it. Yet when the question is asked, "If you could choose between the following two approaches, which do you think is the better penalty for murder: the death penalty or life imprisonment with absolutely no possibility of parole?" only 53 percent favor death while 44 percent favor life without parole.[5]

Supreme Court Justice Sandra Day O'Connor spoke to the Minnesota Women Lawyers organization in July 2001, shortly after McVeigh's execution. What made headlines was her statement that "the system may well be allowing innocent defendants to be executed."[6] Her remarks were significant since she has supported the death penalty and has been the key vote in several cases. Justice O'Connor noted that ninety death-row inmates have been exonerated since 1973.

According to news stories, Justice O'Connor emphasized the importance of effective counsel, noting that in Texas, for example, defendants with appointed counsel are 28 percent more likely to be convicted than those who can hire their own attorneys and are 44 percent more likely to receive death if convicted. "Perhaps it's time to look at minimum standards for appointed counsel in death cases," she said. Timothy McVeigh's was the most highly publicized death penalty case in memory, and the FBI forgot to turn over three thousand pages of documents to McVeigh's counsel. Many wondered what mistakes are routinely made with far less famous death row defendants, especially those without adequate counsel.

And after the exoneration of more than a dozen death-row inmates over the years in Illinois (through DNA testing), Republican governor George Ryan declared an indefinite moratorium on executions in that state. Then in January 2003, as his last act in office, Ryan commuted the sentences of 167 prisoners who had been sentenced to death. In his statement, Ryan said, "Our capital system is haunted by the demon of error: error in determining guilt and error in determining who among the guilty deserves to die. What effect was race having? What effect was poverty having? Because of all these reasons, today I am commuting the sentences of all death row inmates."[7]

The use of the death penalty in the United States has attracted international attention. During President Bush's first trip to Europe in June 2002, anti–death penalty demonstrations were held in several European cities. The International Court of Justice has ruled against the United States in several cases involving the execution of foreign nationals, saying that the executions violated international law because the United States had not informed consular officials of the arrest and conviction of their citizens.

Declining public support, growing international outrage, and deeper legal and political questioning may all add up to more serious and sustained efforts to end the death penalty in the United States. President Bush's strong advocacy of capital punishment and record number of executions in Texas may actually be helping to focus the issue.

Only God can ultimately judge a person's soul, but it is the right and the responsibility of society to punish crime and not just rehabilitate the criminal—especially when the crime is as horribly violent as McVeigh's. When someone takes other lives so deliberately, they should be deprived of any normal life themselves. The only adequate alternative to the death penalty is life without parole. If a convicted murderer repents, as some have suggested McVeigh might have some day, it would be an occasion to celebrate the grace of God, but not to free the killer. The murderer's penance would then be to minister to other prisoners on the inside. Support for the death penalty in the United States is beginning to decline (even though McVeigh was an exception), and moratoriums on capital punishment are emerging in some states. That will only continue if death penalty opponents are willing to take "murder I" just as seriously as those who support capital punishment.

Several years ago, I joined with a group of religious, political, and civil rights leaders in signing a letter to President Clinton urging him to issue a moratorium on federal executions. The letter, issued by the Citizens for a Moratorium on Federal Executions, noted a Justice Department study in which the attorney general acknowledged that "minorities are over-represented in the federal death penalty system." The letter went on to say, "Unless you take action, executions will begin at a time when your own Attorney General has expressed concern about racial and other disparities. . . . Such a result would be an intolerable affront to the goals of justice and equality for which you have worked during your Presidency." The letter concluded, "We are asking only that you prevent an unconscionable event in American history—executing individuals while the government is still determining whether gross unfairness has led to their death sentences."[8]

The U.S. Catholic Bishops then took the next step by urging President Clinton to commute all federal death sentences. Bishop Joseph Fiorenza noted that he had signed the letter urging a moratorium. He continued, "Today, I write on behalf of the country's Catholic bishops to reiterate that call and urge something more fundamental [than a moratorium]. In this Jubilee Year, we ask that you commute the sentences of all 31 people awaiting executions in federal prisons."[9] President Clinton announced that he had decided to stay one federal execution for six months but pointedly emphasized that he was not commuting the sentence or halting all federal executions.

According to Bureau of Justice statistics, at the end of 2003, there were 3,374 people on death rows in thirty-seven states.[10] And the killing continues. In the last four years, nearly three hundred people have been executed.

As Bishop Fiorenza said, "Despite their horrible crimes, the men and women on our nation's death rows are often themselves poor and forgotten. . . . We believe that we have other means to keep society safe from murderers—means that demonstrate a respect for life and ensure that innocent people will never be put to death."

It's called a consistent ethic of life, and it should not be selective. It is selective morality, imposed in partisan political ways, that has so divided us politically. How can our affirmation of "a culture of life," as Pope John Paul has called for, serve to bring us together across ideological and political battle lines? And even more important, how can the precious gift of life be protected and preserved? We are all needed for that battle.

CHAPTER 19

Truth Telling About Race

America's Original Sin

TELLING THE TRUTH about race in America has always been a difficult thing. I remember my early questions as a teenager in Detroit about why we lived the way we did in "white Detroit" when life in "black Detroit" seemed so different. I had started to read the newspapers and listen to the news on radio and television, and I kept hearing about people without jobs, enough to eat, or decent places to live—all in black Detroit—about crime and lots of black men in jail. So a fourteen-year-old innocently asked why. The answers I got are like the answers I've heard ever since: "You're too young to ask those questions." "When you get older, you will understand." "We don't know why it's that way either, but it always has been." The only honest answer I ever received was from a church elder who told me, "Son, if you keep on asking those questions, you're going to get into a lot of trouble." I did, and his prediction proved to be true.

My questions eventually took me into the inner city, to jobs working alongside young black coworkers, and, most critically, to the black churches. There I learned about "two nations, separate and unequal," as the famous Kerner Commission concluded after looking into the causes of the urban riots that tore cities apart in the hot summer of 1967, most devastating, in my hometown of Detroit.

Years later, *Sojourners* magazine put out a study guide on racism that was used in communities across the country, in weekly sessions often cosponsored by black and white churches. I wrote the opening article in the study guide, which we entitled "America's Original Sin." The piece began, "The United States of America was established as a white society, founded upon the genocide of another race and then the enslavement of yet another."

We named racism theologically, as America's "original sin." I've never received so many letters in response to an article, or especially to an opening line. Some attacked the sentence as "outrageous," while others thanked me for being so "courageous." I was puzzled. The controversial line was neither outrageous nor courageous, but rather was simply a statement of historical fact, just a description of how the United States of America was established.

America's original sin has affected most everything about our nation's life ever since. Slavery and the subsequent discrimination against black people in America is of such a magnitude of injustice that one would think national repentance and reparations would be called for. But neither has ever come. Even "apologizing" for this great sin has proved to be quite controversial.

In the spring of 2000, Representative Tony Hall of Ohio called on Congress to apologize for slavery.[1] That's right, apologize. Say we were wrong; say we're sorry. On a sunny day in June, I stood with Tony Hall and Representative John Lewis of Georgia, along with a handful of other clergy, near the east steps of the U.S. Capitol as Hall made the historic announcement. Tony Hall is known for asking not what's in it for him, but rather what's the right thing to do. And he discovered, to his surprise, that the U.S. government had never apologized for slavery. In recent years, Hall pointed out, there have been many public apologies. Congress has apologized to Japanese Americans for imprisoning them during the Second World War and to native Hawaiians for helping overthrow the Kingdom of Hawaii a century ago.

Other countries have also apologized. Britain's prime minister apologized to the Irish people for failing to relieve the nineteenth-century potato famine, East Germany apologized for the Holocaust, and Japan's emperor apologized for atrocities in Korea. And in recent years, many religious leaders and bodies have apologized for slavery—the pope, the Southern Baptists, the Methodists, and the National Association of Evangelicals. The state of

Florida has both apologized and paid reparations to victims of a race riot in the city of Rosewood.

Hall said he was amazed when his research in the Library of Congress revealed that the U.S. government had never apologized for enslaving millions of people, tearing them from their families, killing, torturing, beating, and brutalizing fellow human beings. He said that an apology provides the foundation for a new beginning and that such an apology should not be "wasted." Restoration, even some kind of restitution, and genuine healing should follow, he said.

When Hall introduced a similar bill into Congress three years before, he was "taken aback" by the hate mail he got. But he persisted. "Apologizing is humbling," he said, "but the United States should be big enough to admit its mistakes." He honestly admitted that the resolution will not "fix" the lingering legacy of slavery but that reconciliation always begins with an apology, and this one could be the start of some still-needed healing between the people of this country. The white member of Congress from Ohio said, "This apology will not solve all of the problems, but it will begin new progress on issues that still divide Americans. It's never too late to admit a wrong and to ask forgiveness. In giving those our nation has wronged the dignity of this honest admission, we might all enjoy some measure of healing. And it will set the right example for our children." The resolution again languished in committee and eventually died. Once again, the U.S. government couldn't even say it was sorry for slavery and discrimination against its black citizens.

We teach our children to say they're sorry when they've done something wrong. What do we think black children, and white children, make of our failure to even say we're sorry for slavery and racism, let alone to make any restitution? Hall said we should set the right example for our children.

Yes, our children. I am very conscious of how the discovery of racism in my city, nation, church, and community of friends and family deeply affected my own childhood and teenage years. Now I am a father, raising two young boys in a diverse urban neighborhood in Washington, D.C. Watching my own children's reactions to their discovery of America's great racial offense has been instructive.

When I was a little boy, I would go downstairs to the basement and listen to old Broadway musicals such as *Oklahoma!*, *South Pacific*, *My Fair Lady*, and the movie *Around the World in Eighty Days* on my parent's hi-fi. To this day, I could sing all those songs for you, but you wouldn't want me to do that.

South Pacific was one of my favorites, and there was one song I liked but didn't really understand. It was about a young American Navy sailor and a beautiful Pacific Island girl who fell in love but whose relationship wasn't accepted by the people around them. The song was titled "You've Got to Be Taught." I remember the lyrics: "You've got to be taught to be afraid of people whose eyes are oddly made, or people whose skin is a different shade, you've got to be carefully taught. You've got to be taught before it's too late, before you are six or seven or eight, to hate all the people your relatives hate, you've got to be carefully taught."[2]

The song was about racism, but I didn't understand then what that was.

I see that same lack of understanding in my sons. Even before my son Luke was two years old, he loved to go for walks with his mom and dad: to the park, to the subway, to a restaurant, to church, or especially to a party. And what he always wanted to do was meet people. Luke greeted nearly everybody he met or passed. There are still many spots around town where he's made great friends who love to see him coming through the door.

"Hi, hi, hi," he would say to everybody. "Hi man, hi boys, hi lady," he exclaimed with his winning smile, and if he'd heard somebody's name before, he would be almost sure to remember it. Walking to the subway, Joy and I noticed how Luke was saying "Hi" to everyone—black, white, Latino, Asian, male, female, young, or old. It was obvious that Luke saw no differences that mattered to him. Everybody got his happy greeting. His only preference was for people who talked back to him, and most did.

It made me sad that, in ways both subtle and direct, his society would start teaching him about racism. Of course, we hoped that we could teach him too—about the need to overcome it. But it is a marvelous thing to watch a little white boy walk through streets that pulse with America's diversity and want to meet everybody. I've watched my son's technique carefull, because it seems easily to break down racial barriers, sullen looks, and angry stares. I've paid close attention because, after all, you've got to be carefully taught.

Luke is in kindergarten now in a Washington, D.C. public school—a good school and quite diverse. Joy and I were especially pleased when his teacher and class made so much of black history month in February. Luke is now getting the same things at school that we teach him at home—books about Martin Luther King Jr. and the civil rights movement, for example. At a big kindergarten/first grade performance, each child portrayed a major figure from black history. My five-year-old stood up and proudly proclaimed, "My name is Langston Hughes. I am a poet. Among my works are the *Weary Blues*. I am an important member of the Harlem Renaissance." His teacher took Joy aside one day after school and told her that Luke had learned all the other kids' parts too and would whisper helpful prompts if they forgot their lines.

Every night my son came home with lots of questions about black people and white people, about the Montgomery bus boycott, or about Martin Luther King Jr. and his family. Luke's basic question is the same one I asked a long time ago. "Why?" He just can't figure out why people would do such terrible things to each other. His class learned a song about Rosa Parks that he likes to sing joyously, "If you miss me at the back of the bus, and you can't find me nowhere, come on up to the front of the bus; I'll be riding up there!" Then he starts to dance and sing at the top of his voice, "I'll be riding up there, I'll be riding up there; come on up to the front of the bus, I'll be riding up there!" "Daddy," he told me one night, "I wouldn't drink out of those drinking fountains either if my friends couldn't." The sheer injustice of it all bothers him. At the end of black history month, Luke announced to his parents, "I am going to be just like Dr. Martin Luther King Jr., except I will have a different name and a different skin." Of course, nothing could have pleased his mom and dad more.

But Luke's fundamental question remains "Why?" It's always been the right question. Truth telling about race requires that we understand and confront the purposes that racism plays in American life, even today. And in America, racism has always had a purpose. Slavery, of course, carried an enormous economic profit, as did subsequent racial discrimination—first free and then cheap labor for building a new nation and a growing economy. But the purposes went even deeper. Race has always been a dividing distraction in American life. By offering white people an "other" to blame and resent for their own problems, attention to the real problems of wealth and inequality in America have always been distracted and diverted.

For all the attention given to Howard Dean's presidential campaign of 2004, the media completely missed one of Dean's most significant speeches and, in fact, one of the most important speeches ever given about race in America by a white politician. The fact that Dean's speech really made racial-political history was almost universally overlooked by the mainstream press. But the *Black Commentator,* a publication for black journalists, said, "Howard Dean's December 7 [2003] speech is the most important statement on race in American politics by a mainstream white politician in nearly 40 years. Nothing remotely comparable has been said by anyone who might become or who has been President of the United States since Lyndon Johnson's June 4, 1965 affirmative action address to the graduating class at Howard University."[3]

The article went on, "For four decades, the primary political project of the Republican Party has been to transform itself into the White Man's Party. Not only in the Deep South, but also nationally, the GOP seeks to secure a majority popular base for corporate governance through coded appeals to white racism. The success of this GOP project has been the central fact of American politics for two generations—reaching its fullest expression in the Bush presidency. Yet a corporate covenant with both political parties has prohibited the mere mention of America's core contemporary political reality: the constant, routine mobilization of white voters through the imagery and language of race. Last Sunday, Howard Dean broke that covenant."

In that speech, Dean said, "In 1968, Richard Nixon won the White House. He did it in a shameful way—by dividing Americans against one another, stirring up racial prejudices and bringing out the worst in people. They called it the 'Southern Strategy,' and the Republicans have been using it ever since. Nixon pioneered it, and Ronald Reagan perfected it, using phrases like 'racial quotas' and 'welfare queens' to convince white Americans that minorities were to blame for all of America's problems. The Republican Party would never win elections if they came out and said their core agenda was about selling America piece by piece to their campaign contributors and making sure that wealth and power is concentrated in the hands of a few. To distract people from their real agenda, they run elections based on race, dividing us, instead of uniting us."

Dean's speech in Columbia, South Carolina, did what truth telling about race in America has always done, and that is to draw the straight line that connects racism to poverty. The *Black Commentator* quoted from Lyndon Johnson's Howard University speech, which they compared to Dean's. Johnson said, "Negro poverty is not white poverty. Many of its causes and many of its cures are the same. But there are differences—deep, corrosive, obstinate differences—radiating painful roots into the community, and into the family, and the nature of the individual. These differences are not racial differences. They are solely and simply the consequence of ancient brutality, past injustice, and present prejudice. They are anguishing to observe. For the Negro they are a constant reminder of oppression." Most black Americans hear those words as the truth.

The black journalists called Johnson's speech "a defining moment" and likened it to Dean's South Carolina remarks. "Not since Lyndon Johnson vowed to harness the power of the federal government to redress the historical grievances of Black America has a potential or sitting President made such a clear case against racism as a political and economic instrument—and even Johnson failed to indict corporate interests, or anyone in particular, for wielding race as a political weapon. Howard Dean points the finger straight at executive boardrooms, and directly implicates members of his own party in the coded conspiracy."

In South Carolina, Dean said, "Every time a politician uses the word 'quota,' it's because he'd rather not talk about the real reasons that we've lost almost 3 million jobs. Every time a politician complains about affirmative action in our universities, it's because he'd rather not talk about the real problems with education in America—like the fact that here in South Carolina, only 15 percent of African Americans have a post–high school degree."

Johnson pursued a vigorous program of "affirmative action" to achieve black progress and to overcome past injustice. More recently, Bill Clinton, though more comfortable with black people and culture than any previous American president, was himself more comfortable with purposeless "national conversations" on race that never went anywhere rather than with a substantial strategy to redress historic grievances.

Dean was presenting whites (especially working-class whites) in the South and everywhere else with the real choice they have always had: to vote for their own economic self-interest or to vote for their bosses' interests while blaming black people for their problems. In American history, "populist" and "progressive" movements only became possible when middle- and low-income whites make that choice, not only to vote their own interests but to actually ally with working-class blacks against the wealthy and powerful interests that were aligned against them both. The *Black Commentator* quoted Reverend Jesse Jackson as saying, "The big fight in this state [South Carolina] should be trade policy and the Wal-Martization of our economy. . . . The challenge is to get South Carolina to vote its economic hopes and not its racial fears." Most low-income Americans are white, and "they work every day. They work at Wal-Mart without insurance. They work at fast-food places. They work at hospitals where no job is beneath them, where they don't have insurance, so they can't afford to lay in the beds they make. . . . The challenge for South Carolina is to move from racial battleground to economic common ground to moral high ground."

But while racism has always had an economic motive, its consequences touch every area of our lives. We have been fractured by race, divided into communities that, although living in the same cities, inhabit different countries. Finally, it is our souls that have been fractured by race. Another Republican president, Abraham Lincoln, fought for the freedom of black Americans and famously said, "A house divided against itself cannot stand." But today's Republicans win elections through strategies of racial division. Yet the Democrats, the party of Johnson and Kennedy, have long taken black voters for granted. So black Americans have few political choices today. One party uses them as a wedge issue to draw working-class whites, while the other depends on their support without offering substantial policy proposals to redress their grievances.

But as Howard Dean said in his pivotal South Carolina speech, "In America, there is nothing black or white about having to live from one paycheck to the next. Hunger does not care what color we are. In America, a conversation between parents about taking on more debt might be in English or it might be in Spanish, worrying about making ends meet knows no racial identity. Black children and white children all get the flu and need the doctor. In both the inner city and in small rural towns, our schools need good teachers. When education is suffering in lower-income areas, it means that we will all pay for more prisons and face more crime in the future. When families lack health insurance and are forced to go to the emergency room when they need a doctor, medical care becomes more expensive for each of us. When wealth is concentrated at the very top, when the middle class is shrinking and the gap between rich and poor grows as wide as it has been since the Gilded Age of the 19th Century, our economy cannot sustain itself."

Stagnant wages affect both black and white workers, and neither is responsible for the policies that create them. Dean said that "Americans, black and white, are working harder, for less money, with more debt, and less time to spend with our families and communities." He reminded his audience that a majority of children living in poverty in America are white—eight million of a total thirteen million. The former presidential candidate said, "It's time we had a new politics in America—a politics that refuses to pander to our lowest prejudices. Because when white people and black people and brown people vote together, that's when we make true progress in this country. Jobs, health care, education, democracy, and opportunity. These are the issues that can unite America. The politics of the 21st century is going to begin with our common interests."

The best part of Dean's speech for me (as a new father) was this: "If any politician tries to win an election by turning America into a battle of us versus them, we're going to respond with a politics that says that we're all in this together—that we want to raise our children in a world in which they are not taught to hate one another, because our children are not born to hate one another. We're going to talk about justice again in this country, and what an America based on justice should look like—an America with justice in our tax code, justice in our health care system, and justice in our hearts as well as our laws. . . . The politics of race and the politics of fear will be answered with the promise of community and a message of hope."

Dean's campaign faltered; perhaps it was the right message but the wrong messenger. But the South Carolina speech remains a milestone in American politics. Telling the truth about race is an unusual thing.

I've often said that race is a social construction, having no basis in biology, religion, or philosophy. But what a social construction race has been—affecting virtually everything else in our social, political, and personal lives. And now, the increasing diversity of American life just makes it all more complicated.

Restitution to African Americans for slavery and discrimination has never been done. Indeed, almost all the racial progress made in America has come at African American initiative—through various movements for black legal and civil rights. Whites, rather than supporting proactive racial justice, have mostly been grudging in their acceptance of black progress—even the most famous white political supporters like John Kennedy and Lyndon Johnson had to be pushed to make black civil rights their priority. Blacks have led; whites have eventually acquiesced, with a few white leaders showing real courage by doing the right thing. With notable individual exceptions, real white leadership on racial justice in America has been largely absent.

Even after major victories such as *Brown vs. Board of Education,* the Supreme Court decision that outlawed "separate but equal" in the nation's schools, American education is more segregated than ever at the decision's fiftieth anniversary. Despite the critical accomplishments of the civil rights law of 1964 and the crucial voting rights act of 1965, many white Americans still resist black gains. And without any national repentance and restitution for slavery and discrimination, many black Americans still feel very unwanted and unwelcomed in their own country.

And now, the largest racial "minority" in America has become Hispanic, surpassing African Americans, with Asians also rising sharply in the population. Without morally and economically needed restitution, black progress still lags far behind whites in income, education, and health. Hispanics also lag behind the white majority, with Asians making ground quickly, utilizing strong family and community structures along with easier white acceptance.

Both Hispanics and Asians now legitimately demand minority recognition and rights, which only makes blacks feel, understandably, pushed aside as the nation's historic minority. I've seen it in my own mostly African American neighborhood in Washington, D.C., where blacks resist Hispanic demands for more progress, often using the same language of resentment that whites have used toward them. In supermarket grocery lines, black D.C. residents complain about Hispanics wanting "something for nothing" and accuse their new neighbors from the South of not working hard enough and getting undeserved special treatment from the government. Asians come in for even more harsh treatment, often as small business owners operating in black neighborhoods where they don't comprehend the culture. African American/Korean tensions, for example, have been especially high in American cites like Los Angeles.

I recall being called in to speak at a tension-filled town meeting in Harrisburg, Pennsylvania, after a black teenager had shot and killed an elderly Korean shop owner during a robbery. The community threatened to explode in interracial accusation and conflict. But when I rose to speak, I saw the front row in the assembly room with the slain Korean's family and the community's African American leaders sitting side by side in grief, solidarity, and commitment to reconciliation. But such leadership is too often lacking when urban communities disintegrate into racial hatred and warfare.

The hope of political solidarity between "people of color" in America has often proved to be a wistful and painful myth. Urban elections pit blacks against Hispanics against Asians, with white politicians often able to exploit racial divisions. And again, powerful economic forces (read: white) are also able to use such tensions to divide racial groups against one another and prevent the kind of interracial labor organizing, for example, that might achieve better wages and benefits for low-income people. Yet the best organizing campaigns in the country persist in the commitment to interracial coalitions and still offer the best hope for the future.

One of the groups most directly affected by these racial dynamics is black youth, especially African American young men. I see it every day and feel a mounting concern. Many black youth simply feel left out and left behind in America. Without any real restitution and successful efforts to include them in the society (the civil rights gains achieved by their parents' generation do not substitute for white initiative), they now feel asked to step aside for other minorities. Being especially vulnerable to the alluring sirens of an affluent society, black youth are also subject to a growing resentment for being denied the good life. So, too many elect to take a fast track to the rewards of affluence through drug trafficking and other illegal activities, and much of that comes with a high risk of violence and gang involvement. Too many young black men, in particular, just refuse now to take the slow and gradual path to middle-class stability. Rather they feel "owed" more, they want it all, and they want it now, which is precisely the attitude that makes white society most unsympathetic to black youth.

A good friend of mine, an African American male in his late fifties, is a successful cab driver in Washington, D.C., with a regular clientele of dependable customers. He tells me how hard it is to persuade a new generation of black men to take up cab driving for a good living. They don't want to be drivers for white people, he says, and want economic success to come much more quickly.

The white majority little understands, or even cares about, the great tragedy of how the black anger our society generates in African American youth becomes both self-destructive and a dangerous obstacle to success. This is why most black parents in America worry about their sons and why young black men in America are becoming an endangered species.

But truth telling about race also suggests the responsibilities those parents and other black leaders have to teach the most basic moral values to their children, to directly counter the false and dangerous consciousness of the violent, hedonistic, and consumerist society their children are being dangerously tempted by. Just because white society has yet to do the right thing about race in America is no excuse for the victims of that neglect not to do the right things in their own lives. That insight is always key for an oppressed people and is essential for eventually overcoming their oppression. The drug dealers, gangbangers, and rawest rappers simply must not be acclaimed as role models for the next generation of black youth. Rather it is the persistent adherence to core spiritual values that will finally achieve social justice, economic equality, and personal freedom.

And as far as "spiritual values" are concerned, it may be that religious congregations, in both minority and white communities, will play a major role in ultimately achieving racial justice and reconciliation. Those congregations are still the most universally present institutions in most communities, especially poor communities, and are the natural keepers of the values a society (and the poor) most need to remember.

I've written in other places about the extraordinary role black churches have played in overcoming horrendous youth violence in cities like Boston and the key contributions that Christian, Muslim, and Native American spiritual leaders have made to gang peace "summits" around the country.

But more unexpected is the surprising new focus on racial "reconciliation" and even racial justice among white evangelicals in America. These are the same people who once told me that "Christianity has nothing to do with racism" and who pushed me out of my little evangelical church at a tender young age over the issue of race. But nonetheless, there is more conversation about race in many evangelical circles today than in the more predictable mainline churches, many of whom seem to have run out of energy on the subject. What seems to be emerging now is something more significant than merely evangelical blacks and whites hugging and singing "Kum Ba Ya" together.

While many evangelical and Pentecostal churches remain overwhelmingly white and have yet to give up substantial power to people of color, much of their growth is coming in poor communities of color, both at home and abroad. That has already changed the political dynamics in traditional organizations like the National Association of Evangelicals. The NAE was unable to speak out in support of the U.S. war in Iraq (which they historically might have done) precisely because their member churches of color were mostly opposed to the war. The Evangelical Covenant Church, of very white Scandinavian origins, now sees its primary growth in the African American community and even in Africa itself. Under the leadership of its president, Glenn Palmberg, all of the denomination's pastors now make a pilgrimage to many of the historic sites of the civil rights movement and to forgotten places of extreme poverty and hunger still present in America today. Claiming its historical roots as "an immigrant church" the Evangelical Covenant now embraces the multicultural future of America as its own future too, making it, in my opinion, the most interesting church in America today. And out of the Covenant's emerging multicultural identity is coming a powerful and prophetic commitment to social justice and peace.

All that affirms what the *Sojourners* study guide said twenty years ago: that racism is a theological offense, a spiritual test, and not merely a political and economic issue. It was, after all, a primary theme of Paul's missionary journeys, the reconciling of two peoples—Gentiles and Jews—in the body of Christ. He proclaims in 2 Corinthians 5:18 that "God, who reconciled us to himself through Christ, . . . has given us the ministry of reconciliation." As America moves into a decidedly multicultural future, Paul's command and example become increasingly relevant and challenging.

To rise to the challenge, the racial conversation in America must become more and more honest if it is to take us anywhere. The discussion must be truth telling, and it must be genuinely two-way and now multi-way. That is, we must be able to confront one another, even as we seek to reconcile. Indeed, there really is no reconciliation without confrontation. Whites will continue to bear the responsibilities of privilege; those who benefit from oppression are obligated to change it. Yet historically oppressed racial minorities have their own responsibilities too—for their values, behaviors, and choices. While race affects much of our national and communal lives, it doesn't determine everything. Whites must be willing to see and acknowledge where race clearly affects benefits and outcomes. And blacks, along with other minorities, must not always "play the race card," especially when other factors are clearly involved. Truth telling will be required from all of us. But the rewards of our truth telling could be great, and we really have no other moral choice.

CHAPTER 20

The Ties That Bond

Family and Community Values

THE FOX NETWORK's hottest new show had just aired. *Temptation Island* brought four "committed" couples to a tropical paradise to meet twenty-six "fantasy singles." The idea was to test their relationships by "dating" these beautiful strangers who are there to seduce them. In the tradition of *Who Wants to Marry a Millionaire* and *Survivor,* this newest installment of what's now called "reality television" hoped to attract and make millions.

It did. Carefully selected scantily clad women and men frolicking in the surf, cuddling in hot tubs, and spending the weekend together in cozy cottages produced both high ratings and advertising dollars.

But as a veteran *Washington Post* columnist said to me over lunch one day, "Isn't that what we usually call 'prostitution' when they pay people to have sex with strangers?" Well, yes it is. Will they? Won't they? That became the titillating question. And we all got to watch. Actually, everybody gets prostituted here.

I've often wondered how the Fox Network gets away with preaching conservative politics and family values in their news and commentary shows while producing such a sleazy lineup of "entertainment" programs. On the other hand, my lunch companions were wondering why liberals who consistently advocate for social justice seem to always pass on making critical comments about shows like *Temptation Island*. Maybe they're afraid they will sound like the religious Right or something. Or maybe they really do want to separate personal morality from social commitments. Remember the Clinton White House?

Well, we decided over lunch that it was time for a new option—one that would treat both *Temptation Island* and child poverty as morally offensive. Infidelity, betrayal, broken relationships, and casual sex are undermining the health and integrity of our society. For Fox to offer all of that as exciting and voyeuristic entertainment is pretty despicable. And to do it all for money just compounds the offense. In the worldview of the television networks, everything is a commodity—including our relationships, our bodies, and our values. And everyone is for sale.

We spent the rest of lunch talking about how to build a movement that could bring our best personal and social morality together. The real enemy here isn't sex, but rather the commodification of everything—turning all values into market values, gutting the world of genuine love, caring, compassion, connection, and commitment for what will sell, for example, on a television show.

In a rare moment of truth, the show itself flashed ahead to later episodes and showed a few of the contestants in real despair. One said, "I feel like I've sold my soul." Another added, "Now the fun is over, and I'm paying for it." Finally, a third one concluded, "Now I'm in hell." Exactly.

Super Bowl Sleaze

I'll admit it, I like a good football game, and this turned out to be a very good one, after a boringly slow start. But what everybody was talking about the next day was the baring of Janet Jackson's right breast. Justin Timberlake's little grope and tear-off of Jackson's bustier was the grand finale to their sex-simulating dance to a song called "Rock Your Body," which ends with the romantic line, "I gotta have you naked by the end of this song." And that's just what he did. But that was only the crude climax to what *Washington Post* television critic Tom Shales called "the Super Bowl of Sleaze."[1]

The rest of the MTV-produced half-time show had lots more bumping and grinding, crotch-grabbing rappers, and background girls tossing off their wardrobes to "I'm getting so hot, I wanna take my clothes off."

Then there were the commercials, often creative and funny at Super Bowls but this year featuring horse flatulence, a trained dog that bites men (again) in the crotch to steal their beer, a monkey leering at a girl's breasts and suggesting they go upstairs, and several ads for erectile dysfunction. Bud Light clearly won the night's award for the most stupid, crude, and banal ads, while only Homer Simpson seemed to offer any healthier and funny alternative fare.

My five-year-old son Luke was playing with his friends in another room, but walked in just in time to see a spot for *Van Helsing,* an upcoming and not yet rated horror film that featured very disturbing and graphic images of horrific violence. Seeing the fanged monsters leering at us through the screen literally stopped him in his tracks.

Because of substantial public outcry, CBS issued some unconvincing apologies about how surprised they were at Janet's bare bust (in this first public revelation of the Jackson family values) and insisted they had been to all the rehearsals and never seen it before, while the impressive young Timberlake preserved both his intelligence and integrity by blaming the whole thing on a "wardrobe dysfunction," which as we all know, happens all the time in Hollywood. There was certainly a lot of dysfunction going on here, but it has to do with much more than outfits.

You want to know why people join the religious Right? It may have less to do with wanting to take over the country than being desperate to protect their kids from the crass trash and degrading banality that media conglomerates like Viacom (which owns both CBS and MTV) seem to think is just fine family entertainment for Super Bowl night. Fortunately, my kids were in bed before the half-time show, but next year we may just go with *Mary Poppins* in the other room.

Again, some people think that only right-wing conservatives care about such moral pollution. Wrong. Most parents that I know, liberal or conservative, care a great deal about it, as do most self-respecting women and men. It defies ingrained stereotypes to suggest that a healthy, moral consistency applies to personal and sexual ethics as well as to social and political values. Its time to break out of those old ideological shibboleths and forge a unified front against the amoral corporate greed that violates all our ethics—personal and social—by creating a system that sells beer and breasts in the same advertising plans just to make a buck.

I don't recommend joining the religious Right, but I think it's time to confront the pornographic corporate profiteers—that is, the conglomerates that own and control all of our media networks. Tell them that you're not a member of the religious Right, not a puritan, and not afraid of sex (but actually think it's great when there is the lifelong commitment to match its significance). Especially if you're a liberal progressive type, tell them that. And if you're a parent, ask them if they have any kids and if they would want them watching the low-life ads and actors they put on TV during the Super Bowl. Tell them to put it on cable, where the people who want to watch Jackson and Timberlake hump each other can pay for it. Tell them that you're angry. Tell them that they're not entertaining, interesting, or even sexy. Tell them that they're just pathetic. Tell them that their soulless and mindless "entertainment" won't sell anymore, at least not to you. And tell them to keep their garbage away from your kids.

But while I can get as mad as anybody else about the cultural pollution that surrounds us, the real family values questions go much deeper. The really important issue is parenting.

The Politics of Parenting

Exit polls following the 2004 election showed that 57 percent of married voters supported President Bush, while 42 percent voted for Senator Kerry. Conversely, 58 percent of unmarried voters supported Kerry and 40 percent voted for Bush.[2]

The "marriage gap" and "family values" culture wars have separated liberals from conservatives for several decades now. But the deep divide between married and single voters may go deeper than the typical issues of morality often raised by the political and Religious Right. They may instead go to the increased pressure on parents across the political spectrum. Yet it is a battle that liberals may continue to lose unless they both acknowledge the legitimacy of the "moral" challenges facing families while also changing the terms of the debate.

The legitimacy of the family values debate has been demonstrated in the clear links that have been made between the problem of family breakdown and the social ills of youth delinquency and crime, drug use, teenage pregnancy, welfare dependence, and the alarming disintegration of civic community, especially (but not exclusively) in poor neighborhoods.

Whatever headway the Democrats were making on renewing a discussion on personal responsibility and family values was almost completely undermined by Bill Clinton's impeachment scandal over his affair with Monica Lewinsky. Thanks in part to Bill Clinton's lack of moral compass, the Republicans have succeeded in claiming to be the arbiters of moral and family values.

A January 2001 Pew Research Center study found that 49 percent of Americans trusted the Republicans to improve moral values, whereas only 26 percent trusted the Democrats to do so.[3] In a recent Democracy Corps survey, 57 percent of voters associated the Republicans with "personal responsibility," compared with 25 percent who associated the term with the Democrats. In the same survey, the Republicans had a 20-point advantage on "commitment to family" and a 15-point advantage on "shares your values."[4]

I met Anna Greenberg when I was teaching at the Kennedy School of Government at Harvard. Anna has done extensive research on the question of "values," which has included serious attention to the perspectives of diverse religious communities.

One of the most important issues Greenberg has focused on is the "politics" of marriage. Greenberg says, "It would be easy for Democrats to dismiss this Republican 'morality' advantage as an artifact of the influence of fringe religious voters. But it is evident in a broad swath of voters. And a large portion of these voters are married. But does this mean that married Americans are hijacked by the Christian Right? Hardly. A more likely story is that when people get married and have children, their new experiences alter their political concerns. First, parents often return to the religious fold after dropping off from worship when they left their childhood home for work or college. Second, people with children have concerns that did not occur to them when they were childless. Parents worry about their kids' exposure to sex and violence on television, in music, and on the Internet. They worry about school safety and the impact of peer pressure on taking drugs or committing violence."

Single adults, and especially men, just don't think about these issues as much as married people do. In one survey Greenberg cites, "only 16 percent of single men thought that sex and violence on television were a very serious problem, compared with 47 percent of married men. By comparison, 42 percent of single women and 56 percent of married women thought they were a problem."

Most of these data support the contention that what is wrong with families, in the perception of most Americans, is lack of responsible parenting. From violence-torn inner cities to suburban Littleton, Colorado, where upper-middle-class students shot down their classmates at Columbine High School, a lack of parental supervision and discipline is at the heart of the problem.

And, most tellingly, Greenberg reports her respondents saying that "the biggest challenge facing parents is insufficient time and commitment to teach children respect for rules and responsibility." That finding is confirmed by many other national surveys showing that the pressures of parenting are as important as any of the other worries that most Americans have these days.

Such findings indicate that the answer to the "family values" crisis may not be a return to traditional roles for men and women and combating gay marriage, as the Right suggests, but rather in supporting the critical task of parenting—culturally, morally, and economically. Here again, both the Right and the Left are failing us. The Republican definition of family values, which properly stresses moral laxness but ignores the growing economic pressures on all families, simply doesn't go deep enough. Similarly, the Democrats are right when they focus on economic security for working families but wrong when they are reluctant to make moral judgments about the cultural trends and values that are undermining family life.

Stagnant wages, the loss of health care and other job benefits, the rising cost of housing, and the demands of multiple jobs for low-income workers are all an assault on family time and values. But even in middle- and upper-middle-class households, the growing pressure for two incomes has led to the "latch-key syndrome" that forces young people in families up and down the income ladder to look after themselves.

Greenberg reports that "people are working more hours to keep up, but we have no real support for child care or family leave, some of the biggest issues facing working families. This means people are working longer hours and spending less time at home. Little wonder Americans are feeling stressed about their ability to spend time with their children." Perhaps her key insight is this: "Americans make the connection between moral decline, the challenge of parenting, and the economic squeeze." Americans do, but the political parties don't.

Being new parents ourselves, with two young boys, Joy and I talk to lots of other parents. Our experience suggests that mothers and fathers across the political spectrum now regard parenting in America as a countercultural activity. When I make such a statement on the road, all the parents in the audience begin nodding their heads—whether they are political liberals or conservatives. The values of a materialistic and hedonistic culture are clearly arrayed against our raising children with moral and spiritual values. Objecting to the "spiritual formation" of the television and media culture is no longer an issue that is liberal or conservative. It is a commitment that now bonds parents together no matter what their political or religious affiliations are—and could provide a powerful constituency for holding our society more accountable to the needs of families and the well-being of children.

As a progressive Democrat herself, Greenberg offers a refreshing challenge to her political kindred spirits: "There is no reason for progressive Democrats to shy away from addressing these issues—they are not antithetical to a progressive agenda, and they speak to the core values of the majority of the American public. Democrats should not hesitate to criticize the proliferation of sex and violence in the media. Parents across the political spectrum are concerned and want help in shielding their children from these influences. After all, studies by groups such as the Kaiser Family Foundation find that the majority of television programming contains some sexual content but little discussion of sexual responsibility. Likewise, Democrats should not shy away from talking about the importance of parenting—regardless of people's familial arrangements—and its influence on early childhood development, behavior in school, economic well being, and children's health. And

Democrats, regardless of their own faith or lack of it, should express respect for the religious choices of others and for the role faith-based organizations can play in families and communities. After all, all religious communities, not just religiously conservative groups, have a stake in the quality of our communities."

Joy and I also find that the tremendous economic pressures of our corporate-driven society take their toll on parents of all our kid's schoolmates, from low-income single moms to the households of two-parent professional couples (though the pressures on the former are for survival, while the latter are often from keeping up with the rat race of a consumer culture). Perhaps the key to the politics of parenting will be to connect the moral agenda to the economic agenda. How do we support the values that most parents want for their children by also supporting their ability and time to actually teach and nurture the next generation?

At Harvard, I also met Theda Skocpol, whose book *The Missing Middle* proposes a new political agenda centered on "the struggles of working mothers and fathers, especially those with modest means." She also wants to "better support parents" in a society that undermines and devalues families. Skocpol argues that "most Americans worry about both the material circumstances *and* the moral climate for family life today."[5] Again, that connection may be the key to a new politics of parenting. She believes the country would be receptive to progressive policies for parents: expanded and intergenerational use of Social Security, ensured child support, affordable child care, a steadily expanding Medicare, and universal access to paid family leave. Skocpol aims to strengthen parents so they can do a better job in supporting their children and have more time to offer moral guidance. She agrees with those who speak of the central role of parenting in a healthy society, and she asks how we can provide social supports for parents who carry so much responsibility.

In a critical article entitled "Adding Values," Anna Greenberg and her father, Stanley Greenberg, claim that "more progressives have begun to understand that the future progressive agenda must focus not just on people's work lives but on their family lives." They conclude, "Progressives also need to rediscover the family. For some three decades, the left neglected to affirm the centrality of the family, especially of two-parent households. The left must give itself permission to recognize the benefit of two-parent families to children. Democrats need to affirm the value of such family structure, even as they redouble their efforts on behalf of single-parent families, whose battles are tougher."[6]

The battle for good parenting must not be a politicized one; it is too critical for the future of our children and too important for any definition of a good society. Indeed, the need to support parents in their deepest desires to be responsible parents could be one of the few areas left where we could and should find a broad political consensus and a crucial common ground. "Conservative" and "progressive" parents alike have many of the same worries and concerns and could unite across their political dividing lines to combat the polluting and undermining cultural influences that put good kids and good values at risk. Indeed, when it comes to parenting, most of us are conservative, and most of us would support a progressive pro-family agenda if one were to be laid out. This is one issue, the passion to protect our kids—and other people's children—that could really bring us together.

The Controversy over Gay Marriage

Family and "family values." Seldom has the importance of an issue been so great, and seldom has the political manipulation of it been even greater. The bonds of family that tie together our most important and intimate relationships and provide the critical framework for raising the next generation are too important to be reduced to mere symbols in our most partisan and bitter political wars. Yet that is precisely what has happened. Liberals and conservatives, Democrats and Republicans are all responsible for politicizing the questions of family life and thus contributing to the disastrous weakening of this most basic and important institution.

Each of us, across the political spectrum, must now recognize the crisis of family life in America. The statistics are overwhelming and simply demand our attention. Nearly half of all marriages end in divorce, 1.5 million women a year are assaulted by their current or former husbands or boyfriends, one in three children are born outside marriage (even in poor communities), and so on. Family ties and relationships are growing weaker at an alarming pace, with disastrous consequences—especially for children. And the consequences are also clear for poverty; delinquency and crime; sexual promiscuity; education and employment; physical, emotional, and mental health; spiritual well-being; and social pathologies that transmit themselves intergenerationally. Family breakdown figures into a wide variety of other social problems as a primary causal factor.

But as soon as the issues of family breakdown are raised as a public issue, a very ideological debate quickly ensues about the ideal forms of family life that various cultural, religious, or political constituencies wish to advance. What is soon lost are the "facts" of family life and how those facts can be either weakened or strengthened.

There have been many forms of family in history, the most common being the extended family where blood relationships across generations create strong webs of social networks and living arrangements, which provide security and stability for all its members, and multiple role models guide the young. The extended family is arguably the most historically common form of family life and is undoubtedly the form of family operative in ancient biblical cultures, both Jewish and Christian. Of course, forms of extended family life, some stronger and some weaker, continue today as most families have networks of broader relations.

But the more modern phenomenon of "the nuclear family" is what most today refer to as family and is the particular result of a modern industrial and consumer society. Because they have fewer resources to draw upon, nuclear families are more subject to many pressures and tensions that can be more easily absorbed in the extended family model. Today, many forces undermine nuclear families and thus imperil family life as most people in our cultures experience it. And today, there are new forms of family life emerging: single-parent families, families that skip a generation with grandparents having the primary child-rearing responsibilities, some same-sex couples rearing children, and cohabiting couples of unmarried heterosexual adults who choose not to marry but may also be raising children (often from previous marriages). And more people are divorced, widowed, and otherwise living alone than ever before. All of these forms of family have their impact on children.

In the fall of 2003, the Massachusetts Supreme Court ruled that gay and lesbian couples in that state are legally entitled to marry, thereby entitling them to the same "legal, financial, and social benefits" as heterosexual couples.

That decision set off a great controversy that played a prominent role in the 2004 election debate. The city of San Francisco and other jurisdictions began issuing marriage licenses to same-sex couples, and President Bush announced his support for a constitutional amendment to stop gay marriages—which were then performed by the thousands in both San Francisco and Massachusetts. The issue was joined. The topic now seems destined to continue as one of the most controversial issues in America for some time to come.

Over the past decade, this "family values" question has become very difficult and has been polarized by both the religious Right and the cultural Left. To move forward, we must simply refuse the false choices offered by both sides.

As we have already noted, the Left has often misdiagnosed the roots of our present social crisis, mostly leaving out the critical dimension of family breakdown as a fundamental component of problems like poverty and violence. These issues are not important just to the religious Right and are not simply bourgeois concerns. We do need to rebuild strong and healthy two-parent families. And we desperately need more strong male and female role models in both "nuclear" and extended families.

Today, family breakups, broken promises, marital infidelity, bad parenting, child abuse, male domination, violence against women, lack of living family wages, and the choice of material over family values are all combining to make the family norm in America more and more unhealthy. A critical mass of healthy families is absolutely essential to the well-being of any society. That should be clear to us by now, especially in neighborhoods where intact families have all but disappeared.

But the Right has seized upon this agenda and turned it into a mean-spirited crusade. To say gay and lesbian people are responsible for the breakdown of the heterosexual family is simply wrong. That breakdown is causing a great social crisis that affects us all, but it is hardly the fault of gays and lesbians. It has very little to do with them and honestly more to do with heterosexual dysfunction and, yes, "sin." Gay civil and human rights must also be honored, respected, and defended for a society to be good and healthy. It is a question of both justice and compassion. To be both pro-family and pro–gay civil rights could open up some common ground that might take us forward.

Do we really want to deny a gay person's right to be at their loved one's deathbed in a hospital with "family restrictions"? Do we also want to deny that person a voice in the medical treatment of his or her partner? And do we really want all the worldly possessions of a deceased gay person to revert to the family that rejected them thirty years ago, instead of going to their partner of the last twenty years? There are basic issues of fairness here that can be resolved with or without a paradigm shift in our basic definition of marriage.

We can make sure that long-term gay and lesbian partnerships are afforded legitimate legal protections in a pluralistic society no matter what our views on the nature of marriage are. But the question of gay marriage is important; it is a major issue in the religious community, and it is unlikely to be resolved for many years. Many in the churches and the society believe that the long-standing and deeply rooted concept of marriage as being between a man and a woman should not be changed, but same-sex couples should be granted the rights of "civil unions." That's still my own view. For others, only gay marriage fulfills the requirements of equal protection under the law. There are at least three different views being debated in the churches. Most Christians still believe that the sacrament and theology of the church on marriage should not be altered, while others are exploring new rites of church "blessings" for gay and lesbian couples committed to lifelong relationships, and still others want full sacramental inclusion.

The states themselves will ultimately resolve the legal and civil issues through referendums and legislative proceedings. While the issues of gay civil rights are fundamental matters of justice in a democratic society, the legal and ecclesial questions of how to handle the specific marriage issue should not be used to divide the churches, as too many on both the religious Right and religious Left seem eager to do.

In 2004, *Sojourners* published a dialogue between two seminary presidents, Richard Mouw of Fuller Theological Seminary and Barbara Wheeler of Auburn Theological Seminary. One is more conservative on the issues of homosexuality, the other more liberal. Both are members of the Presbyterian Church USA, which like many other denominations, threatens to divide over the issue. But Mouw and Wheeler believe that would be a mistake.

Richard Mouw says:[7]

I hope with all my heart that we can avoid the divorce court. I want us to stay together. I do not have a clear sense of what it would take to avoid what many of our fellow Presbyterians apparently are convinced is an inevitable separation. I do sense, however, a strong need to keep talking. The church, some insist, is not some mere voluntary arrangement that we can abandon just because we do not happen to like some of the other people in the group. God calls us to the church, and that means that God requires that we hang in there with each other, even if that goes against our natural inclinations. I agree with that formulation.

I want with all my heart for this to happen to us in the Presbyterian Church—that we take up our arguments about the issues that divide only after we have knelt and laid our individual and collective burdens of sin at the foot of the cross. Needless to say, if it did happen, I would be surprised. But then, the God whom we worship and serve is nothing if not a God of surprises.

Barbara Wheeler says:[8]

How's this for a model of the church that we are called to become: "They confessed that they were strangers and foreigners on the earth." What if instead of denying our estrangement, or bemoaning it, we embraced it as a gift from God?

A church that contains members we think strange, even barbaric, is a healthier setting for us, for our formation as Christians. We like to think that a church of our kind, one that excludes those who believe incorrectly and behave badly by our lights, would be a better school for goodness than the mixed church we've got. It is not necessarily so. Familiarity and affinity breed bad habits as well as virtues.

The last and most critical reason for all of us Presbyterian strangers to struggle through our disagreements is to show the world that there are alternatives to killing each other over differences. As long as we continue to club the other Presbyterians into submission with constitutional amendments, judicial cases, and economic boycotts, we have no word for a world full of murderous divisions, most of them cloaked in religion.

I believe that the protection of gay civil rights should be a bottom line in the debate, and I try to say that especially in very conservative evangelical circles. On the issues of gay marriage, our *Sojourners* magazine has not taken a "position" but rather an "approach" of civil dialogue between Christians who are committed to justice and compassion but who have different understandings of how best to resolve the question of gay marriage. It is our view that while gay civil rights is a fundamental justice issue, the controversies over gay marriage and the ordination of gay bishops, and so on, should not be seen as "faith breakers." The church is going to have to learn to stay together and talk about these things until we find some resolutions together. Many feel that legal protections can and should be extended to same-sex couples, without necessarily changing our whole definition of marriage. But one could also argue that gay civil marriage is necessary under "equal protection." One could also argue for church blessings of gay unions. I think all those are strong points, even if the churches are unlikely to change their whole theology and sacrament of marriage itself. But this is the good and necessary dialogue. And in the meantime, the church must stand up for gay and lesbian people under attack and must welcome them into the community of faith.

Evangelical leader Tony Campolo offers one proposal. While taking the conservative view that the Bible does not affirm homosexual practice, Campolo denounces homophobia and supports gay civil rights. But he believes that the legal issue of marriage should be separated from the church—as it used to be. Clergy should no longer pronounce marriage blessings "by the authority vested in me by the State of Pennsylvania." That should be left to the civil authorities, and both heterosexual and gay couples should apply to the state for the civil right—be it civil marriage or civil union. Then religious couples, heterosexual or same-sex, could approach the church of their choice to seek a religious blessing. Most churches will only provide marriage ceremonies for the union of a man and a woman. But there are a growing number of local congregations that are open to offering blessings to same-sex couples. It is a solution that might satisfy most, as it preserves the Christian conscience of diverse positions. But for it to work, those congregations that open up to gay couples should not be cut off from their denominations, and neither should more traditional churches be pressured to accept gay marriage by liberal voices in their own denominations.

Conservative Christians should be careful not to draw their primary line in defense of family at the expense of gay couples who want to make a lifelong commitment, instead of standing prophetically against the cultural, moral, and economic forces that are ripping families apart. And liberal Christians should not just argue for gay marriage on the grounds of human rights, but rather should probe more deeply into the theological, biblical, and sacramental issues that are also at stake.

When conservatives seem to suggest that the future of western civilization is at stake in the battle over the legal status of same-sex couples, they seriously overstate the issue. Likewise, when liberals say that resolving the legal issues surrounding gay unions is morally equivalent to the issues of racism, apartheid, and the Holocaust, they make the same mistake. Solutions to the definitional legal issues for gay couples will eventually be found, but this is not the ultimate moral confrontation of our time, as partisans on both sides have tried to make it. It will take many years before we are able to better understand and resolve the many complicated factors surrounding the issue of homosexuality. When we do, we should be able to look back and feel good about the way we conducted our dialogue and our relationships in the process.

I wrote during the 2004 election campaign that unless people of faith insist that the biblical imperatives for social justice, the God who lifts up the poor, and the Jesus who said "blessed are the peacemakers" were brought into the political debate, the "values questions" and "moral issues" in the election would be restricted to the Ten Commandments in public courthouses, marriage amendments, prayer in schools, and abortion. Many in the media continue to treat issues like gay marriage as the only values and moral issues in an election campaign.

But in a national poll done in early 2004, an overwhelming percentage of voters saw it another way. The poll, commissioned by the Alliance to End Hunger and Call to Renewal, was conducted by a leading bipartisan polling group.[9] Those polled were a representative sample of likely voters—Democrats and Republicans, liberals and conservatives—and were diverse in racial and faith background. The poll asked, "The question of values is sure to be important to many voters this November. As you decide your vote for president of the United States, which of the following would be more important to you: hearing a candidate's position on gay marriage or hearing a candidate's plan for fighting poverty?" An overwhelming majority of voters said that they would rather hear a candidate's plan for fighting poverty (78 percent) than a candidate's position on gay marriage (15 percent). The polling also showed that even in the midst of budget belt-tightening, voters want Congress to strengthen anti-hunger programs. Traveling around the country as I do, I was hopeful about the result of the poll, but I didn't expect that 78 percent would see poverty as such an important values question.

Apparently, thirteen million children still living in poverty is indeed a moral issue for most Americans. And apparently, the pundits are misreading and misrepresenting the people about what the most important moral issues really are. "Hunger and poverty are on the rise in our country and this poll confirms that voters want to hear more from political leaders about real solutions to these serious problems," said Reverend David Beckmann, president of Bread for the World and a founding member of the Alliance to End Hunger.

We must show that people of faith are united in believing that millions of children and their parents living in poverty is a "family issue" that our political debate must address. We must articulate the moral issues of social justice, or others will define the values questions in much more narrow ways. The poll tells me that people are ready to hear another view. We have to make sure that they do.

Similarly, the controversies over gay marriage may not be as important as the deeper ethical issues of war and the emerging theology of war in American politics. Duke Divinity School professor Richard B. Hays wrote in an important article, "Let us stop fighting one another, for a season, about issues of sexuality, so that we can focus on what God is saying to the church about our complicity in the violence that is the deepest moral crisis of our time. And let us call the church to fasting and prayer in repentance for the destruction our nation has inflicted upon the people of Iraq."

And he appealed to his own church: "In the United Methodist Church, we say nothing about the horrifying violence in Iraq, while at the same time we exhaust ourselves going around in circles debating issues of sexuality."[10]

The Culture Is Broken

Ultimately, I suspect the legal definitions we use for gay and lesbian relationships will have less impact on our society than the definition of community that shapes our culture.

Is the wildly successful *Survivor* television series just a game, or is it another sign that our culture is broken? I was on vacation, doing some beach reading, when the popular culture swooped down to drop another bomb. I was finally getting to read *Tuesdays with Morrie,* the *New York Times* bestseller about a very successful, but unhappy, sportswriter who learns life's greatest lessons from his former college professor who was dying with Lou Gehrig's disease.[11] On one of his weekly visits to see his old mentor, Mitch Albom asks Morrie why so many people are living unhappy lives. Morrie says, "Well, for one thing, the culture we have does not make people feel good about themselves. We're teaching the wrong things. And you have to be strong enough to say if the culture doesn't work, don't buy it. Create your own."

No sooner had I read the sage's words when a copy of *USA Today* landed in my lap with the cover story, "Tonight, There'll Be Just One Survivor."[12] I have to admit I had missed all this. But I read that CBS was expecting upwards of forty million people to watch the finale to the highest rated summer series in television history, and advertising was selling for $600,000 for a thirty-second slot. As most of America apparently knew, sixteen "contestants" had been stranded on a desert island in the South China Sea for thirty-nine days and had been successively voting one another off with the goal of getting down to a "sole survivor" who would win $1 million. It had become what the media calls a "cultural phenomenon."

The newspaper cover story said, "The show's genius is its seeming replication of the underbelly of corporate America: The innocents who joined *Survivor* as a lark, relishing the adventure, have been summarily dispatched by the schemers, conniving to shape the game's outcome by joining forces." Now the sixteen were down to four for the final episode. And the article concluded, "Each now knows the formula for success—cunning and manipulation win out over likeability—employed by all four remaining contestants." An expert on national trends added, "Pop culture dominates our society, and it's not to be taken lightly."

So I just had to watch, since I had all this vacation time on my hands anyway!

It was fascinating. The show opened with a very spiritual dilemma, the announcer saying, "Our goal was to work together to build a new world but compete against each other to become the sole survivor and to win $1 million. It was to become the ultimate human experiment." I'll say.

A series of "alliances" had worked well in getting rid of almost everybody else, but now it was "all for ourselves," said one of the four remaining survivors. "What's it like sleeping next to your enemy?" said another. Wonderful human comments were now flowing freely. "I'm here for me, and everyone else can kiss my ass. I don't give a crap about anyone on this island." That was day 38.

Sue, the truck driver, got voted out, and the last three put mud all over themselves, walked over hot coals, and then had to stand for hours with one hand on a totem pole of sorts—"attached to an idol," one survivor insightfully remarked. Rudy, the former Navy Seal, had a seeming lapse of concentration and let go of the pole, and soon they were down to two—Kelly, the river rafting guide, and Rich, the corporate trainer who had earlier gained viewer fame by going naked on camera.

Kelly was far superior to Rich in physical ability, discipline, and stamina, but it was Rich who walked away with the million dollars. Kelly was just too conflicted. "I've had a lot of moral issues," she said at a critical point. "There have been some moral low points, not-so-proud moments that don't belong in my life." Rich, as even Kelly pointed out, "never pretended he wasn't scheming and conniving. He played the game from the beginning. I respect that." And that's what Rich said over and over again in his final speech. It was all "strategy." The ultimate winner acknowledged there were some "sincere interactions" and "maybe starts of friendships," but "it's about who played the game better." The sole survivor actually did poorly on a quiz asking what they all had learned about the lives and families of the other contestants. Instead, he was "observing" people's strategies and concluded, "I just want the million." The last seven contestants who were ejected became the final jury to vote the winner. In the end, they concluded it wasn't the best or most deserving person up there, but merely the "least objectionable."

Did the end justify the means? It did on the island. The final four were praised for "doing what they had to do," and Rich for doing it best. He got the "alliance" going, which proved so decisive in whittling the number down to the few. And in the end, he was the only one left standing. I guess the corporate trainer has been watching those fun-loving mergers, which have become such a game in the business world. Kelly admitted she had too many "identities" going on the island. Even Rich said his only regret was trusting people too much.

One earlier contestant put it most bluntly: "I couldn't have been that ruthless. I can't even lie. My God, I'm glad I got out early. At least I can live with myself. That's the important thing." Yes, but she won't get the money and all those potential corporate contracts. By the end of the show, several "survivors" were already in the advertisements, drinking Bud Light or talking on cell phones on the beach. "It was a game that definitely parallels our regular lives," said the announcer. I wondered why nobody pointed out that all the white men on the jury voted for the white man. How's that for paralleling regular life?

Survivor II was already in the making, and fifty thousand contestants had applied. But the deeper question is whether the ethics of *Survivor,* do really mirror our lives, our economy, and our politics. Can you really build a better world with no ethic of the common good, but only one of strategic self-interest? The *Survivor* series has appeared in many other incarnations, perhaps culminating in the latest version featuring Donald Trump offering his final and most ruthless contestant a dream job in one of his companies and gleefully telling all the others, "You're fired!" Is "surviving" to win the million dollars or work for the ultimate capitalist, Donald Trump, to be the bottom line of our society?

Morrie would scream "No!" And remember, the book about his spiritual wisdom is also a best-seller. Maybe some people are hungry for the new bottom line he's talking about—love, caring, family, friendship, service, and community. Could Morrie become the sole survivor?

Perhaps the biggest adjustment my English wife has had to make in coming to America is the definition of community used in her new country. American culture is much more individualistic than her familiar British context. In everything from the role of government, to the experience of her parish church, to the solidarity of her old neighborhood—even across racial and class lines—the British are more communal than we are. They have a deeper sense of belonging to one another and a responsibility for one another. No British politician, for example, whether Labour or Tory, would challenge the basic nature of a health-care system that regards medical care as a mutual social obligation rather than a commodity you may or may not be able to afford. The Anglican churches from which Joy comes often look like the neighborhoods they are rooted in, rather than being commuter centers for like-minded homogeneous people who travel miles away from their homes to worship.

America used to be more like that, too, but becomes less and less communal every day. It is a harsh and cruel individualism that is now being forced upon us by our corporate, media, and political culture. And that is not good for family or community values. In fact, it destroys them. Yet family and community values remain primary religious values, even if they are disappearing as American values. To reestablish that spiritual sense of community in our churches, our neighborhoods, and even in our national politics is a counter-cultural activity indeed—but one also in the best traditions of an older America that must now be urgently renewed.

It is indeed the strength and health of the bonds between us that are so key to our future. There are no more important bonds than those between parents and children, and it's time we achieved a political consensus about that. Strengthening marital fidelity, commitment, and longevity should also become a key bipartisan agenda. This includes encouraging healthy, monogamous, and stable same-sex relationships—which religious conservatives should be careful not to pit themselves against, regardless of how such relationships are ultimately defined. And the bonds between individuals and families that we call *community* are absolutely essential if we are to protect the key religious and political concept of the common good. Protecting these ties that bond could be the beginning of the new political agreements we are in such desperate need of today.

PART VI

Spiritual Values and Social Change

CHAPTER 21

The Critical Choice

Hope Versus Cynicism

> Do not remember the former things, or consider the things of old, I am about to do a new thing: now it springs forth, do you not perceive it? (Isaiah 43:18–19)

> And the one who was seated on the throne said, "See, I am making all things new." (Revelation 21:5)

First, Two Family Stories—About Dancing

My wife was dancing with the archbishop—and in front of twelve thousand people! Joy Carroll had just con-celebrated the Eucharist at the Sunday worship service of the United Kingdom's Greenbelt music and arts festival with the new archbishop-elect of the Church of England, Rowan Williams. After we finished the wonderful liturgy, the musicians on the main stage struck up some lively Celtic music that got the whole crowd dancing. I was grinning at our four-year-old son, Luke, dancing away with his ten-year-old cousin, Steven, when somebody tapped me on the shoulder and said, "Look, Joy's dancing with the archbishop!" The next day's *London Times* carried the picture with a caption that read, "The Archbishop dancing with the 'Vicar of Dibley.'" (Joy was the script consultant and role model for this hit British comedy show and is known in England as the "real" Vicar of Dibley! In her book, *The Woman Behind the Collar,* Joy tells the compelling story of the ordination of women in the Church of England and her own pilgrimage as a woman called to ministry.)[1] I'm told it's been a long time since British teenagers were heard chanting the name of an archbishop: "Rowan! Rowan!" The Welshman and Oxford scholar is not necessarily known for his dancing but rather for his theological intellect, social conscience, poet's heart, and deep faith. Just a few weeks earlier, Williams said a war with Iraq would not be a "just war." Still in his early fifties, he brings new energy to the Church of England in its theology, spirituality, and commitment to social justice.

Greenbelt is a festival of music, arts, faith, and politics that has attracted people of all ages in Britain for nearly thirty years. I've often been a speaker there, and it is where Joy and I first met. The days are full of lively discussions, and the nights are alive with music. The most serious topics in the church and the world are regularly dealt with at Greenbelt, but most who attend would testify to having their spirits lifted (and even having a great deal of fun) at these annual events over the August bank holiday weekend. Under big tents with thousands of people, with lots of families and kids camping in their own tents nearby (often in the mud of England's wonderful weather), Greenbelt is known for inspiring hope (and even dancing).

Here's the second story. "Happy New Year! Happy New Year! Happy New Year!" That's what my two-year-old son, Luke, said to everyone who entered the airplane on our return flight home after a New Year's retreat our family attended several years ago. I always enjoy watching expressionless faces turn into smiles when Luke starts talking to strangers. "He must be the official greeter," one of the passengers remarked with a big grin.

Luke just had a great time on New Year's Eve. Joy and I had hoped to use him as an excuse to go to bed early, but Luke danced until midnight at a big party with a live band that was just too much fun. Twirling, jumping, hopping, swinging, waving his arms, and even doing a little break dancing on the floor, this boy didn't want the party to end. The countdown to midnight fascinated him, the kisses and hugs all around were a welcome surprise, and then everybody started yelling "Happy New Year!" He thought that was so cool, he kept saying it to everybody for the next three days.

And I loved watching the responses he got. Maybe it was the freshness and utter sincerity of the traditional greeting from the mouth of a little child with sparkling eyes that turned busy travel faces into broad smiles. And Luke would keep saying it until the person said it back!

Or maybe it was because we really want to believe things can be new again at the start of another calendar year, even though our adult minds make us cynical about it. But isn't that the promise of Isaiah, John's Revelation, and the whole biblical story? Our faith says that new beginnings are possible—always and in every circumstance. Yes, in the midst of our relationships, in the course of our work, in the health of our communities, and even in the history of our nations—change is always a real prospect. That's the promise of faith.

And believing that is what makes the change possible. We are in the midst of great transition, in the politics of the country, perhaps in the dynamics of the economy, in the state of a world deeply mired in both poverty and conflict, in the battle against terrorism, in the struggle for justice, and in many of our personal lives. If we expect the worst, we'll probably get it. But if we can be as open to the new opportunities as we are to all the potential dangers, who knows what could happen?

I enjoyed seeing people's responses to Luke's "Happy New Year" greeting. It literally lit up their faces. And maybe it reminded them of their hope for things really being new—a hope just beneath the surface for most of us too busy to stop and wish each other a happy New Year.

The Real Battle

When I was growing up, it was continually repeated in my evangelical Christian world that the greatest battle and the biggest choice of our time was between belief and secularism. Everything was under attack from "secular humanists," and believers had to defend whatever was sacred and significant from the onslaught. That kind of thinking led to the rise of the religious Right, which I have challenged many times in this book. But I have also been critical in these pages of what I call "secular fundamentalism" among too many of our liberal elites who seem to have an allergy to spirituality and a disdain for anything religious. In particular, they have such a visceral reaction to the formulations of the religious Right that they make the mistake, over and over again, of throwing all people of faith into the category of right-wing conservative religion. That mistaken practice has further polarized the debate over religion and public life in America and has even deepened the impression among many Christians that the real battle is indeed between belief and secularism.

But as I have argued in this book, the answer to bad and even dangerous religion is not secularism, but better religion. And the best religion to counter the religious Right is prophetic faith—the religion of the prophets and, of course, Jesus.

Prophetic faith does not see the primary battle as the struggle between belief and secularism. It understands that the real battle, the big struggle of our times, is the fundamental choice between cynicism and hope. The prophets always begin in judgment, in a social critique of the status quo, but they end in hope—that these realities can and will be changed. The choice between cynicism and hope is ultimately a spiritual choice, one that has enormous political consequences.

First, let's be fair to the cynics. Cynicism is the place of retreat for the smart, critical, dissenting, and formerly idealistic people who are now trying to protect themselves. They are not naive. They tend to see things as they are, they know what is wrong, and they are generally opposed to what they see. These are not the people who view the world through rose-colored glasses, the ones who tend to trust authority or who decide to live in denial. They know what is going on, and at one point, they might even have tried for a time to change it. But they didn't succeed; things got worse, and they got weary. Their activism, and the commitments and hopes that implied, made them feel vulnerable. So they retreated to cynicism as the refuge from commitment.

Cynicism does protect you in many ways. It protects you from seeming foolish to believe that things could and will change. It protects you from disappointment. It protects you from insecurity because now you are free to pursue your own security instead of sacrificing it for a social engagement that won't work anyway.

Ultimately, cynicism protects you from commitment. If things are not really going to change, why try so hard to make a difference? Why become and stay so involved? Why take the risks, make the sacrifices, open yourself to the vulnerabilities? And if you have middle-class economic security (as many cynics do), things don't *have* to change for *you* to remain secure. That is not intended to sound harsh, just realistic. Cynics are finally free just to look after themselves.

Perhaps the only people who view the world realistically are the cynics and the saints. Everybody else may be living in some kind of denial about what is really going on and how things really are. And the only difference between the cynics and the saints is the presence, power, and possibility of hope. And that, indeed, is a spiritual and religious issue. More than just a moral issue, hope is a spiritual and even religious choice. Hope is not a feeling; it is a decision. And the decision for hope is based on what you believe at the deepest levels—what your most basic convictions are about the world and what the future holds—all based on your faith. You choose hope, not as a naive wish, but as a choice, with your eyes wide open to the reality of the world—just like the cynics who have not made the decision for hope.

Dancing in South Africa and Beyond

The former South African archbishop Desmond Tutu used to famously say, "We are prisoners of hope." Such a statement might be taken as merely rhetorical or even eccentric if you hadn't seen Bishop Tutu stare down the notorious South African Security Police when they broke into the Cathedral of St. George's during his sermon at an ecumenical service. I was there and have preached about the dramatic story of his response more times than I can count. The incident taught me more about the power of hope than any other moment of my life. Desmond Tutu stopped preaching and just looked at the intruders as they lined the walls of his cathedral, wielding writing pads and tape recorders to record whatever he said and thereby threatening him with consequences for any bold prophetic utterances. They had already arrested Tutu and other church leaders just a few weeks before and kept them in jail for several days to make both a statement and a point: Religious leaders who take on leadership roles in the struggle against apartheid will be treated like any other opponents of the Pretoria regime.

After meeting their eyes with his in a steely gaze, the church leader acknowledged their power ("You are powerful, very powerful") but reminded them that he served a higher power greater than their political authority ("But I serve a God who cannot be mocked!"). Then, in the most extraordinary challenge to political tyranny I have ever witnessed, Archbishop Desmond Tutu told the representatives of South African apartheid, "Since you have already lost, I invite you today to come and join the winning side!" He said it with a smile on his face and enticing warmth in his invitation, but with a clarity and a boldness that took everyone's breath away. The congregation's response was electric. The crowd was literally transformed by the bishop's challenge to power. From a cowering fear of the heavily armed security forces that surrounded the cathedral and greatly outnumbered the band of worshipers, we literally leaped to our feet, shouted the praises of God and began . . . dancing. (What is it about dancing that enacts and embodies the spirit of hope?) We danced out of the cathedral to meet the awaiting police and military forces of apartheid who hardly expected a confrontation with dancing worshipers. Not knowing what else to do, they backed up to provide the space for the people of faith to dance for freedom in the streets of South Africa.

I very often tell the story of the inauguration of Nelson Mandela, a decade later, which I was blessed to attend. I have never seen history change so much in a single day as on that day. I asked Bishop Tutu (the virtual master of ceremonies for the birth of a new nation), if he remembered the morning in St. George's and what he had said. His cherublike face lit up with a big grin (the characteristic smile the world associates with the Nobel Peace Prize–winning bishop). He remembered. "Bishop," I told him with tears running down my face, "today, they've all joined the winning side!" Indeed, you could hardly find, on that day, any white person in South Africa who hadn't *always* been against apartheid.

Desmond Tutu is a believing Christian, and it was his Christian belief that kept him going during so many difficult times. I think it is significant that many leaders in the most oppressive circumstances find themselves going more deeply into their faith, becoming more personal about their relationship to God, and even more orthodox in their religious convictions. Martin Luther King Jr. tells the famous story about his "kitchen table" conversion in the middle of the night after his home had been firebombed. He tells of how his liberal and intellectual faith became much more real and concrete as he cried out to God in fear and distress. Ideas about religion would no longer be enough for King, as his life and that of his family sleeping in the next room were now at stake. God had to be personal now and very real.

I remember a conversation with friends who were biblical scholars from a more liberal tradition than mine. We shared so much. They too saw the gospel as offering a fundamental alternative to the political idolatries and oppressions of our age. But they questioned me on my theology concerning the resurrection of Christ. "Your belief in the resurrection is more literal than ours," they said. "Ours is more metaphorical. But we come out at the same place politically. Is that alright?"

It is at this point in the conversation that the liberal/evangelical debate about the resurrection usually begins. But that debate is often a mostly intellectual one, with heady arguments flying back and forth, and usually ends up quite unresolved. Their sincere question prompted a different response in me. I simply asked a question back: "In the heat of South Africa's oppression and the heart of apartheid's despair, do you think that a merely metaphorical resurrection would have been enough for Archbishop Desmond Tutu? It wouldn't have been for me." Mere intellectual debates aren't enough when it comes to faith. It is what we face in our real lives and in the real world that has the most capacity to deepen our faith.

At the very end of Mandela's magnificent speech, another incredible moment occurred. I had already abandoned my seat among the international guests on the stage in order to stand with my South African activist friends who had hoped so long and paid so dearly for this day. All of a sudden, we heard a distant rumble. The crowd of 150,000 South Africans looked in the direction of the sound that was quickly increasing in volume. Soon military helicopters were plainly visible, and not far behind them were fighter planes. The South African Air Force was rapidly approaching. You could feel the crowd of 150,000 South Africans almost shrink away from the now roaring sound in reflexive anxiety. No, this could not be. They wouldn't do this! Not now, not after all this. No, that's too horrible to imagine. But what are they doing? Would the white South African military, even at this last moment, try to undo this new government, this South Africa, this new hope? Those were the unspoken thoughts I could literally feel in the crowd, which had been so jubilant a few moments before, and they spoke to the cold fear that gripped my own heart.

All of a sudden an extraordinary thing happened. The lead planes began to release a visible stream of colors—green, red, and yellow–in the shape of the new South African flag. The South African Air Force (the white guys) was doing a flyover of the inauguration ceremonies in honor of their new president. It was amazing.

It reminded me of another incident a few days before when some friends picked me up at the airport for the inauguration activities. On our way into Johannesburg, we rounded a corner and I spotted a squad of the South African Security Police. Having once been interrogated by these dangerous thugs, I instinctively shouted, "Quick, let's get away!" My South African friends smiled at me and said, "Don't worry. They're ours now." Who would have ever believed? And that's just the point. We have to believe.

I've simply seen too much not to believe—not to live and act in hope. In addition to seeing South Africa set free, here are just ten more things I've seen and had the privilege to be a part of that give me great hope.

As a teenager in the 1960s, I saw a civil rights movement, led by the black churches, suffer enormous violence but lead the nation to a history-changing civil rights act in 1964 and a democracy-changing voting rights act in 1965.

I've seen former gang leaders on American inner-city streets literally standing between warring factions ready to attack with their hands up in the air shouting "No, no, no! Don't do this!" And they didn't.

I've seen campaigns of nonviolent direct action topple generals in Latin America, the dictatorship of Ferdinand Marcos in the Philippines, and the bloody regime of Slobodan Milosevic in Serbia.

I've seen people's movements bring down the Communist governments of the Soviet Union and Eastern Europe, and I sat glued to my television set as gleeful Germans tore down the Berlin Wall brick by brick.

I've seen North American religious people stand at the edge of Nicaraguan villages, holding prayer vigils within clear and easy rifle range of

U.S.-backed "contra rebels" in the nearby hills who suspended their attacks against local civilians because of the "outside" presence. And I saw the U.S. State Department blame the churches when the Reagan administration lost the final vote that cut off funding for the "contra war" and decided not to invade the country because of a Pledge of Resistance that promised civil disobedience signed by almost one hundred thousand—mostly people of faith.

I've seen the religious community become the animating core of the 1980s movement to halt the nuclear arms race and help alert the nation to the nuclear danger.

I've seen a prophetic minority, but significant number of churches, try to deal seriously with the proposition that racial reconciliation requires racial justice.

I've seen churches and their leaders from across the theological and political spectrum come together on the issue of poverty and perhaps lay the foundations for a new movement of religious conscience for economic justice.

I've seen courageous Christian lawyers and activists act to protect and rescue vulnerable women and children from the brutal and violent forces of the world's sex-trafficking industry.

I've seen wealthy Christian businessmen leave behind lucrative careers and dedicate themselves to lives of service to the poor and, even better, use their skills to help create sustainable economic development in the developing world.

Now at the beginning of a new century and millennium, I see a new generation of young Christian activists coming of age and committing themselves to seeking social justice, preserving the environment, acting for peace, and trying to live consistent lifestyles—often in ways that require significant sacrifice and risk taking.

The Story of Hope

The fuller story of hope is finally one of many stories. Here are a few you may not have heard.

Our new field organizer for Call to Renewal was working the room at a conference of leaders of local councils of churches and interfaith organizations from around the country. I saw how pleased these slightly aging ecumenical leaders from mostly mainline Protestant churches were to have such an impressive young Christian woman among them, particularly one so passionate about the mission of overcoming poverty.

But as they got to know Christa Mazzone, they learned she is not the predictably liberal social gospel Christian one used to find at such conferences. Instead, this twenty-four-year-old Christian activist quietly talks about what "the Lord" is doing in her life and prayerfully considers what God might have in mind for her vocation.

Christa is an evangelical Christian, one who could never separate her faith in Jesus Christ from her commitment to social justice. Actually, it would never even occur to her. She is exemplary of a new generation of evangelical Christians for whom social activism is the natural outgrowth of personal faith. I first met Christa when I was teaching at North Park University, which, like a growing number of evangelical Christian campuses, has built social justice into the curriculum—even more so than secular universities.

I often have the opportunity to participate in gatherings of church leaders that bring evangelicals and Pentecostals together with Catholics, orthodox, and mainline Protestants. Quietly emerging in the United States is the kind of broad interdenominational and cross-confessional table that America has rarely seen, now called Christian Churches Coming Together in the USA (CCT-USA). One clear consensus among the church leaders is on the centrality of the issue of poverty. During one of the initial proceedings, one of the nation's most prominent Pentecostal leaders asked me if I thought evangelical and Pentecostal Christians were developing a deeper social conscience. I could tell that he thought so and hoped that I did too. When I told him I absolutely agreed that leaders from his tradition, especially the younger generation, were undergoing a real social transformation on issues of compassion and justice, a big smile broke out across his face.

The conventional wisdom still says that liberal Christians have a social conscience and evangelicals do not, preferring instead to focus only on the personal morality of issues such as abortion and homosexuality. The media in particular keep that perception alive.

But the big story that most of the press (including the religious press) continues to miss is how much that reality is changing. On at least three key social issues—poverty, race, and the environment—evangelicals are exhibiting a growing conviction and conscience. In local congregations, poor neighborhoods, and legislative halls, a new evangelical activism and advocacy is emerging.

One area where that new evangelical social conscience is clearly on the rise is on the environment, or the stewardship of creation, as many Christians would name it. One sign of the new Christian insurgency on ecology was the highly controversial campaign launched in 2002 that asked the provocative question, "What would Jesus drive?" A challenge to the nation's addiction to SUVs and their air-polluting emissions is not something evangelical Christians would have been expected to lead. But they did. As I survey the list of new Christian organizations and campaigns that focus on environmental stewardship, I observe that most of them have been founded by evangelicals—young evangelicals.

Sojourners has focused on the environment and the increasing Christian activism—much of it evangelical—that is rising up to offer new leadership. It may well be that only theology—good theology—can save the Earth now. And the fact that a new generation of Christians is offering an environmental social conscience is a sign of hope indeed.

I remember a week I spent in the Pacific Northwest. The *Portland Oregonian* story about Call to Renewal and our town meeting said, "Imagine attending a town meeting on ending poverty and coming away feeling hopeful instead of hopeless."[2]

I also spoke to two thousand high school students who were holding a mock political convention and said that unless their generation helps create a "new politics" in this country, all our future political conventions were destined to be "mock" ones. The popular wisdom says young people don't care about public life. The popular wisdom is wrong. Breaking the death grip of money over politics and dealing with real issues like child poverty in the midst of prosperity are things they really care about.

Seattle was the site of hopeful meetings with a cross section of church and community leaders. But the most moving experience for me was a speech given by Richard Stearns at World Vision's annual Washington

Forum. Rich is the new president of World Vision, a Christian relief and development organization. Before coming to World Vision, he was CEO of Lenox China. What does a successful businessman who sold fine china know about the poor, some critics asked when Stearns was named World Vision's new leader?

In a talk he titled "A Letter to the American Church," he cited letters to some of the earliest churches from the biblical book of Revelation. Stearns eloquently described the lives of two very different churches in the world today—one an affluent suburban American congregation, and the other a poor rural African faith community. You could tell he knew both churches very well. In Rich's story, the pastor of the poor church came to visit the rich church and was so excited to finally "meet the Christians who can help my people." But as he rose to the beautiful pulpit and began to speak his heart, the visiting preacher realized that nobody was listening and finally that these Christians couldn't even see or hear him. The poor Christian pastor was invisible in their midst. Desperate, he even followed them into the parking lot, pleading with them as they drove away in their luxury cars and SUVs. For his letter to the American churches, Stearns took every line from the Scriptures themselves so that these would be God's words, not just his.

I have another title for this talk about the church's responsibility to the poor, given by a former CEO from one of the nation's major corporations. I'd call it "The Rich Young Ruler Who Didn't Turn Away." It seems that those infected with HIV/AIDS, especially in sub-Saharan Africa (where the pandemic especially strikes the poorest of the poor), have become Stearns's abiding passion, another wholly unpredictable commitment from a suburban churched evangelical businessman. Rich Stearns continues to provide me with an astonishing sign of hope.

Lessons for Life

Sometimes the greatest stories of hope are the ones closest to home. It's often been my own best friends who have offered the hope that helps keep me going. In 2002, I participated in a memorial service for Chuck Matthei in Providence, Rhode Island. Five hundred people from across the country came together to remember, celebrate, and say good-bye to one of the most remarkable people that most of us had ever known. Chuck was a founding board member of Sojourners and, for me, a beloved friend for twenty-five years.

Chuck was the best and clearest thinker I ever knew on the critical issues of land, labor, and capital and especially on the moral values of economic life. He was the consummate "moral economist," after the manner of his mentor, Mahatma Gandhi.

Chuck had been fighting thyroid cancer for many months and was only fifty-four when he died. I was in Belfast when I heard that Chuck had taken a turn for the worse and likely had only a few days left to live. I got to him as quickly as I could and was blessed to spend the last hour of Chuck's life on earth with him. He couldn't speak anymore, but he was responsive as we had our last talk together. One of the things that most preoccupied Chuck at the end of his life was some "life lessons" he wanted to pass on to people who were very important to him. I promised I would help him do that, and it was Chuck's lessons that became the substance of my contribution at the memorial service. They came from my notes of an extended conversation we had in Boston a few months earlier. Here is the summary of the lessons that I passed on.

1. *Choice.* You can always make a choice. No matter what is done to you, no matter what people think, you have choices. Illness is like being in jail. You can think about what you've lost—but you'll go crazy. Or you can say, Here I am, what do I do? There is never a time when you can't make a meaningful decision. You can choose to be grateful.
2. *Integrity/dignity.* There is the "soul" or, as the Quakers say, "that of God in every person." This is key. Bernadette Devlin once said, "To keep your dignity, you may have to give up everything else." Integrity is who you are when your eyes are wide open. It is to make the best and most thoughtful choice. They can't take your dignity away from you. They can make you pay a price for it, but they can't take it away. My advice is to maintain an identity and dignity that are profound, while humble. The first two lessons represent opportunities—choice and integrity/dignity. The third is an obligation.

3. *Open your eyes.* Augustine said, "Love God, and do as you please." He did not say, "Do as you please." Look around you. Who do you see? I would say, "Open your eyes, then do as you please." Find God in the other, in what Christians call "the least of these." Decide what you want with your eyes wide open—depending on who is at the table. See what's there. Put yourself in context. Walk across the street. See who is around. See as much as you can see. Then make a decision.
4. *Relationships.* If you exercise your ability to choose, maintain your integrity, look at the other and include them in your field of vision, then you will see the web of relationships with others and the whole creation. Then you will value a spirit of welcome and the practice of compassion.

Before he died, Chuck said, "I have no fear of loneliness. People who are facing death, especially untimely, often feel two things. They feel alone, and they feel regret. I am surrounded by family and friends. And I have no regrets. Sure, I would change some things—if you wouldn't you haven't learned anything. But I have no regrets." In our last phone call from Northern Ireland, Chuck could barely talk. But the words he struggled to repeat to me, over and over, were "family . . . friends." In Chuck's life I saw again that it is the myriad acts of kindness that keep us alive. In history, it's not the will to power, wealth, or sex that ultimately matters—but rather the will to meaning. We are not alone, even though the culture teaches us to fear. If you do these things, Chuck reminded us, the world can be a good and welcoming place. These were Chuck's lessons for life.

Matthew's Homeless Church

I remember a homeless young man I met in my course at Harvard's Kennedy School, of all places. I had spoken at a conference on homelessness at the Massachusetts Statehouse, and this young man was there. In the small discussion groups, he met some of my students, who were speaking about our class. It sounded interesting to him, so Matthew started coming. Every week he would show up on time and quietly sit at the back, paying close attention.

One night, I invited him to join several students and me for dinner afterward. In conversation around the table, he began to open up to us about how a workplace conflict had left him without a job and eventually homeless.

We all noticed that Matthew always carried around a large cardboard box, which he would carefully set down next to him. Exactly what was in that box? After the last night of class, Matthew came with several of us to a Call to Renewal organizing meeting at the Divinity School. Afterward, as we were in the refectory for refreshments, I looked over and saw that Matthew had opened his box and placed the contents on one of the tables. People gathered around to view a beautifully crafted model of a church made from white cardboard. All along the outer walls of the steepled church were the words of the prophets and the sayings of Jesus, beautifully written in Matthew's own hand—almost like calligraphy. Over the front door, Christ's words appeared, "Come to me, all you who are heavy laden, and I will give you rest." Right beneath the words was a door, closed shut with a little padlock. The message was clear.

Matthew told me how he made his church and protected it from the elements. He smiled and said, "Sometimes, people like the church so much they offer to keep it in their apartment for me, so it doesn't get damaged." With a twinkle in his eye, he added, "But they don't make the same offer to me. Only to my church."

Some would not see a story of hope in Matthew's church, but I do. Matthew understood, better than most church people, what the teachings of the prophets and Jesus were and really meant. He believed in their power and their hope. Why else would he carry that beautiful miniature church around with him everywhere he went? When he spoke of it, Matthew wasn't bitter, but rather sad. He was sad that the church people didn't quite get it. Yet his small act of consistent witness was a way of hoping they someday would.

The Power of Reconciliation

Bill Bright was the founder and president of Campus Crusade for Christ, an evangelical student organization on campuses around the country. A conservative businessman from southern California, Bill became the ultimate evangelist. Motivated, above all else, by the Great Commission, he wanted to reach every person on the planet for Christ "in this generation." But his special burden and passion was to reach young people. He was also known to be very conservative politically and, on virtually every single issue, usually found himself in agreement with the right wing of the Republican Party. Concerned about the "moral degeneration" of America, he wanted America to come back to God, which for him meant an ultraconservative political agenda. Bill and I were both evangelical Christians but clearly disagreed on a whole range of political issues.

In 1976, he got involved in a project to get conservative Christians organized politically. With John Conlan, a far-right congressman from Arizona who had presidential ambitions, and a whole range of conservative Christian groups, a plan was formulated to mobilize evangelical prayer and cell groups for political purposes. It was, in fact, the first attempt to create a religious Right in American politics—several years before the eventual founding of successful groups like the Moral Majority and the Christian Coalition.

Several people who had gotten wind of this new evangelical political initiative contacted us at *Sojourners* and suggested we do an investigative story on the plan. We decided to investigate and found a very ambitious political effort under way. It became the most extensive investigative project we had ever undertaken, and the story took several months to complete.

In April 1976, it became a major cover story for *Sojourners* called "The Plan to Save America." The story got picked up across the nation and in several local newspapers, including in Arizona. The more people heard about the effort to "politicize" Christian prayer and fellowship groups, the more many didn't like it. There was a clandestine feel to the political organizing effort that also turned many off when it was finally exposed. A few years later, much more overt political initiatives to create the religious Right that we know today would prove far more successful than this awkward first attempt.

But Bill Bright was publicly embarrassed by our story and the whole experience. He blamed us for that, and me in particular. Though we had been scrupulously careful to back up every fact in the story with at least three sources, Bright angrily denounced me. We invited him or the others involved to respond, both before and after the article was published, but they never chose to. Because we also differed on almost every political issue, from Vietnam and foreign policy to economic and domestic issues, a bitter and public polarization grew up between Bill Bright and me.

The "bad blood" continued for many years. I remember a particularly painful moment at a dinner for evangelical leaders the night before the annual Presidential Prayer Breakfast in Washington, D.C., one year. In the personal introduction time, Bright again went on the attack with me in a very public way, calling me a "liar." A shouting match erupted between Bright and my defenders, and I feared a fistfight would break out.

We didn't see each other for many years, and the bitterness seemed to subside over time. More than two decades had passed when Bill Bright and I found ourselves at yet another leaders' dinner the night before another Presidential Prayer Breakfast. People were milling around before the meal, and I saw him across the room. I swallowed hard and headed in his direction. He turned to shake another friend's hand and obviously didn't recognize me after so long (we had both gotten a bit older). I introduced myself, and he became quiet. I said, "Bill, I need to apologize to you. I was in a hotel several months ago and knew you were also there. I should have come to your room, knocked on your door, and tried to mend the painful breach between us after all these years. I didn't do that, and I should have. I'm sorry."

The now old man named Bill Bright melted into my arms and rested his head on my shoulder. Then he said, "Oh Jim, we need to come together. It's been so long, and the Lord would have us come together." We both had tears in our eyes and embraced for a long time. Then Bill said, "Jim, I'm so worried about the poor, about what's going to happen to them. You're bringing us together on that, and I want to support you." I was amazed and humbled. We agreed to get together soon to spend more time.

A few months later, Bill and I were again, coincidently, at the same hotel, a wonderful place in Daytona Beach, Florida, called El Caribe, where an extraordinary Christian laywoman named Mary Ann Richardson has turned a business into a ministry, serving both local and national Christian groups across the country. She has been a good friend to both Bill and me for many years. I called Bill and told him I was on the floor below. We agreed to a walk on the beach together the next morning.

It was a wonderful stroll along Daytona Beach. We shared our own conversion stories, which we had never done before, our callings and dreams for our respective ministries, and how we might be more connected. Bill Bright, the conservative businessman from southern California then astounded me. He said, "You know Jim, I'm kind of a Great Commission guy." I smiled and nodded my head. "And I've discovered that caring for the poor is part of the Great Commission. Because Jesus instructed us to 'teach the nations to observe all the things I have commanded you.' And Jim, Jesus certainly taught us to care for the poor, didn't he? Caring for the poor is part of the Great Commission!" said Bill Bright. When we got back to El Caribe, Bill asked if we could pray together. We sat down and grasped each other's hands. First praying for each other, we also prayed for each other's ministries. Bill Bright prayed for me and for the work of Call to Renewal and *Sojourners*. When we were finished, he said he wanted to raise some money for our "work of the Lord."

In the months that followed, Bill began to get sick; he was now over eighty years old. I kept track of how he was doing. Then one day I got a letter—from Bill Bright. Here's what it said:

> My Dear Jim,
>
> Congratulations on your great ministry for our Lord. I rejoice with you. An unexpected gift designated to my personal use makes possible this modest contribution to your magazine. I wish I had the means to add at least three more zeros to the enclosed check.
>
> Warm affection in Christ.
>
> Yours for helping to fulfill the Great Commission each year until our Lord returns.
>
> Bill

Inside the letter was a check for $1,000.

As I was reading Bill's letter, my colleague Duane Shank walked into my office. "Did you hear?" he asked. "Bill Bright just died." We looked at the postmark on the letter and compared it to the news reports of Bill's death. We concluded that writing me this letter was one of the last things that Bill Bright did on earth. Bill sent a $1,000 gift to the magazine that had exposed his most embarrassing moment more than thirty years before, as an affirmation of the ministry of another Christian leader whom he once regarded as his enemy. I couldn't hold back the tears, and I can't again as I write down this story for the first time. The experience of my relationship with Bill Bright has taught me much about the promise and power of reconciliation. I will never deny the prospect of coming together with those I disagree with—ever again. It is, indeed, the power of the gospel of Jesus Christ to break down the walls between us. Thank you, Bill. I will never forget you.

Our Interfaith World

Our world and the United States, in particular, are becoming increasingly pluralistic religiously. How do we navigate the new waters of religious diversity in America and beyond? I remember a prayer vigil that Call to Renewal held in the Rotunda of the U.S. Capitol, when fifty Christian leaders came to protest legislation we found very detrimental to poor families and children. The night before the action, I got a call from Rabbi Michael Lerner, who asked, "Can you use a Jewish rabbi?" Sure, I replied, you can read the Isaiah text!

The scene the next morning was dramatic indeed. Dressed in clerical garb, pastors and priests, lay leaders and seminary professors, religious sisters and organizational heads—and one rabbi—stood in a circle to pray and read from the Scriptures. "Woe to you legislators!" read the words of Isaiah, "who write oppressive statutes, to turn aside the needy from justice and to rob the poor of my people of their right." The sergeant at arms of the National Capitol Police offered our "call to worship" when he proclaimed over his bullhorn, "You are not allowed to pray in the Rotunda of the United States Capitol!" That was all we needed, and we settled in to pray.

One by one, we were handcuffed and led away by the police. Rushed off to jail, we began many tedious hours of processing and waiting until they finally let us go (pending trial dates) in the middle of the night. But we had almost twelve hours in jail together with no phones, conference calls, meetings, deadlines, or to-do lists. For most of us, the break in the routine was glorious, as was the fellowship with our brothers and sisters. When do people like us get to spend so much time with one another?

I smiled as I looked around the massive jail cell and focused on what seemed to be a very vigorous conversation going on between Baptist preacher Tony Campolo and Jewish rabbi Michael Lerner. I moved closer to hear. They were discussing Christology, the theological topic concerning the nature and identity of Jesus Christ. "Who was Jesus?" is still perhaps the most provocative religious question in our interfaith world. Tony and Michael were deeply engaged together. Was Jesus the very Word of God made flesh or a great Jewish prophet? They clearly didn't agree and were not likely to convince each other anytime soon.

But the amazing dialogue taught me something very important. Perhaps the best place to discuss theology is in jail, after you have all been arrested for acting on your faith. Act faithfully, then talk about why you have faith and what your faith means. Michael always tells me he wants Christians to be the best Christians they can be and then find ways to cooperate together with Jews around the biblical vision of justice.

I don't believe in "interfaith" services and activities where everybody is afraid to say anything that someone in the congregation might not agree with. Don't change or water down who or what you are for your interfaith brothers and sisters. Don't be afraid to express your faith clearly and compellingly. Act on your faith, then reflect together.

And the best place for our theological seminars might well be in jail!

Dare to Dream

The students in my class at the Kennedy School of Government were a sign of hope to me. That's why I have kept going back to Harvard to teach. The course is called "Faith, Politics, and Society." At the end of each class, I write them all a letter on the difference between vocation and career and encourage them in their decision to act in hope. Here is the letter I wrote to my students at the completion of one of our semesters:

Dear Class Members:

You are a bright, gifted, and committed group of students. There are probably many people who tell you about your potential, and they are right. You are people who could make a real contribution to the movement for social and economic justice that we dreamed about this semester.

In that regard, I would encourage each of you to think about your vocation more than just your career. And there is a difference. From the outside, those two tracks may look very different or very much alike, but asking the vocational question rather than just considering the career options will take you much deeper. The key is to ask why you might take one path instead of another—the real reasons you would do something more than just because you can. The key is to ask who you really are and want to become. It is to ask what you believe you are supposed to do.

Religious or not, I would invite you to consider your calling, more than just the many opportunities presented to graduates of the Kennedy School. That means connecting your best talents and skills to your best and deepest values; making sure your mind is in sync with your soul as you plot your next steps. Don't just go where you're directed or even invited, but rather where your own moral compass leads you. And don't accept others' notions of what is possible or realistic; dare to dream things and don't be afraid to take risks.

You do have great potential, but that potential will be most fulfilled if you follow the leanings of conscience and the language of the heart more than just the dictates of the market, whether economic or political. Rather than merely fitting into systems, ask how you can change them. You're both smart and talented enough to do that. That's your greatest potential.

One of you told me as we said goodbye last night, "When I started this course, I was cynical. I'm not cynical anymore." Nothing could make me feel better about teaching this course. Cynicism really comes out of despair, but the antidote to cynicism is not optimism but action. And action is finally born out of hope. Try to remember that.

I first went to Harvard as a Fellow at the Center for the Study of Values in Public Life, based at the Divinity School. It was a wonderful year of residence in Cambridge for Joy and me at the beginning of our marriage, and our son Luke was born in Boston. When the regular seminars I convened on faith and politics attracted students from across the university and the wider Boston community, I knew the topic had struck a chord among a new generation of young people. When invitations started coming from other schools at Harvard, it became clear that the interest in religion and public life was much broader here than I expected.

My first invitation to speak at Harvard's Kennedy School of Government was sponsored by the Christian Fellowship and the Progressive Caucus—two groups that had never worked together before. The room was packed with both religious and "secular" students who were hungry to discuss the connections between politics and moral values. Another invitation followed from the Institute of Politics at the Kennedy School to speak at one of their regular IOP Suppers, which normally bring people from the worlds of government and journalism to discuss the hottest political topics. Apparently, faith-inspired movements to overcome poverty had become one. More venues at other schools, churches, and organizations around Harvard and Boston opened up throughout the year, and I was continually amazed at the turnouts and responses from a wide variety of students to these conversations about spirituality and social change.

I could feel myself moved by the students. They were the next generation, and they became the highlight for me at Harvard and the reason I've kept coming back. And there has begun a steady stream of Harvard students and graduates who come to work at Call to Renewal and write for *Sojourners*. The focus for so many of the young people I've met is how to put their faith into action in a world in desperate need of both justice and hope.

The remarkable year at the Divinity School led to an invitation to teach a course at the Kennedy School for the next two years, titled "Faith, Politics, and Society." Our class drew up to ninety students from most of Harvard's graduate schools and undergraduates from many majors and departments. We also attracted students from more than twenty countries and from diverse racial and religious backgrounds. Our topics ranged from domestic to international issues and included how faith-inspired movements and initiatives were helping to change America and the world.

But what inspired me most were the questions my students were asking. By halfway through the course each year, my individual conversations with students were less about the final paper topic and more about what they were going to do with their lives. I learned that mentoring was as central to teaching as lecturing and that when you have the opportunity to help a young person shape his or her life choices and direction you are treading on very holy ground. My last lecture always focused around the crucial difference between career and vocation.

And the choices these students are making never cease to amaze me: running for Congress on an agenda of overcoming poverty, going to medical school to do health care reform, building youth networks to find solutions to HIV/AIDS in Africa, applying MBAs to managing nonprofits seeking social change.

It was the students who encouraged me to return to Harvard again as a Fellow at the Institute of Politics. The topic for my weekly study group was activism, spirituality, and social change, and the student response indicated the interest for this generation in the "soul" of politics. My five student "liaisons" were freshmen, sophomores, and juniors who evidenced a hunger for public service and a new kind of politics rooted in moral values. That last dinner had each one sharing the next steps they hoped to take.

It is this next generation that keeps me trekking back and forth to Harvard and other campuses across the country. They are shaping a vision for going public with their faith and hopes, whether they are religious or not. And they give this middle-aged activist a lot of renewed energy.

The Nice Guys Finished First

Hope isn't always just about ending apartheid, ending global poverty, and other matters of life and death. In the midst of tax cuts for the rich and no child tax credits for the poor, weapons of mass destruction missing in Iraq, eroding civil liberties, and Code Orange as a way of life, sometimes it's good to talk about something uplifting: basketball. Remember the comments earlier in this chapter about fun? One thing I've learned after many years of social activism is that you have to learn to enjoy the world while you're busy trying to change it. And some consistent sources of enjoyment are necessary for a balanced and, yes, hopeful life.

For those of you who don't follow the game, the San Antonio Spurs defeated the New Jersey Nets, on Father's Day 2003, to win the National Basketball Association championship. The Spurs are a team full of nice guys, in stark contrast to the thug and flash role models who have come to dominate the NBA in recent years. Who says nice guys finish last?

It was the final game in the remarkable career of David Robinson, the fourteen-year veteran San Antonio center—and what a way to go out. Robinson had been a league Most Valuable Player, won Defensive Player of the Year, made the All-Star team fourteen times, and was named among the best fifty NBA players of all time. But most important, David Robinson is known as one of the best people ever to play the game. Motivated by a strong Christian commitment, Robinson spent $9 million to start a school for inner-city kids (and didn't even name it after himself!). He and his wife, Valerie, run a faith-based foundation that addresses the physical and spiritual needs of families, and he created "Mr. Robinson's neighborhood," a section of special seats reserved at all Spurs home games for needy kids and their parents.

In the final game and throughout the playoff series, the thirty-eight-year-old Robinson made a mighty contribution to winning his second NBA title. He scored key points, snatched critical rebounds, blocked shots, and raced down the court on his old legs to make a dramatic steal that turned pivotal game 4 around. After the sweet victory, the scene on the court of an emotional David Robinson holding his three sons with his own father at his side—on Father's Day—was a rare and welcome picture of "family values" in big-time sports.

The Most Valuable Player in the championship series, and the season MVP, was Tim Duncan, Robinson's companion at power forward. Duncan was utterly dominant during the entire series, with his consistent spectacular play and team leadership, yet at the same time, he evidenced a humble spirit almost never seen in professional sports today. A humble NBA basketball star? Duncan is a quiet, even self-effacing, just-get-the-job-done kind of guy. But get it done he does. In the last game, Duncan accomplished the nearly unbelievable—"a quadruple double," meaning he reached double figures in four categories: points, rebounds, assists, and blocked shots. Game announcer and Hall of Fame center Bill Walton just kept repeating that Duncan was simply the best player in basketball today. (Take that Shaq, Kobe, and Iverson!)

Duncan doesn't trash talk, chest bump, strut his stuff, go into tirades for the camera, or embarrass his city off-court. Instead, he just plays good, smart, team-oriented, and amazing basketball and brings a big breath of fresh air for all of us who love the game but have gotten depressed by players with no work ethic, no respect, no character, and lots of commercial endorsements.

The whole San Antonio team seems to embody the ethos of Robinson and Duncan, showing that good habits and attitudes can be as infectious as bad ones. While, of course, staying focused on Iraq, the federal budget, and faith-based initiatives that bring a prophetic political witness, I watched every game of the championship series. It proved to be a welcome break and lovely diversion from life's intensity these days. It brought together two great things: good basketball and good values. What a combination.

Culture does shape politics, and spiritual values are central to that. Role models, in particular, are hugely important today, especially in shaping the character of young people. Perhaps the best thing the Spurs did that year was to beat the mighty Los Angeles (Hollywood Hype) Lakers who have gotten used to cruising through the season and winning the playoffs. But in 2003, the postgame interviews were not with oversized prima donnas droning on endlessly about themselves, but with polite and modest superstars giving the credit to their teammates, a great game, and even to God. Sweet.

Predictions for the New Millennium

Everyone says don't make predictions, so I can't resist. Of course, America missed any real chance to celebrate the coming of the new millennium in a significant way as other people were doing around the world. Many of us were too busy worrying about Y2K, stocking up on cans of Spam, or booking ourselves into expensive Las Vegas parties awash with celebrities. In England, every community received a grant to improve or create something new in their public common space to mark the millennium. We didn't do anything like that in America.

We could have done so much more. The nation could have used the historic occasion to candidly acknowledge the deep injustices that attended the founding and formation of our country—Native American displacement and genocide, slavery, racial and gender discrimination, and labor exploitation—then gratefully celebrate progress made in civil rights and women's enfranchisement and commit ourselves to fulfilling the promise of our democracy. We could have celebrated the richness of American literary, musical, and artistic expression by teaching young people to value books and culture over mindless materialism. Churches could have marked the two-thousandth birthday of Jesus by asking their members to examine seriously how his teachings might really be applied to our lives and society in the new millennium. Another missed American opportunity. But let's at least make some predictions for the new millennium.

1. Faith in the new millennium will be defined much more by action than by doctrine.
2. At the same time, religious fundamentalism will continue to rise in the face of moral decline.
3. Bible study will continue to grow in popularity among a wide variety of people.
4. Prayer will be even more important to people than it is now.
5. The religious Right will lose control of the discussion of religion and public life and other voices will be heard.
6. The secular Left will give up its hostility to religion and spirituality, or it will die.
7. The Spice Girls won't be remembered, and Martin Luther King Jr. will.
8. Family values (meaning what's good for parents and their kids) will be embraced by people across the political spectrum.
9. Women in leadership in every area of life will become a given.
10. Overcoming poverty will become the great moral issue as we move into the new millennium.
11. The unfinished agenda of racism will be impossible to ignore in the face of increasing diversity.
12. Internet pornography will quietly undermine people's lives and relationships, if there are no restraints.
13. Nelson Mandela's stature will grow as a role model for moral integrity and spiritual discipline, while most of our other political leaders will be forgotten.

14. Some liberals will get the values questions right, and some conservatives will really care about poor people.
15. So a new option will emerge: conservative in personal values, radical for social justice. The number of spiritual progressives will grow.
16. A new alliance across political lines between parents of all stripes will take on the moral pollution of the culture by Hollywood, the Internet, and the corporate advertising world.
17. Old ecumenical structures will gradually dissolve in favor of new tables that bring together evangelicals, Pentecostals, Catholics, mainline Protestants, and historic black churches along with Latino and Asian congregations.
18. The abortion rate will continue to decline as moral concern grows and practical alternatives spread.
19. The challenge of pluralism will replace the challenge of secularism as many diverse religious and spiritual traditions have to learn to live with one another.
20. Sexual restraint, fidelity, and integrity will make a comeback as the results of "sexual freedom" are understood and rejected.
21. More parents will choose good books over mindless and soulless television.
22. Those who don't will produce children who are increasingly mindless and soulless.
23. The enormous and growing gap between the rich and the rest of us will finally be recognized as a real problem for both democracy and religion, shaking up our two-party politics (which is really only one party of the very rich and powerful).
24. The movement that started with the Jubilee 2000 campaign will ultimately succeed in eliminating the unpayable Third World debts.
25. Fair trade will become as important to us as free trade.
26. Nuclear weapons will become a big issue again, but the real question is whether anything will really be done about them until a city is incinerated.
27. Human rights will replace national sovereignty as the key international issue.
28. Wealthy countries will become inundated with immigrants unless the North/South economic divide is faced.

29. Something like the Marshall Plan, which rebuilt vanquished nations after World War II, will be created for the developing world.
30. Billy Graham will be remembered with more respect than all the presidents he knew.
31. More and more affluent families will get off the pressure train and adopt simpler lifestyles.
32. More churches will throw their arms around at-risk kids, but it won't be enough unless the whole society puts children first.
33. Faith-based organizations will become critical partners in forging new social policy, but will tell government that they can't solve poverty by themselves.
34. The need for prophetic religion will grow.
35. More and more people will ask why we're spending more for cosmetics, pet food, and ice cream than on making a decent and dignified life possible for the world's poorest people.
36. Television will get worse, and more people will decide they don't want their reality to be like reality TV.
37. Radio will become more and more important as an alternative media.
38. The Internet will further isolate the poor, and the Internet will help create greater democracy—raising the question of whether those two trends ultimately are reconcilable.
39. In the Catholic Church, we'll have married and women priests, and the importance of lay and female leadership will continue to grow.
40. The churches finally will not divide over homosexuality.
41. The concept and discipline of the Sabbath will see a great comeback in the lives of overworked and overstressed people.
42. Violence will be a culture-wide issue, not just an inner-city problem.
43. Peacemaking and conflict resolution will be regarded as being among our most highly valued skills.
44. We will have to learn much more about forgiveness and reconciliation if we are to heal the violence.
45. All our media will be owned by two or three corporate conglomerates unless an effective movement rises up to stop the trend and restore a genuinely democratic public discourse.

46. Wal-Mart will sell us everything unless we act strategically as consumers to restore a genuinely free and diverse marketplace and as citizens to support a revitalized labor movement.
47. The Left will decide, as the conservatives already have, that ideas are important and will begin to offer some better ones.
48. Having fun will become more important.
49. Raising children will be seen as the most important thing.
50. Hope will be the most essential ingredient for social change.

EPILOGUE

We Are the Ones We've Been Waiting For

One of the best street organizers I ever met was Lisa Sullivan. Lisa was a young African American woman who earned the trust of urban youth in Washington, D.C., and around the country. Lisa was from D.C., a smart kid from a working-class family who went to Yale and earned a PhD. With early jobs in major national foundations and nonprofit organizations, Lisa felt called back to the streets and the forgotten children of color who had won her heart. With unusual intelligence and entrepreneurial skills, she was in the process of creating a new network and infrastructure of support for the best youth-organizing projects up and down the East Coast. But at the age of forty, Lisa died suddenly of a rare heart ailment. At her graveside, Marian Wright Edelman and I held each other and wept for the loss of one of the most promising leaders of her generation, a friend to us both. Lisa had worked with Marian's Children's Defense Fund and was on the board of Sojourners.

Lisa's legacy is continuing though countless young people whom she inspired, challenged, and mentored. But there is one thing she often said to them and to all of us that has stayed with me since Lisa died. When people would complain, as they often do, that we don't have any leaders today or would ask where the Martin Luther Kings are now, Lisa would get angry. "We are the ones we have been waiting for!" she would declare. Lisa was a person of faith. And hers was a powerful call to leadership and responsibility and a deep affirmation of hope.

Lisa's words are the commission I want to use to conclude this book. It's a calling the prophets knew and a lesson learned by every person of faith and conscience who has been used to build movements of spiritual and social change. It's a calling that is quite consistent with the virtue of humility because it is not about taking ourselves too seriously, but rather about taking the commission seriously. It's a commission that can only be fulfilled by very human beings, but people who, because of faith and hope, believe that the world can be changed. And it is that very belief that changes the world. And if not us, who will believe? After all, we are the ones we have been waiting for.

Notes

Introduction

Why Can't We Talk about Religion and Politics?

1. Exit poll at www.cnn.com.
2. E. J. Dionne, ". . . He Didn't Get," *Washington Post,* November 5, 2004.
3. U.S. Conference of Catholic Bishops, *Faithful Citizenship,* November 2003, p. 2.

Chapter 1

Take Back the Faith: Co-opted by the Right, Dismissed by the Left

1. David Morgan, "Bush Seeks to Mobilize Religious Conservatives," Reuters, July 1, 2004.
2. David Brooks, "How to Reinvent the GOP," *New York Times Magazine,* August 29, 2004.
3. Poll by Tom Freedman, Bill Knapp, and Jim McLaughlin for the Alliance to End Hunger and Call to Renewal, February 2004.

Chapter 3

Is There a Politics of God? God Is Personal, but Never Private

1. Jim Wallis, "A Pastor Takes on Jesse Ventura," *MSNBC Online,* October 13, 1999.

Chapter 5

How Should Your Religious Faith Influence Your Politics?

What's a Religious Voter to Do?

1. Pew Research Center for the People and the Press, "The 2004 Political Landscape Evenly Divided and Increasingly Polarized," November 5, 2003.
2. www.cnn.com/election/2004.
3. John Kerry, "Remarks to the 124th Annual Session of the National Baptist Convention," September 9, 2004.
4. Cal Thomas and Ed Dobson, *Blinded by Might* (Zondervan, 1999)
5. "Lift Every Voice: A Report on Religion in American Public Life," a special report by the Pew Forum on Religion and Public Life, December 2001; "Religion and Politics: Contention and Consensus," Pew Forum on Religion and Public Life, July 24, 2003.
6. E. J. Dionne, "The Lieberman Litmus Test," *Washington Post,* September 5, 2000.
7. Joseph Lieberman, speech at the Fellowship Chapel, Detroit, August 28, 2000.

Chapter 6

Prophetic Politics: A New Option

1. "Pat Robertson: God Says Bush Will Win in 2004," AP/*Fox News,* January 2, 2004.

Chapter 7

Be Not Afraid: A Moral Response to Terrorism

1. Thomas Merton, "The Root of War Is Fear," October 1961.
2. Gallup poll on response to 9/11.
3. *Meet the Press,* September 23, 2001.
4. "But Then, Nothing, Nothing Justifies Terrorism," *Al Ayyam,* September 17, 2001.

Chapter 8

Not a Just War: The Mistake of Iraq

1. Dana Milbank and Claudia Deane, "Hussein Link to 9/11 Lingers in Many Minds," *Washington Post,* September 6, 2003.
2. Rose Marie Berger and Jim Rice, "The Burden of Truth," *Sojourners,* November–December 2003, p. 22.
3. Peter J. Gomes, "Patriotism Is Not Enough," *Sojourners,* January–February 2003, p. 20.

4. Iraq military casualties, U.S. Department of Defense: http://www.defenselink.mil/news/casualty.pdf. Iraq Coalition Casualty Count: http://icasualties.org/oif. Civilian casualties: http://www.iraqbodycount.net.
5. Steven Mufson, "There Are 87 Billion Reasons to Revisit Those Tax Cuts, Mr. President," *Washington Post,* September 14, 2003.
6. Jeffrey D. Sachs, "A Better Use for Our $87b," *Boston Globe,* September 13, 2003.
7. Patrick Graham, "'We've Had a Lot of Experience of US Weapons,'" *Observer,* May 2, 2004.
8. Seymour M. Hersh, "Torture at Abu Ghraib," *New Yorker,* May 10, 2004.
9. Amnesty International, "Iraq: Amnesty International Reveals a Pattern of Torture and Ill-Treatment," http://web.amnesty.org/pages/irq-torture-eng.
10. Tom Regan, "US General: Abu Ghraib Abuse Coverup," *Christian Science Monitor Online,* May 3, 2004, http://www.csmonitor.com/2004/0503/dailyupdate.html.
11. "Iraq: ICRC Explains Position over Detention Report and Treatment of Prisoners," International Committee of the Red Cross, May 7, 2004, http://www.icrc.org/web/eng/siteeng0.nsf/iwplist322/7ee8626890d74f76c1256e8d005d3861.
12. "Interview of the President by Al Arabiya Television," May 5, 2004, http://www.whitehouse.gov/news/releases/2004/05/20040505-2.html
13. "Unilateral Attack Would Be 'War of Aggression,' Says Vatican Official," Zenit News Agency, March 11, 2003.
14. "Military Intervention in Iraq Would Be a Crime, Says Vatican Official, Zenit News Agency, March 17, 2003.
15. "Pope to Ask Bush for Radical Shift in Policy, Says Cardinal Laghi," Zenit News Agency, May 13, 2004.

Chapter 9

Dangerous Religion: The Theology of Empire

1. Joseph Nye, "Ill Suited for Empire," *Washington Post,* May 25, 2003.
2. Jay Bookman, "Now in Open, 'Empire' Talk Unsettling," *Atlanta Journal Constitution,* May 8, 2003.
3. Project for the New American Century, "Rebuilding America's Defenses: Strategy, Forces and Resources for a New Century," September 2000; Project for the New American Century, "Statement of Principles," June 3, 1997.
4. Lawrence F. Kaplan and William Kristol, *The War over Iraq,* Encounter Books, 2003.
5. The second Gore-Bush presidential debate October 11, 2000: http://www.debates.org/pages/trans2000b.html.

6. Joe Klein, "Blessed Are the Poor—They Don't Get Tax Cuts," *Time*, June 2, 2003
7. David Frum, *The Right Man* (New York: Random House, 2003).
8. George W. Bush, "State of the Union," January 28, 2003.
9. Deborah Caldwell, "George Bush's Theology: Does President Believe He Has Divine Mandate?" *Religion News Service*, February 12, 2003.
10. Michael Duffy, "The President, Marching Along," *Time*, September 1, 2002.
11. Martin Marty, "The Sin of Pride," *Newsweek*, March 10, 2003.
12. "George W. Bush's Religious Rhetoric," *Christian Century*, March 2003.
13. Tony Carnes, "The Bush Doctrine," *Christianity Today*, May 2003.
14. Joe Klein, "The Blinding Glare of His Certainty," *Time*, February 18, 2003.
15. Alan Cooperman, "Bush's Remark About God Assailed," *Washington Post*, November 22, 2003.
16. Jane Lampman, "New Scrutiny of Role of Religion in Bush's Policies," *Christian Science Monitor*, March 17, 2003.
17. George F. Will, "No Flinching from the Facts," *Washington Post*, May 11, 2004.
18. Archbishop of Canterbury Rowan Williams, "John Mere's Commemoration Sermon," St. Benet's Church, Cambridge, April 20, 2004.
19. Hanna Rosin, "When Joseph Comes Marching Home," *Washington Post*, May 17, 2004.
20. Jane Lampman, "New Scrutiny of Role of Religion in Bush's Policies," *Christian Science Monitor*, March 17, 2003.
21. William M. Arkin, "The Pentagon Unleashes a Holy Warrior," *Los Angeles Times*, October 16, 2003.
22. R. Jeffrey Smith and Josh White, "General's Speeches Broke Rules," *Washington Post*, August 19, 2004.

Chapter 10

Blessed Are the Peacemakers: Winning Without War

1. Peter Ackerman and Jack DuVall, "With Weapons of the Will," *Sojourners*, September–October 2002, p. 20.
2. John Howard Yoder, *The Politics of Jesus*, Eerdmans, (2nd edition), 1994, p. 204.
3. Walter Wink, *The Powers That Be* (New York: Galilee/Doubleday, 1998), p. 159.
4. Gerald Schlabach, "Just Policing, Not War," *America*, July 7, 2003.
5. Hauerwas, Wink & Lederach—Jim Wallis, "Hard Questions for Peacemakers," *Sojourners*, January–February 2002, p. 29.
6. April 2002 Islamic conference on terrorism speech.

7. "Kuala Lumpur Declaration on International Terrorism," adopted at the Extraordinary Session of the Islamic Conference of Foreign Ministers on Terrorism, April 1–3, 2002, http://www.oic-oci.org/english/fm/11_extraordinary/declaration.htm.
8. William Orme, "Anti-terrorism Drive Idling," *Los Angeles Times,* April 16, 2002.
9. Kofi Annan, "Address to the General Assembly on Terrorism," October 1, 2001. http://www.un.org.pk/latest-deve/hq-pre-011001.htm.

Chapter 11
Against Impossible Odds: Peace in the Middle East

1. "Fatalities in the al-Aqsa Intifada, 29 Sept. 2000–15 October 2004," B'Tselem, Israeli Information Center for Human Rights in the Occupied Territories, http://www.btselem.org/english/statistics/index.asp.
2. Rabbi Brian Walt and Rabbi Arthur Waskow, "Replanting Trees, Rebirthing Peace: Tu B'shvat in Israel and Palestine," Shalom Center, February 1, 2002.

Chapter 12
Micah's Vision for National and Global Security: Cure Causes, Not Just Symptoms

1. "Radio Address by the President to the Nation," August 31, 2002. http://www.whitehouse.gov/news/releases/2002/08/20020831.html
2. Project for the New American Century, "Rebuilding America's Defenses: Strategy, Forces and Resources for a New Century," September 2000, p. 14.
3. Martin Luther King Jr., "A Time to Break Silence," Riverside Church, New York City, April 4, 1967.
4. Bono, remarks at Africare's Annual Bishop Walker Awards Dinner, October 24, 2002, Washington, D.C., http://www.data.org/archives/000229.php.
5. Chancellor of the Exchequer Gordon Brown speech to the Federal Reserve Bank, New York, November 16, 2001.
6. Micah Challenge: http://micahchallenge.org/overview/.

Chapter 13
The Poor You Will Always Have with You?
What Does the Bible Say about Poverty?

1. Stewart Burns, *To the Mountaintop: Martin Luther King Jr.'s Sacred Mission to Save America, 1955–1968,* (San Francisco: HarperSanFrancisco, 2004).

2. Ken Medema, "Coal Black Jesus," in Jim Wallis and Ken Medema, "Let Justice Roll," *30 Good Minutes,* December 16, 1990. http://www.30goodminutes.org/csec/sermon/medema_3410.htm.

Chapter 14

Poor People Are Trapped—in the Debate About Poverty: Breaking the Left/Right Impasse

1. Poverty stats: http://www.census.gov/hhes/www/poverty03.html.
2. "Out of Reach 2003: America's Housing Wage Climbs," National Low Income Housing Coalition, September 2003, http://www.nlihc.org/oor2003/data.php?getmsa=on&msa%5B%5D=denver&state%5B%5D=CO.
3. "President Announces Welfare Reform Agenda," February 26, 2002, http://www.whitehouse.gov/news/releases/2002/02/20020226–11.html
4. "Key Facts About American Children," Children's Defense Fund, August 2004.

Chapter 15

Isaiah's Platform: Budgets Are Moral Documents

1. "Shields and Brooks," *PBS News Hour,* January 31, 2003, http://www.pbs.org/newshour/bb/political_wrap/jan-june03/sb_1–31.html.
2. Susan Pace Hamill, "An Argument for Tax Reform Based on Judeo-Christian Ethics," *Alabama Law Review,* 54:1 (fall 2002), http://www.law.ua.edu/pdf/hamill-taxreform.pdf.
3. Governor Bob Riley, "It's Time to Choose," Address to the People of Alabama on the Occasion of the Extraordinary Session of the Alabama Legislature, May 19, 2003, http://www.governorpress.alabama.gov/pr/sp–2003–05–19-session.asp.
4. "Governor Tries Religion to Solve Deficit," CBS Evening News, June 25, 2003.
5. David Firestone, "Tax Law Omits Child Credit in Low-Income Brackets," *New York Times,* May 29, 2003.
6. Jonathan Weisman, "Senators Rush to Propose Expanded Child Tax Credit," *New York Times,* June 3, 2003.
7. David Firestone, "DeLay Rebuffs Move to Restore Lost Tax Credit," *New York Times,* June 4, 2003.
8. David Cay Johnston, "Very Richest's Share of Income Grew Even Bigger," *New York Times,* June 26, 2003.
9. David Firestone, "Dizzying Dive to Red Ink Poses Stark Choices for Washington," *New York Times,* September 14, 2003.
10. Poverty stats: http://www.census.gov/hhes/www/poverty03.html.

11. "Out of Reach 2003: America's Housing Wage Climbs," National Low Income Housing Coalition, September 2003, http://www.nlihc.org/oor2003/.
12. Iraq casualties, U.S. Department of Defense, http://www.defenselink.mil/news/casualty.pdf.

Chapter 16

Amos and Enron: What Scandalizes God?

1. "Transcript: O'Neill on Enron Mess," Fox News, January 13, 2002, http://www.foxnews.com/story/0,2933,42952,00.html
2. Judy Olian, "Trusting Big Business," Penn State Smeal College of Business," http://www.smeal.psu.edu/news/releases/jul02/trusting.html; USA Today/CNN/Gallup poll results, July 9, 2002; http://www.usatoday.com/news/2002-07-09-poll.htm.
3. "Military, Police Top Gallup's Annual Confidence in Institutions Poll," Gallup News Service, June 19, 2003, http://www.gallup.com/poll/releases/pr030619.asp.
4. Sean Gonslaves, "WTO Protesters Appear Prophetic," *Seattle Post-Intelligencer,* July 16, 2002.
5. Richard Reeves, "Democrats: Loyal Losers or Disloyal Winners?" Universal Press Syndicate, January 22, 2004, http://www.uexpress.com/richardreeves/?uc_full_date=20040122.
6. David Batstone, *Saving the Corporate Soul,* Jossey-Bass, 2003, p. 2.
7. Estate tax fact sheets, United for a Fair Economy, http://www.faireconomy.org/estatetax/factsheets/index.html.
8. William H. Gates Sr. and Chuck Collins, "Tax the Rich?" *Sojourners,* January–February 2003, p. 36.

Chapter 17

The Tipping Point: Faith and Global Poverty

1. Chancellor of the Exchequer Gordon Brown, speech at a conference on "Making Globalisation Work for All—The Challenge of Delivering the Monterrey Consensus," February 16, 2004.
2. Joseph Kahn, "Congressional Leadership Agrees to Debt Relief for Poor Nations," *New York Times,* October 18, 2000.
3. Bachus floor speech.
4. David Beckmann, "Jubilee Begins with Me," *Sojourners,* July–August 2000, p. 47.
5. Susan Dominus, "The Way We Live Now: Questions for Bono," *New York Times Magazine,* October 8, 2000.

6. Charlotte Denny and Larry Elliott, "Bank Works for Change, Its Chief Insists," *Guardian,* September 27, 2000.
7. Romilly Greenhill and Elena Sisti, "Real Progress Report on HIPC," Jubilee Research/New Economics Foundation, September 2003.
8. Christian Aid's Trade Justice Campaign: The Basics, http://www.christainaid.org.uk/campaign/trade/basics.htm.
9. Equal Exchange: http://www.equalexchange.com.
10. Pura Vida mission statement, http://www.puravida.coffee.com.
11. Fair trade criteria: http://www.transfairusa.org.
12. Celeste Kennel-Shank, "Java Justice," *Sojourners,* April 2004, p. 8.
13. Ten Thousand Villages: http://tenthousandvillages.com.
14. International Fair Trade Association: http://www.ifat.org.
15. Millennium Development Goals: http://www.un.org/millenniumgoals/.
16. Jim Wallis, interview with Clare Short, "Lies Leaders Tell," *Sojourners,* January 2004, p. 21.
17. World Vision on HIV/AIDS.
18. Kofi Annan, interview with BBC, November 28, 2003, http://news.bbc.co.uk/2/hi/africa/3245014.stm.
19. Adam Taylor, "The Devil's in the Details," *Sojourners,* May–June 2003, p. 18.
20. Taylor, "The Devil's in the Details."
21. Millennium Development Goals: http://www.un.org/millenniumgoals/.

Chapter 18

Abortion and Capital Punishment: A Consistent Ethic of Life

1. Joseph Cardinal Bernardin, "A Consistent Ethic of Life: An American-Catholic Dialogue," Gannon Lecture, Fordham University, December 6, 1983.
2. U.S. Conference of Catholic Bishops, *Faithful Citizenship,* November 2003, p. 11.
3. Bud Welch, "A Father's Urge to Forgive," *Time,* June 16, 1997.
4. Lois Romano and Dan Eggen, "For McVeigh's Victims, Different Paths to Peace," *Washington Post,* April 15, 2001.
5. Jeffrey M. Jones, "Support for the Death Penalty Remains High at 74 Percent: Slight Majority Prefers Death Penalty to Life Imprisonment as Punishment for Murder," Gallup News Service, May 19, 2003.
6. Associated Press, "O'Connor Questions Death Penalty," July 3, 2001.
7. Jeff Flock, "'Blanket Commutation' Empties Illinois Death Row Incoming Governor Criticizes Decision," CNN, January 13, 2003, http://www.cnn.com/2003/law/01/11/illinois.death.row/.

8. Citizens for a Moratorium on Federal Executions, letter to President Bill Clinton, November 20, 2000, http://www.federalmoratorium.org/cmfeletter-topresident.htm.
9. Bishop Joseph Fiorenza, "Bishops' Head Urges President Commute Death Sentences of 31 Federal Prisoners," December 4, 2000, http://www.nccbuscc.org/comm/archives/2000/00–301.htm.
10. Bureau of Justice Statitistics, http://www.ojp.usdaf.gov/bjs/cp.htm.

Chapter 19

Truth Telling About Race: America's Original Sin

1. Representative Tony Hall, press release, "Hall Calls on Congress to Apologize for Slavery," June 19, 2000, http://www.directblackaction.com/tony_hall.htm.
2. Oscar Hammerstein II, "You've Got to Be Taught," lyrics from the 1949 musical *South Pacific* by Richard Rodgers & Oscar Hammerstein II, Copyright © 1949, Renewed.
3. "Dean Makes Racial-Political History," *Black Commentator,* December 11, 2003, http://www.blackcommentator.com/68/68_cover_dean.html.

Chapter 20

The Ties That Bond: Family and Community Values

1. Tom Shales, "Incomplete!" *Washington Post,* February 2, 2004.
2. http://www.cnn.com/election/2004.
3. "Clinton Nostalgia Sets in, Bush Reaction Mixed," Pew Research Center for the People and the Press, January 11, 2001.
4. Anna Greenberg, "The Marriage Gap," *Blueprint Magazine,* July 12, 2001.
5. Theda Skocpol, *The Missing Middle* (New York: W.W. Norton, 2001).
6. Anna Greenberg and Stanley B. Greenberg, "Adding Values," *American Prospect,* August 28, 2000.
7. Richard Mouw, "Why the Evangelical Church Needs the Liberal Church," *Sojourners,* February 2004, p. 14.
8. Barbara Wheeler, "Why the Liberal Church Needs the Evangelical Church," *Sojourners,* February 2004, p. 15.
9. "Voters See Poverty and Hunger as the Real Moral Issues in the Presidential Race," Alliance to End Hunger and Call to Renewal press release, February 27, 2004, http://www.religionnews.com/press02/pr030104.html.
10. Richard B. Hays, "A Season of Repentance," *Christian Century,* August 24, 2004, p. 8.

11. Mitch Albom, *Tuesdays with Morrie* (Doubleday, 1997).
12. Gary Levin, "Tonight There'll Be Just One 'Survivor' Susan, Kelly, Richard or Rudy?" *USA Today,* August 23, 2000.

Chapter 21

Hope Versus Cynicism: The Critical Choice

1. Joy Carroll Wallis, *The Woman Behind the Collar,* Crossroad, 2004.
2. Nancy Haught, "Social Activist Will Use Forum to Tell the Truth About Poverty," *Portland Oregonian,* April 26, 2000.

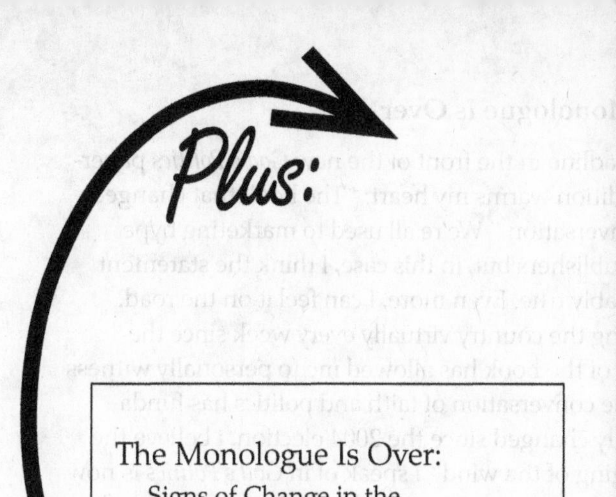

Plus: Insights, Interviews, and More

The Monologue Is Over

The headline in the front of the new *God's Politics* paperback edition warms my heart: "The book that changed the conversation." We're all used to marketing hype from publishers but, in this case, I think the statement is probably true. Even more, I can feel it on the road. Traveling the country virtually every week since the launch of the book has allowed me to personally witness how the conversation of faith and politics has fundamentally changed since the 2004 election. I believe the "changing of the wind" I speak of in *God's Politics* is now actually happening—in the religious community, in the broader society, in the media discussions of religion and public life, and even in politics.

The monologue of the Religious Right is finally over, and a new dialogue has begun! At last, the loudest and most vitriolic religious voices are no longer the only ones, and the Religious Right no longer controls the agenda. Global and domestic poverty, the environment, HIV/AIDS, and the ethics of war are all now clearly on the agenda of the faith community, alongside the critical issues of family values and the sacredness of life—all being discussed in deeper and broader ways. I feel very grateful that *God's Politics* has helped that happen.

The three great hungers in our world today are for *spiritual integrity, social justice,* and the *connection* between the two. That's exactly what this book has helped to do—to build that bridge, and people are responding. A book on faith—and, even more amazingly, faith and politics—jumped to number two on Amazon its first week, and then onto the *New York Times* bestseller list for four months. The book tour for *God's Politics* quickly became a whirlwind, and then a "movement" tour.

Almost right from the start, I realized something important was happening, and that it was about more than a book. *God's Politics* became the right book at the right time, and revealed what was already there waiting to be expressed. Many people of faith felt their voice was not being heard in the national debate over faith and politics and found something to point to. I soon realized the large numbers of people were not just coming to hear my voice, but also to express *their* voice.

During the year I traveled to sixty-four cities (some more than once), doing one hundred thirty major events, speaking face-to-face to more than one hundred thousand people, and reaching millions more through almost a thousand interviews. The book had struck a chord and it was reaching beyond the usual audience to ordinary people in America—people who are eager to apply their faith or values to politics and are ready to get involved. From the first week, book signings turned into town meetings and bookstore events into revivals. At churches, universities, and conferences these town meetings involved Christians of every kind, Jews, Muslims, those who don't call themselves "religious" but rather "spiritual," and even those who are agnostic about faith but believe that "moral values" should shape our public life.

As I have traveled, I have seen the hunger for a deeper understanding of the relationship between faith and politics from persons of all ages. I spoke on thirty-three campuses—Evangelical, Jesuit, public and private universities, and seminaries. At all of them, students came out by the thousands, stayed for discussion sessions that lasted for hours, and were full of questions about how to get involved. I often did an "e-mail altar call," and many eagerly wrote down their e-mail addresses to sign up for a new movement.

But there were also high-school-age young people and even middle schoolers who showed up. In Minneapolis, I greeted a girl waiting with her book to be signed. I asked her how old she was. "I'm eleven years old," she replied. I stopped the line to talk to her and asked, "What did you get from tonight?" The very poised and articulate little girl said, "Well, I think we are just going to have to change the world." I asked her further, "And who is going to do that?" She thought for a moment and then told me, "I suppose people like me." So when I spoke to my Harvard students the next week, I said if they wanted to lead in this movement they better move quickly because a whole new generation is coming up fast behind them!

The book tour got lots of media coverage, but the television appearance that really reached this generation of young people was an interview on *The Daily Show with Jon Stewart.* We had a good time on the show, with lots of humorous banter back and forth, but also with real substance. And it was great to see a youthful audience whooping and hollering for progressive religion, especially when I said that I didn't think Jesus's top priorities would be a capital gains tax cut and the occupation of Iraq.

When I traveled to an event in my home town of Detroit, a young African American front desk clerk in her early twenties proudly told me, "You have an upgrade!" When I asked who had given me the upgrade, she replied, "Well, I did. I saw *Jon Stewart* and bought the book. It was awesome!" At a later event, a seventeen-year-old came up and said, "I just wanted to shake your hand. You're the only Christian I see on TV and don't throw up afterward!"

Then the e-mails came, from thousands of young people who said things like, "I lost my faith because of television preachers, bad religious fundraising, pedophile priests, bishop cover-ups, and White House theology." And many said, "I didn't know you could be a Christian and care about poverty, the environment, or the war in Iraq." I was amazed by the response. How has this happened? How has faith been so misrepresented and even stolen in the public arena? The popular response to *God's Politics* revealed the enormous number of people who believed their faith had been hijacked by the Religious Right and were now ready to take it back.

Something else very moving happened at many stops. Parents told me how their sons or daughters had lost their faith and left the church. One father said, "But my son saw you on TV and got the book. He just wrote his mother and me to tell us that he is finding his way back to faith." There was a tear in that dad's eye when he told me. I heard many stories like that, about sons and daughters, husbands and wives, and even parents who hadn't been to a church in many years now taking a fresh look at faith and how it applies to the social issues they cared about. Reading the book was actually bringing families back together around the issues of faith and social justice.

Also encouraging is what many pastors tell me about the new book studies for *God's Politics* in their local congregations. First, that they are some of the best attended book studies they've seen and that they are bringing together people across the political spectrum to talk for the first time about how their faith relates to the world. A prominent leader of the so-called mega-church movement called me personally to say that he had commended *God's Politics* to many of the pastors in his large network because it was so "balanced" and "thoughtful." After I spoke at the annual pastors' conference of a major evangelical denomination, the organization's president told me it was one of the few events that had ever united the "right" and the "left" in his church.

Along with the evangelicals coming out to seek a broader social conscience, I met Catholics at many stops along the way who don't want to be spoken for by a handful of bishops who tell them they can vote on only one issue—abortion—and that they must ignore the rest of Catholic social teachings—teachings I wish Protestants had. Abortion is a crucial moral issue over which we need a much deeper discussion on all sides, but single-issue voting is not in the mainstream of Catholic social teaching. The "consistent ethic of life," at the heart of Catholic social teaching, could be a key to America's political future.

Respect for life extends to our environment. In early 2006 the National Association of Evangelicals (NAE), began to speak up on the issue of "creation care." Leaders of the NAE began to sound like biblical prophets, describing their experiences of "epiphany" and "conversion" on the issue. In 2004, the NAE adopted an important social policy statement "For the Health of the Nation: An Evangelical Call to Civic Responsibility," which included a principle titled "We labor to protect God's creation." The NAE began to participate at many major climate change and environmental meetings—both domestically and internationally—and a series of press stories about the new evangelical environmentalists appeared. The NAE constituency is a significant part of the Republican base, and the new environmental concern did not go unnoticed by the White House.

Then, in January 2006, the Religious Right reared its head. A letter addressed to the NAE and signed by twenty-two of the Right's prominent leaders said, "We have appreciated the bold stance that the National Association of Evangelicals has taken on controversial issues like embracing a culture of life, protecting traditional marriage and family." But it went on to say, "We respectfully request, however, that the NAE not adopt any official position on the issue of global climate change. Global warming is not a consensus issue." It was a clear effort to prevent the NAE from taking a stand on environmental issues and even to veto the whole effort. Five years ago, so powerful a group of conservative

Christian leaders probably could have tamped down this new evangelical effort that served to broaden the range of "moral values" and issues of biblical concern. But not this time.

On February 9 a full-page ad appeared in the *New York Times* with the headline: "Our commitment to Jesus Christ compels us to solve the global warming crisis." The striking ad announced the "Evangelical Climate Initiative" and was signed by eighty-six prominent evangelical leaders, including the presidents of thirty-nine Christian colleges. The "Evangelical Climate Initiative" is of enormous importance and could be a tipping point in the climate change debate.

Our values as people of faith involve other moral issues that demand deeper discussion. Chapter fifteen of *God's Politics* is subtitled "Budgets are Moral Documents." In 2005 Sojourners and Call to Renewal used that slogan for a campaign to tell our political leaders that budgets reflect the values and priorities of a family, church, organization, city, state, or nation. A budget tells us who and what are most valued by those making it. We said that the question that America's religious communities must ask of any budget is what happens to the poor and most vulnerable—especially what becomes of the nation's poorest children in these critical fiscal decisions.

Many people of faith concluded that the administration's proposed budget was morally unacceptable. It projected a record deficit, promised to make permanent tax cuts benefiting mostly the wealthiest, and made cuts in vital programs and services for low-income people—all in a time of war. We said a budget that places the costs and burdens of fiscal responsibility mainly on the poor while further benefiting the rich only asks for sacrifice from those who can least afford it—and that was a moral offense.

Before long, that language of "budgets as moral documents" and the phrase itself were being picked up and used not only by churches, but by political leaders and the media, with newspaper headlines such as "One More 'Moral Value': Fighting Poverty." Clearly, the influence of the faith community was being felt. Words like "Christian" and "religious groups" were now associated with words like "poverty," "low-income families," and "economic justice," instead of just "abortion" and "gay marriage." Poverty's moral and political urgency was seen by more and more people as a natural outcome of faith.

Then came Hurricane Katrina, and pictures from New Orleans that stunned the nation and exposed the stark realities of who suffered the most, who was left out even before they were left behind, who was waiting in vain for help to arrive, and who faced the most difficult challenges of recovery. From the reporters covering the unprecedented disaster to ordinary Americans glued to their TVs, a shocked and even outraged response was repeated: "I didn't realize how many Americans were poor." And the faces of those who suffered most also revealed the still persistent connection between race and poverty in America.

But only a few months later, as Congress began to finish its budget work, cuts were again on the agenda. After all that Hurricane Katrina revealed, how could we construct the budget at the expense of our poorest and most vulnerable citizens? Among the expected cuts was one to food stamps. But many Republican senators on the Agricultural Committee responsible for nutrition programs were from states where faith-based organizations and religious leaders actively pressed their lawmakers with calls, e-mails, letters, and recess visits not to cut crucial programs for the poor. Those religious voices were heard and food stamps were restored. The story in the *Atlanta Journal-Constitution* read, "Under pressure from religious service groups, Senate Agricultural Committee Chairman Saxby Chambliss of Georgia reversed course and announced he won't seek a $574 million cut that the White House wants in the federal food stamps program."

Still, in the early hours of the morning before leaving for their Thanksgiving break, the House passed a budget that cut $50 billion, including essential services for low-income families. Funding for health care, foster care for neglected children, student loans, and enforcing child support orders all fell to the ax. So we issued an "altar call" to come to Washington. On a bitterly cold December day, we brought our message to the steps of the Cannon House Office Building. After powerful prayers and preaching on the steps and a press conference that seemed more like a revival, 115 pastors and leaders were arrested. Many of those who took part in the nonviolent civil disobedience were Christians who live and work alongside poor people every day. We sounded like a choir (and a good one at that) as we sang Christmas carols while being arrested, handcuffed, put into buses, and taken to a large holding facility to be processed. Our vigil in Washington was followed that evening by more than seventy vigils in more than thirty states and was covered in at least one hundred newspapers across the nation and on the television networks. The budget passed the House by only two votes, and Vice President Cheney had to fly back to Washington to cast the tie-breaking vote in the Senate.

The struggle for a moral budget—and a more decent society—of course didn't end with this latest legislative battle. More cuts are slated for 2007, including another attempt to cut food stamps. We remain committed to a budget that does not harm people in poverty, but in fact supports and empowers them. The biblical prophets tell us that the truest test of the "righteousness" of any society is how it treats its most vulnerable citizens.

In the time since *God's Politics* was published, public and congressional opinion on the war in Iraq has reached a tipping point. The continued violence (now more than twenty-four hundred Americans killed, nearly twenty thousand maimed and wounded, tens of thousands of Iraqis dead) and the cost (now more than $330 billion) have opened a new debate.

On the book tour, I met with military veterans from Iraq, the first Gulf war, and Vietnam about how we are again forgetting the warriors we send to fight, die, and be changed forever after they come home. I was persuaded that the churches must become the welcoming places that these forgotten soldiers really need.

God's Politics opposed the war in Iraq because that war was based on deception, manipulation, and fundamental miscalculation. The subsequent American occupation has become more the catalyst than the solution to a violent and bloody insurgency. The truth in Iraq is that the warring factions are engaged in a civil war, which the occupation is only making worse. Military leaders give a much grimmer picture of the prospects for success in Iraq than the political (and mostly civilian) architects of the war, who repeat tired mantras about "staying the course."

The Bush administration needs to be content with having removed Saddam Hussein from power, which it now claims was the real goal of the invasion. It must give up the control over Iraq that the neoconservatives who run American foreign policy have long desired—despite the administration's rhetoric about democracy. And it must finally open up the situation for genuine international participation in securing Iraq's security and reconstruction by disavowing permanent military bases or a proprietary interest in oil profits. Only that kind of international involvement will ultimately achieve security and stability in Iraq and eventually end the deadly and costly insurgency.

As I pointed out at the beginning of this essay, the most important new development I saw as I traveled the country with *God's Politics* was a new movement of moderate and progressive religious voices challenging the monologue of the Religious Right. An extremely narrow and aggressively partisan expression of right-wing Republican religion has controlled the debate on faith and politics in the public square for years. But that is no longer true.

As you'll read in this book, during the 2004 election campaign Jerry Falwell and Pat Robertson virtually said Christians could vote only for George W. Bush. Many Christians and people of faith responded by reminding America that "God is not a Republican . . . or a Democrat." In 2005 the Religious Right took a new step, saying that supporting the president's judicial nominations was a test of orthodoxy. That was a dramatic and serious breach in the relationship between faith and politics.

In April 2006 major Religious Right leaders hosted "Justice Sunday," a telecast from a mega-church in Louisville, Kentucky. Their message was that those who don't support President Bush's judicial nominees were hostile to "people of faith." Republicans and their religious supporters were openly questioning the faith and religious integrity of their political opponents.

I spoke to more than a thousand people at an alternative event in a Louisville Baptist church that day. I said that behind these activities lies a fundamental assumption by Republican operatives and their conservative religious allies that they own religion in America. They claim that "values voters" in America belong to them, and they disrespect the faith of those who disagree with them. The clear implication of their message is that those who oppose them are not people of faith. There are better words for this than just "politically divisive" or "morally irresponsible." For these are not merely political offenses, they are religious ones. And for offenses such as these, theological terms are better—terms such as "idolatry" and "blasphemy."

A central theme of *God's Politics* is that we should bring our religious convictions about all moral issues to the public square—the uplifting of the poor, the protection of the environment, the ethics of war, the tragic number of abortions in America, the social consequences of family breakdown—without attacking the sincerity of other people's faith or demanding that we should "win" because we are religious. Rather, we must make moral arguments (not special religious claims) and mobilize effective movements for social change that can powerfully persuade our fellow citizens, religious or not, of what is best for the common good.

We can get some historical perspective by looking at how Dr. Martin Luther King Jr. did it—and he was the church leader who did it best. Once, after he was arrested, he wrote the famous "Letter from a Birmingham Jail," addressed to the white clergy who opposed him on the issues of racial segregation and violence against black people. Never once did he say that they were not people of faith. He appealed to their faith, challenged their faith, asked them to go deeper with their faith, but he never said they were not real Christians. If Dr. King refused to attack the integrity and faith of his opponents over such a clear gospel issue, how can the Religious Right do it over presidential nominees and a Senate filibuster? "Justice Sunday" was an attempt to hijack Christianity for a partisan and ideological agenda. The Religious Right was virtually declaring a religious war to give their version of faith religious supremacy in America.

A great deal is at stake in this battle for the heart and soul of faith in America and for the nation's future itself. We must reject the assumption that true Christians must accept only one partisan political position on issues. When either party tries to politicize God or co-opt religious communities, it makes a terrible mistake. We must not allow faith to be put into the service of one political agenda. Rather, we must insist on the deep connections between spirituality, morality, and values with politics while defending the proper boundaries between church and state. We can demonstrate our commitment to pluralistic democracy and support the rightful separation of church and state without segregating moral and spiritual values from our political life or banishing religious language from the public square.

Effects of the new dialogue were made visible in mid-May 2005 at Calvin College. Karl Rove, seeking a friendly Michigan venue for a commencement speech, approached Calvin and offered President Bush as the speaker. But the White House was not counting on the reaction of students and faculty. Rove expected the evangelical Christian college, in the dependable "red" area of western Michigan, to be a safe place. He was wrong.

The day the president was to speak, a letter signed by one-third of Calvin's faculty and staff ran in the *Grand Rapids Press*. Noting that "we seek open and honest dialogue about the Christian faith and how it is best expressed in the political sphere," the letter said that "we see conflicts between our understanding of what Christians are called to do and many of the policies of your administration." The letter asserted that the administration had "launched an unjust and unjustified war in Iraq," "taken actions that favor the wealthy of our society and burden the poor," "harmed creation and [had] not promoted long-term stewardship of our natural environment," and "fostered intolerance and divisiveness and [had] often failed to listen to those with whom it disagrees." On commencement day, according to news reports, about a quarter of the nine hundred graduates wore "God is not a Republican or a Democrat" buttons pinned to their gowns.

This and other events signal a sea change in evangelical Christian politics: the Religious Right is losing control. It has now lost control on the environmental issue; caring for God's creation is now a mainstream evangelical issue, especially for a new generation of evangelicals. But now so is sex trafficking, the genocide in Darfur, the pandemic of HIV/AIDS, and, of course, global and domestic poverty. The call to overcome extreme poverty, abroad and at home in the world's richest nation, is becoming a new "altar call" around the world. I'm still amazed how much the national conversation about faith and politics has already changed as a result.

Our basic message has not changed, but the openness to it in the church, political world, media, and culture has changed dramatically. That is due to many factors: the 2004 election, the heightened role of religion and "moral values" in our political discourse, the reaction among a large number of other people of faith to the Religious Right's hubris and pursuit of power, and, perhaps most important, the essential moral and spiritual character of the most pressing issues our society confronts—the massive nature of global and even domestic poverty, the crisis of the environment, the cost and consequences of war, the selective moralities of both left and right in regard to the sanctity of life, and the breakdown of both family and community.

Family values and the sacredness of life are deeply important—too important to be used as partisan wedge issues that call for single-issue voting patterns that ignore other critical biblical matters. The Religious Right has been able to win when they have been able to maintain and control a monologue on the relationship between faith and politics. But when a dialogue begins about the extent of "moral values" issues and what biblical Christians should care about, the Religious Right begins to lose.

The best news of all for the American church and society is this: the monologue of the Religious Right is over, and a new dialogue has begun. And I am pleased that this book has helped to spark that new dialogue. Many people around the country saw the breakthrough of *God's Politics* as the success of *their* message, *their* hopes, and *their* work. Now when I step up to speak, I look over the crowd and see them looking at each other, and sense the excitement in the room that people feel when they realize they are a part of something much bigger than themselves, something that might be important. So I often begin, "People of faith who are driven to justice and peace have often felt alone. When all the political events and decisions in your country go the other direction from your own convictions, it makes you feel alone. When you never see your own voice and faith represented in the media, it makes you feel alone. Well, my friends, look around tonight, you're not alone anymore."

Jim Wallis
Washington, D.C.
May 2006

SOJOURNERS
faith, politics, culture

CALL to RENEWAL
A faith-based movement to overcome poverty

Jim Wallis is founder and editor-in-chief of *Sojourners,* and convener to *Call to Renewal.*

About Sojourners / Call to Renewal

www.sojo.net / *www.calltorenewal.org*

Sojourners and Call to Renewal share an over-arching statement of mission: To articulate the biblical call to social justice, inspiring hope and building a movement to transform individuals, communities, the church, and the world.

Sojourners, established in 1971, is a nonprofit organization whose mission is to offer a voice and vision for social change. Sojourners attracts a diverse group of evangelicals, Catholic, and Protestant Christians, as well as others who are united on issues of justice and peace.

Call to Renewal is a national network of churches, faith-based organizations, and individuals working to overcome poverty in America. Through local and national partnerships with groups from across the theological and political spectrum, Call to Renewal convenes the broadest table of Christians focused on anti-poverty efforts.

For FREE regular updates, please sign up for the SojoMail e-zine at www.sojo.net.

Sojourners/Call to Renewal
3333 14th St. NW, Suite 200
Washington, DC 20010
Phone: 202-328-8842 or 1-800-714-7474
Fax: 202-328-8757
Website: www.sojo.net and www.calltorenewal.org
E-mail: sojourners@sojo.net and ctr@calltorenewal.org

Sojourners/Call to Renewal Congregational Network

Sojourners/Call to Renewal is launching a new congregational network aimed at empowering, resourcing, and connecting clergy and laity interested in issues of social justice, peace and spiritual renewal. A variety of resources for preaching, teaching, and issue advocacy as well as discounts on Sojourner's/Call to Renewal conferences and events will be included in the network offering. We are excited to be able to provide these tools and would like to offer you the chance to get in on the ground floor. To find out more and to receive announcements about the roll-out of the program, please indicate your interest by going to www.sojo.net/network and provide your contact information.

C4NA

The "Covenant for a New America" is a policy platform intended to move beyond the debate between left and right by seeking to create a common commitment to identify, pursue, and bring about real solutions to poverty.

The Covenant lifts up both personal and social responsibility with policies that address the individual decisions and social systems that trap people in poverty. It identifies policies that move beyond looking solely to charity or only to government. It acknowledges that budgets are moral documents, and budget priorities can help or hurt poor people—and that negative family and cultural values also impact low-income people.

A combination of policies in four major areas will allow the Covenant's vision to engage concretely with the policy-making process. We seek changes that will promote: 1) a living family income for all who work, 2) rebuilding of neighborhoods and communities, 3) strengthening of families and the renewing of culture, and 4) an end to extreme global poverty.

Increasingly, religious leaders from across the theological and political spectrum are standing together against poverty. Concerted efforts to overcome poverty are helping building new common ground and are reinvigorating one of the core purposes of the church: a fundamental commitment to the ones Jesus called "the least of these." The Covenant will allow people of faith across the theological and political spectrum to commit together to recognize the valid concerns of both sides in the political debate, but then move to higher ground by working together to make overcoming poverty a top priority.